Tsuneo Someya (Ed.)

Advanced Combustion Science

With 245 Figures

Springer-Verlag
Tokyo Berlin Heidelberg
New York London Paris
Hong Kong Barcelona
Budapest

Editor in Chief
Professor Tsuneo Someya
Department of Mechanical Engineering, Faculty of Engineering, Musashi Institute of
Technology, 1-28-1 Tamatsutsumi, Setagaya-ku, Tokyo, 158 Japan

ISBN-13: 978-4-431-68230-1   e-ISBN-13: 978-4-431-68228-8
DOI: 10.1007/978-4-431-68228-8

Printed on acid-free paper

Preface

Non–uniform combustion, as encountered in diesel and gas turbine engines, furnaces, and boilers, is responsible for the conversion of fossil fuel to energy and also for the corresponding formation of pollutants. In spite of great research efforts in the past, the mechanism of non–uniform combustion has remained less explored than that of other combustion types, since it consists of many, mostly transient processes which influence each other.

In view of this background, a group research project, "Exploration of Combustion Mechanism", was established to explore the mechanism of combustion, especially that of diffusive combustion, and also to find efficient ways to control the combustion process for better utilization of fuel and the reduction of pollutant emission. The group research was started, after preparatory activity of 2 years, in April 1988, for a period of 3 years, as a project with a Grant–in–Aid for Scientific Research of Priority Area subsidized by the Ministry of Education, Science and Culture of Japan. The entire group of 43 members was set up as an organizing committee of 13 members, and five research groups, consisting of 36 members. The research groups were: (1) Steady combustion, (2) Unsteady spray combustion, (3) Control of combustion, (4) Chemistry of combustion, and (5) Effects of fuels. At the beginning of the project it was agreed that we should pursue the mechanism of combustion from a scientific viewpoint, namely, the target of the project was to obtain the fundamentals, or "know why", rather than "know how" of combustion.

As one of the project activities, world–famous researchers from foreign countries were also invited to stimulate and refine the group research work ; the financial support of the Ministry of Education, Science and Culture and other foundations was given for this. Therefore, the results of the research carried out by the present group are prominently evaluated from the international point of view.

This book is the final collection of the outcome of the activities of the group research project, including several collections of precious work presented by researchers from the USA and Europe. The publication of this book is partially supported by the Grant–in–Aid for Publication of Scientific Research Result. It consists of eight chapters: (1) Structure of turbulent diffusion flames, (2) Modeling of turbulent diffusion flames, (3) Spray formation and combustion, (4) Kinetics, (5) Soot formation fundamentals, (6) Emissions and heat transfer in combustion systems, (7) Fuel effects in combustion systems, and (8) New approaches to controlling combustion. It is our hope that this book will provide a worthwhile understanding of and useful information on combustion science, especially on non–uniform combustion.

The research project, and hence the present book, would not have been possible without the wise selection of the project by the evaluating committee of the Ministry of Education, Science and Culture. Its support for the research group was also indispensable. The editors owe a special debt of gratitude to the late Prof. T. Ishihara, who encouraged the research group members from the very beginning of the project. We also acknowledge deeply the untiring support and competent guidance given by the members of the organizing committee, especially by Professors G. T. Sato, T. Saito, S. Matsuoka, and H. Matsushita.

Tsuneo Someya

Table of Contents

Chapter 1
Structure of Turbulent Diffusion Flames

1.1 Introduction of Turbulent Diffusion Flame Structures

1.1.1 Importance of turbulent diffusion flames

Diffusion flame is the most popular flame formed in the industrial furnaces because of its safety. In this case the fuel and oxidizer are separately supplied into the combustion region and the back fire or the flash back could not occur like a premixed flame. On the contrary to this advantage the diffusion flame has some limitations for higher combustion intensity.

The first one of them is the fact how the molecular size mixing must be achieved between fuel and oxidizer before the combustion takes place. In order to enhance this mixing the practical burners introduce the swirling vanes in the fuel flow and/or the combustion air surrounding the fuel jet.

On the other hand to these kinds of techniques we have no way to control the combustion zone after supplying the fuel and oxidizer. Almost of all the ways of controls we can do by the present combustion technology must be said being concentrated to the burner design. However, once the ignition takes place the flow field, therefore, the macroscopic mixing phenomena and the transportation coefficients of momentum, heat and mass vary so much and the combustion zones become quite different from the ones expected in the cold flow.

The second limitation could be said that the diffusion flame has not the auto-flame-propagation phenomena like the premixed flame. This characteristics could not always be said as the short point, but an burner operator could not expect when the flame would blow off. The mechanisms of flame blow off of the diffusion flame are much more complicated than that of the premixed flame. The mixing of chemical species in the turbulent diffusion flames are started by the macroscale mixing by the large scale eddy motion and transferred to the microscale eddies followed, and finally break down to the molecular level mixing by the molecular motion. The heat transfer to or from the reaction zone by convection and conduction from or to the surrounding media is also done by the same turbulent heat transfer mechanisms. In larger scale flames the radiative heat transfer becomes one of the most important physical factors to maintain the stable combustion. The delicate combinations of these factors are the keys of maintaining the turbulent diffusion flames.

How we could attain higher combustion efficiency and cleaner exhaust combustion gas ? We have to know the above said local and time sequential phenomena including not only the mixing between fuel and oxidizer molecules but also the elementary reactions in these complicated turbulent flame zones.

We have accustomed to think the every phenomena by the time and space averaged values. But, this kind of custom could not be applied to the turbulent combustion regions. We have to correctly know this important phenomena or draw the pictures what are happening within these fields.

1.1.2 History of measuring methods

Figure 1.1 shows the history or the trends of measuring methods for the combustion regions. The measuring methods have been changed by the degree of development of measuring instrumentations. The single point measurements have been applied for long time and are still keeping its popularity. One of the strongest reasons is that this is not always necessary the expensive instrumentations and the other is that the results could be compared with those obtained by the ordinary time averaged turbulent flame theories. We must, however, point out the facts that single point measurements give us only the time sequential information at the measured point, but not the spatially relational ones between even the adjacent points. Therefore the single point measurements produce only the auto-correlation and this gives only the time scales of turbulence.

The simultaneous two point measurements can produce the cross-correlation between the data measured at these two points with changing the distance between them. This gives us the spatial scales of turbulence as the results. If we could take these data simultaneously with some kinds of pictures, a high speed movie or TV image, if it could output the digital information, being much more desirable, we could correlate the signals with the total phenomena surrounding the measured points and understand how fluid moves and mixes with each other, how heat is transported and how reaction takes place.

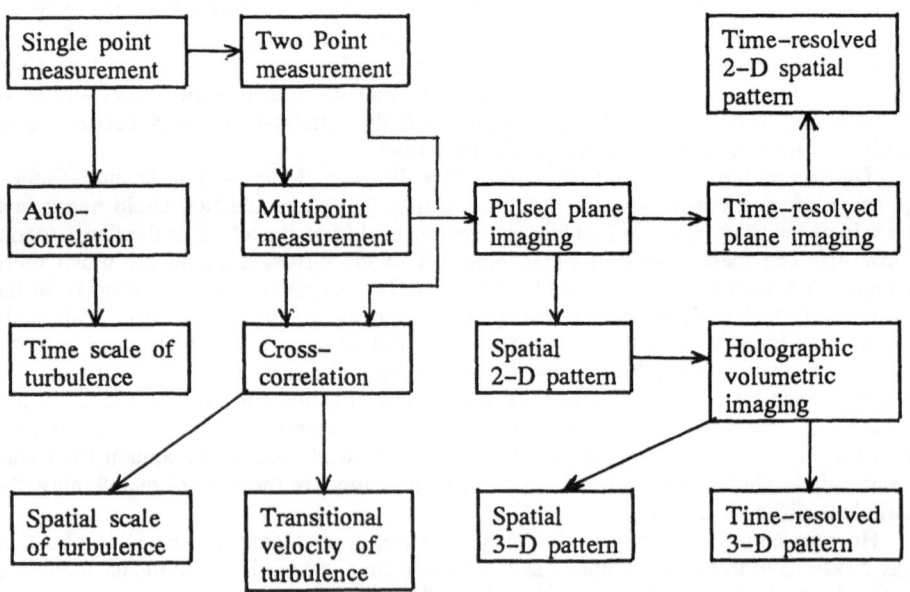

Fig. 1.1   Trend of measuring method

In the early stage of history of measuring device for turbulent flame, touch sensors, such as a fine thermocouple of less than 25 μm in diameter together with an electronic compensator, were used. The measured field are affected somehow by the insertion of this kind of solid material.

In order to eliminate this problem the optical methods like schlieren, shadow photographs and interferometry were adopted. These methods have great advantage compared to the touch sensors because they are free from disturbing the flow fields and reaction regions by the insertion of solid materials. However the images obtained by these traditional methods give the optical information integrated along the optical path, not the local ones. So the quantitative interpretation of obtained pictures rises another problem.

Among these three methods, the interferometry has been developed to the modernized technology of holography to be able to reproduce the three dimensional images from any direction. This method has the unexpected possibilities for the image analyses of the combustion phenomena.

1.1.3 Laser diagnostics

The above said measuring methods are classified as the passive ones as we utilize the phenomena occurred without exciting the energy levels of the molecules there by an applied light source from outside of the flames. On the contrary to those passive methods, the laser optical diagnostics could be said as the active ones because they let the illuminated molecules must be excited up to the upper energy levels and measure the light intensity emitted from the molecules when lose their internal energy with converting into the specific energy of light at a certain wave length by falling down from the upper to lower energy levels. The modern laser sources could produce not only the high intensity coherent light but also the appropriate wave length to excite them up to a certain internal energy level of molecules.

If one point in the flame is focused, we could obtain the information emitted from the point. If the laser has higher emission power, the laser beam can be expanded as a sheet by a special optical setup and we could obtain the two dimensional information.

Optical methods applicable to the flame diagnostics are summarized in Table 1.1. In general the optical phenomena could be treated by the wave function decided by Maxwell equation as [1] ;

$$\{\nabla x(\nabla x)+(\frac{1}{c^2})\frac{\partial^2}{\partial t^2}\}\vec{E}(\vec{r},t)=-(\frac{4\pi}{c^2})\frac{\partial^2}{\partial t^2}\vec{P}(\vec{r},t) \tag{1.1}$$

where $\vec{E}$ denotes the electric field of incident light, c does the light speed, $\vec{P}$ induced electric polarization. Several different monochromatic lights coexist as usual, and we could expand $\vec{E}$ and $\vec{P}$ in the different Fourier components and $\vec{P}$ could be expanded by the power series of $\vec{E}(\omega_i)$ as ;

$$\vec{P}(\omega_i)=\vec{\chi}^{(1)}(\omega_i)\cdot\vec{E}(\omega_i)+\sum_{j,k}\vec{\chi}^{(2)}(\omega_i=\omega_j+\omega_k):$$

$$\vec{E}(\omega_j)\vec{E}(\omega_k)+\sum_{j,k,l}\vec{\chi}^{(3)}(\omega_i=\omega_j+\omega_k+\omega_l): \tag{1.2}$$

$$\vec{E}(\omega_j)\vec{E}(\omega_k)\vec{E}(\omega_l)+...$$

where $X^{(1)}$ denotes the linear susceptibility of the medium, $X^{(n)}$ does the nonlinear susceptibility of nth order and effective to the intensive laser. By the effect of induced linear polarization, the direction of scattered light varies by way of the refractive index. The real and imaginary part in the complex refractive index corresponds to the light dispersion and absorption phenomena, respectively. These phenomena could be explained by the classical model of electron harmonic vibration.

Raman and Rayleigh scattering are emitted by the linear polarization effect related to the first term in Eq.(1.2). The polarization or the vibration of electrical bipolar produces the electromagnetic wave. The polarization phenomena depends on the positions of nuclei of the molecule and it is modulated by the molecular rotation and vibration. The higher mode polarization becomes the weaker with the higher number of n as ;

$$\vec{P}^{(n+1)}/\vec{P}^{(n)} = \vec{E}/\vec{E}_{at} \tag{1.3}$$

where, $\vec{E}_{at}$ at denotes the interatom electric field and is about 3×10^8 V/cm.

In case of the media of gas molecules, which is isotropic and the second term in Eq.(1.2) diminishes due to the inversion symmetry, and the lowest nonlinear susceptibility starts from the third term. CARS and Stimulated Raman Gain/Loss Spectroscopy utilize this third term effect [1].

The types of energy transition between the internal energy levels for Raman scattering family, in which the excited upper level is not specified, are schematically illustrated in Fig. 1.2 [2]. Therefore the wave length of laser as a light source is not specified. This figure shows only the transitions between the vibrational levels. The rotational levels belong to each of them and the internal energy transition between rotational levels occurs simultaneously with the vibrational one followed by the

Table 1.1 Optical methods applicable to combustion measurement

Principles	Method
Geometric optics	Shadow graph, Schlieren photograph
Wave harmonics	Interferometry { Two light beam interferometry; Holography { single pulse / double pulse } } Speckle method
Scattering	Gas { linear — Rayleigh scattering; non–linear { Stokes Raman scattering / Anti–Stokes Raman scattering } } Solid — Mie scattering
Fluorescence	Laser induced fluorescence
Emission	Chemiluminescence

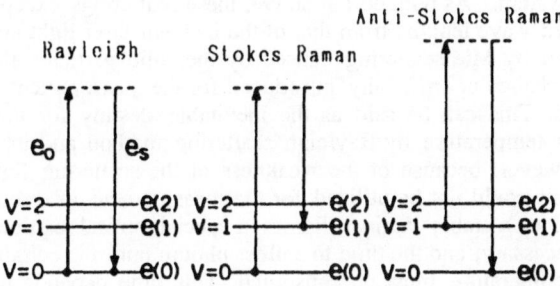

Fig. 1.2 Schematics of internal energy transition

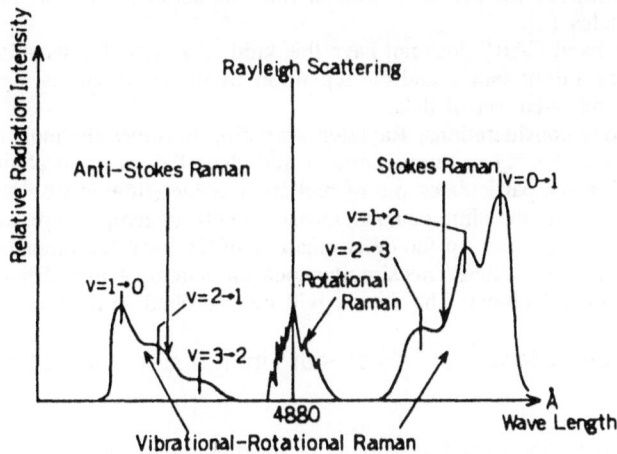

Fig. 1.3 Schematics of spectrum corresponding to the transition shown in Fig. 1.2

rotational energy transition law.

Rayleigh scattering is induced when the transition occurs onto the original energy level from the hypothetical upper level. This says that the Rayleigh scattering irradiates the light at the same wave length as the incident light. On the contrary to this the light irradiated by the Stokes Raman (simply said as Raman) scattering irradiates when the energy transition occurs onto the one upper level than the original one. This transition irradiates the light at longer wave length than the incident one. The Anti–Stokes Raman scattering irradiates the light of shorter wave length as its transition occurs onto the one lower level than the original one as shown in the same figure. This transition system is included in the CARS (Coherent Anti–Stokes Raman Scattering) phenomena. The Scattering spectra by these transitions including the rotational transitions result the complicated ones as shown in Fig. 1.3 [2].

The ratio of light intensities produced by these transitions can be roughly said as 1 for the spontaneous Raman scattering, 10^3 for the Rayleigh scattering and from 10^9 to 10^{12} for the CARS system. As pointed out above, these scatterings except Rayleigh one change their scattered wave lengths from that of the incident laser light and the scattered light are not affected by Mie scattering caused by the solid particles flowing into the measuring point by chance or artificially introduced for the measurement of gas velocity by an LDV method. This can be said as the inevitable destiny for the simultaneous measurement of gas temperature by Rayleigh scattering method and its velocity by an LDV one [3,4]. However, because of the weakness of the scattering light intensity by Raman scattering this could not be utilized for the measurement of turbulent flames. If this is to be applied to a stable laminar flames a special optical equipment such as a photon counter is necessary, and the time to collect photon until the certain level reliable to decide the gas temperature must be consumed. This time depends on the intensity levels of the scattered light, but around ten minutes are necessary for one wave length in case of non–luminous gas flame. If the flame is luminous the light intensity at the measuring wave length from the solid radiation by soot becomes higher than Raman scattering intensity level from the gas and this method could not be adopted. Even if the flame is non–luminous the fuel and oxidizer must be perfectly filtered to let them free from solid particles [2].

On the other hand CARS does not have this kind of trouble, but usually a pulse laser system is used as a light source and the repetition frequency produces another problem to get the time–resolved set of data.

From the above considerations, Rayleigh scattering becomes the most useful method to measure the gas density or temperature in turbulent flames up to about 10 kHz [5]. Complete filtering the particulates out of fuel and oxidizer flow is absolutely necessary in this case. However, the simultaneous measurements of temperature and velocity in turbulent flame are required for the determination of the local turbulent heat transfer in the turbulent diffusion flames. This simultaneous measurement includes a contradictory requests as mentioned above. The details will be described in Section 1.5.

1.1.4 Two–dimensional or sheet–cut imaging of turbulent diffusion flames

As mentioned above the two–dimensional or sheet–cut imaging of turbulent diffusion flames can be said as a new tool to understand more detailed structure than a single point or a simultaneous multi–point measurement. The latter gives more detailed information, but the obtained data are not continuous in space and the information between the measuring points are vacant. In order to solve this problem, the two dimensional continuous information obtained by introducing the laser sheet into the measuring section becomes popular for these few years. Computer imaging technology accelerates this trend and moreover the rapid development of optical image intensifier mountable to a high speed CCD TV camera strongly supports the connection of these powerful instruments. This set of instruments gives various possible applications.

Rayleigh scattering can be as one of the useful method among them to obtain the temperature mapping. The high intensity continuous laser must be used as a light source to obtain the time–resolved long term records. The pulse laser can be a high intensity light source, but because of the limitation of repetition frequency it is difficult to obtain the long term time sequential records. However, it can be a powerful source for the determination of mapping the gas movements using Mie scattering by double exposure technique by the adjacent couple of pulses.

From this conditions, a high power gas laser such as an Ar-ion laser becomes the popular light source among the present lasers to get the high time resolution and long time sequential records. If we could not use a high output power Ar-ion laser as expanding to a sheet light source, scanning the laser light beam quicker than the characteristic time of turbulence and the exposure time for one frame of movie camera or one field of TV camera we could take the two-dimensional Rayleigh scattering photos by the high intensity light source. But, the attention must be paid for these pictures that their images are not taken simultaneously even within one frame of picture. This is analogous to a still picture taken by a focal plane shutter, not by a lens mounted shutter camera.

If the fine particles are uniformly introduced in the flow and taking the Mie scattering picture, we can map the temperature distribution from the intensity irregularity as it corresponds to the gas expansion or shrink by the temperature change. This method can give us the detailed information on turbulent flame structures, but when the steep temperature gradient exists the introduced fine particle is pushed to the colder region because the momentum exchange given by the collision of gas molecules to the particle at a hotter side is larger than that at a colder one. This causes the phenomena that the finer particles could not enter into hotter regions if it is not moved by the big inertia force such as the turbulent transportation. This problem will be fully discussed in Section 1.4.

1.1.5 Turbulent diffusion flame structure and how to attain high combustion efficiency and control pollutants formation

Turbulent diffusion flames are the main combustion situations formed in almost all the practical combustion instrumentations such as internal combustion engines, boilers, incinerators and various furnaces. In all these instrumentations the methods to attain the high combustion efficiency and low pollutants formations are the main concern for both manufacturers and fundamental combustion researchers. In order to realize these two big requests the study on turbulent diffusion flame structure and detailed combustion reactions occurring there becomes very important. The pollutants formed from combustion are summarized in Table 1.2.

Table 1.2 Environmental pollutants generated from the combustion

Phase	Species	Generally Called
Gas	NO, NO_2, N_2O, SO_2, SO_3 CH_4, $C_nH_mO_j$, PAH CO_2 Others	NOx SOx Hydrocarbon
Liquid	High boiling point hydro-carbon droplet, Small water droplet containing poisonous chemicals, Others	Mist, Aerosol, Submicron particulate
Solid	Soot, Solid particulate, Metal oxide, Others	Small particulate, Aerosol, Submicron particulate

Among these pollutants NOx and SOx have been paid the biggest attention to protect the acid rain. Recently N_2O becomes next important earth warming chemical species to CO_2, hydrocarbon and other materials. The formation and destruction of NOx including N_2O as well as NO and NO_2 are the functions of temperature, mixture ratio of fuel and oxygen in case of premixed situation, the diffusion flux of fuel and oxygen into the combustion zone in case of diffusion combustion conditions. These factors are tightly related to the attainment of high efficiency of combustion. However, the suppression of formation of NOx and the realization of high combustion efficiency require the contradictory reaction conditions. The combustion engineers have fought with this problems for these two decades and have developed the suitable technology in some fields. But still remain many unsolved problems, especially in the small scale combustors and high combustion intensity conditions.

From the standpoint of suppression of NOx production, the lower temperature, the richer or leaner mixture, the slower combustion are the more desirable, but for the attainment of higher combustion efficiency all of these conditions are contradictory. For example, in an electric power plant boiler they introduce the combinations of low NOx burners and multistage combustion technology. The basic ideas applied by almost of all the boilers are same as mentioned above. Even in this kind of rather easier combustor the detailed combustion mechanisms are still under the cover, because the size of the combustion zone are so large as impossible to measure the zone directly, and almost of all the explanation could be said as done by expectations. They install large scale burners to the boilers and the Reynolds number based on the burner size becomes in the order of 10^4 or sometimes larger than this. However, large Reynolds number does not always produce a highly turbulent mixing and a better combustion conditions. The combustion reaction occurs at the place where fuel and oxidizer are mixed to form the combustible mixture or the fluxes of these materials to the combustion zones are in appropriate stoichiometric ratio for combustion. The combustion intensity can be decided by the fact how densely and quickly these phenomena occur per unit volume and time in the combustor. They are decided by the order of turbulent scales and intensity at the considering points. Especially from the standpoint of combustion reactions in the turbulent diffusion flames, the microscale turbulence and its decay time has the more important role to the local combustion phenomena. If we could follow the series of instantaneous phenomena of these factors, we would refine more the present combustion technology.

In the modern combustion technology they adopt the complex mixing enhancement by swirling and cross flow, and sometimes the traditional bluff body and step. These technologies are adopted at the primary burner port and/or at the secondary air entrance.

In this Chapter we will describe the structures of various turbulent flames from the fundamental viewpoints to clarify the complicated but very important phenomena to attain the high efficiency of combustion and control the pollutants formation.

(Kazutomo Ohtake)

1.2 Optical Measurement of Flame Structure Analysis

1.2.1 Introduction

As have been mentioned in Section 1.1 the detailed understanding of turbulent flame structure is very important to attain the high combustion efficiency and control the pollutants formation in the combustion systems. Turbulent diffusion flames are roughly categorized into two groups. One is the flame in which the combustion rate per unit volume and unit time is controlled by the aerodynamic turbulent mixing mechanisms because the Reynolds number based on the burner diameter and fuel flow velocity is rather low, and the other is the one in which the burner Reynolds number is high enough in the order of 10^4, and the fuel and oxidizer are well mixed just after the burner exit and the flame behaves like a turbulent premixed flame, so that this kind of flame is categorized as a non–premixed turbulent flame. The latter flames are extensively studied recently by many researchers, typically by Dibble at al [6]. On the contrary to this kind of flames, the flames of former category have been studied for long time [3–5,7,8]. Some used the touching sensor like a fine thermocouple and an ion probe. Laser diagnostics mentioned in Section 1.1 have been used for these two decades and many measuring methods based on the new ideas have been presented. However, almost of all the studies have concerned to the discussions on the local turbulent signals, structures and/or the images, or on the whole flame behaviors. Few studies have correlated these signals with the pictures taken simultaneously.

In this Section we would like to discuss on the turbulent flame structures at the moderate burner Reynolds numbers at around 5×10^3 which are controlled by the aerodynamic and combustion generated turbulence. A Laser Rayleigh scattering (LRS), an LDV, and a high speed video camera are used to measure the flame structures and for the further understanding of controlling factors of turbulent diffusion flames.

1.2.2 Experimental Apparatus and Procedures

The layout of the adopted apparatus is shown in Fig. 1.4 Two Ar–ion lasers were used as the light sources. The gas temperature was measured by the LRS at the wave length of 488 nm injected from one of the lasers. This system allows to measure the temperature fluctuation up to 10 kHz. The other laser which injected two wave lengths of 488 and 514.5 nm simultaneously was used to measure two components of gas velocity by the two–colored LDV system. These systems gave us the information of turbulent heat fluxes of $\overline{u'T'}$ and $\overline{v'T'}$ at the same point.

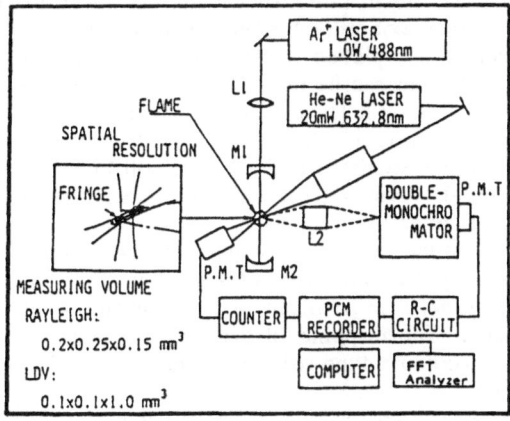

Fig. 1.4   Optical layout for Rayleigh scattering and LDV system

Two high speed video camera (nac 400) was synchronized to take the schlieren and laser sheet Mie scattering pictures at the same time. These pictures were taken simultaneous with the above mentioned temperature and velocity data, and we could correlate the measured signals and the pictures, and understand what were occurring in the studied turbulent diffusion flames.

In order to obtain the macroscale of turbulence two point time–resolved LRS measurement was carried out separate from the above experiments. The distance between these two points onto which two laser beams of different wave lengths of 476.5 and 488 nm were focused, respectively, to collect the LRS signals to determine the gas temperature at each point was varied to get the spatial correlation. All the said data were recorded by PCM recorders and processed by computers and an FFT analyzer.

Furthermore we adopted the connection of a closeup lens, narrow band filter and an image intensifier which were mounted on the high speed video camera to get the information about the fine turbulent structures of the measured flames. The connection of this system and the computer image processor gave us the microscale spatial distributions.

The mixed gas fuel of 62.2% H_2 + 37.8% CH_4 was adopted because this mixed fuel shows the minimum change of Rayleigh scattering cross section before, during and after the combustion compare to that of air. The maximum deviation was estimated as 5% by the calculation of equilibrium concentration.

1.2.3 Results and Discussions

(a) Time scale and normalized rms of temperature fluctuation

In order to study what phenomena could determine the turbulent diffusion flame structures, we mixed nitrogen into the fuel to vary the kinematic viscosity of fuel, which could change the burner Reynolds number with keeping the fuel velocity at the burner exit constant and change the burner exit velocity with keeping the burner Reynolds number constant. By this method we could change the burner Reynolds number from 5,000 to 7,000 and burner exit velocity from 64.7 to 92.3 m/s for the burners of inner diameters of 2 and 4 mm. Turbulent time scale was obtained by the integration of auto–correlation of temperature fluctuation. Thus obtained result for constant Reynolds number for different burner exit velocities is shown in Fig. 1.5 The normalized rms temperature fluctuation is also plotted in the same figure. This result shows that the turbulent diffusion flame zone can be divided into four distinguishing regions by the following characteristics. Region I: The increasing rate of time scale and normalized rms of temperature fluctuation vs. radial distance is slow and almost constant. Region II: Both values steeply increase. Region III: Time scale still increases gradually while the normalized rms suddenly decreases after reaching the maximum value. Region IV: Time scale suddenly drops off after reaching the maximum value and the normalized rms reaches almost zero. All of the tested flames show that the time scales except in Region I decreases with increasing the burner Reynolds number. This tendency coincides well with the fact that the wrinkles in schlieren pictures become finer with increasing the burner Reynolds number. The mapping of these four regions for the whole flame is shown in Fig. 1.6 The maximum time mean temperature appears in Region II near the boundary of Region III.

From these results we can characterize the four flame regions as follows. Region I is the fuel flow region into which the fine islands of high temperature burnt gas and the broken up tiny air islands around which the diffusion flame is formed are entrained and

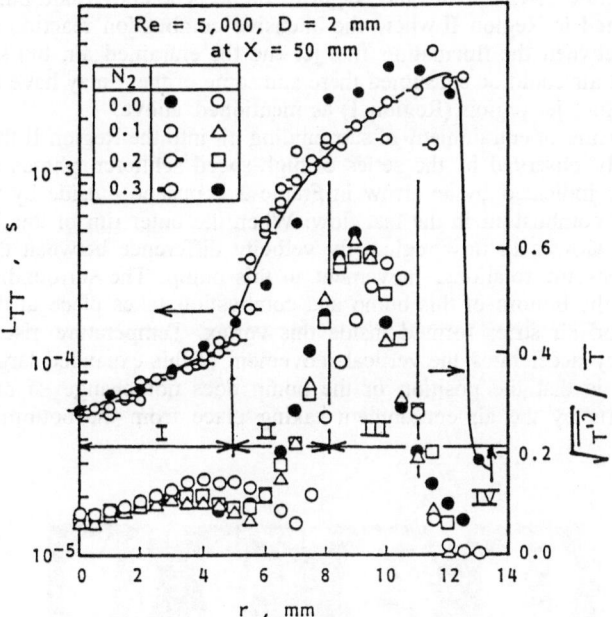

Fig. 1.5 Radial distribution of time scale

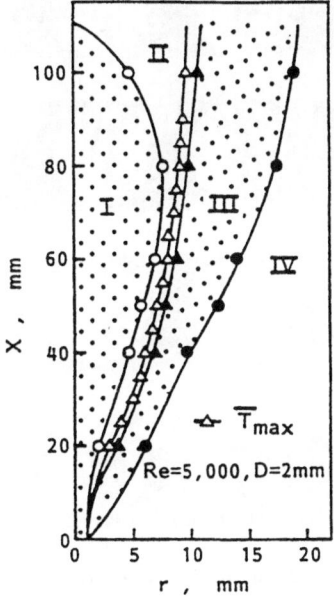

Fig. 1.6 Mapping of characteristic regions

the fuel temperature rises up with the height of flame. The above said burnt gas and air islands are formed in Region II where the intensive combustion reaction takes place at the boundary between the fluctuating fuel jet and the entrained air, but sometimes not all the entrained air could be consumed there and some of them may have the possibility to get into the fuel jet region (Region I) as mentioned above.

The mechanisms of entrainment of surrounding air into the Region II through Region III can be clearly observed by the series of high speed schlieren photos shown in Fig. 1.7 [9]. A bump indicated by an arrow in Frame 2 is pushed outside by the expansion caused by local combustion in the fast flow. When the outer rim of this bump reaches the surrounding slower air flow region, the velocity difference between the central and rim part produces the rotational movement to this bump. The surrounding air is thus entrained from the bottom of this bump and combustion takes place at the boundaries between fuel and air strips formed inside this vortex. Temperature rises up and thus formed buoyancy accelerates the vertical movement of this expanded large vortex. The interesting fact is that the position of the bump does not change so much until the combustion starts by the air entrainment taking place from the bottom of the bump (Frame 11).

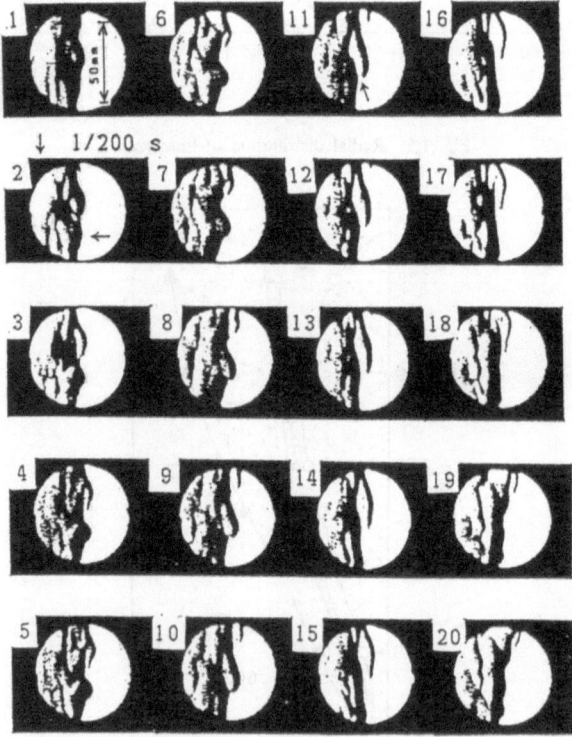

Fig. 1.7 Time–resolved schlieren TV photos. Exposure time:1/10000 sec.
Frame speed:1/200 sec.

As mentioned before, schlieren method gives us the integrated optical information along the light path and does not correspond to the local information. In order to certify the above considerations the corresponded schlieren and laser sheet Mie scattering photos simultaneously taken by the synchronized high speed video cameras are compared in Fig. 1.8 [9]. Both photos are computer image processed ones to make them clearer for the comparison. Not only the vertical sheet but also the horizontal one were adopted to take the cross-sectional cut photos ((c) in Fig. 1.8). The problem occurring in the diffusion flame when the laser sheet Mie scattering method are applied to it will be discussed in Section 1.3, but both photos taken by the different two methods are similar especially the shape of the peripheries between the hot and cold air region. We, however, do not know yet the holes existing in a large vortex appearing in the Mie scattering photo (b) are the islands of the entrained air or the cross-sections of the bending vortex tube of air.

The speed of changing the images of horizontal cut is much higher than that of vertical one. We can follow the change of images from frame to frame for the vertical cut, but it is difficult for the horizontal one. This does not say the fact that the horizontal movement of fuel flow is faster than that of the vertical one, but that the complex construction of fuel jet flow moves vertically in high speed and we could not connect the images in the successive frames like the vertical photos.

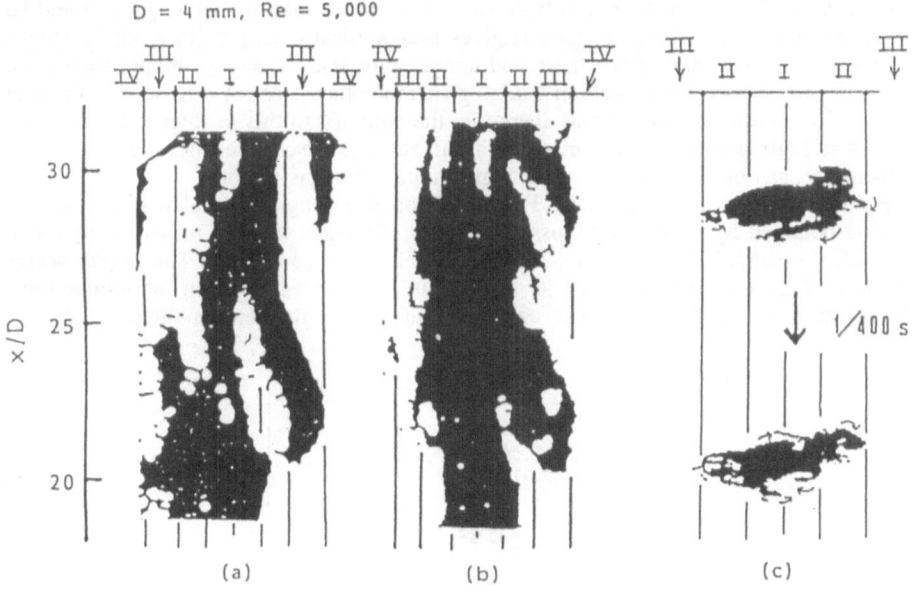

Fig. 1.8 Image processed still photos taken by the different methods at the same instance. (a)Schlieren, (b)Laser sheet Mie scattering fuel flow, vertical cut, (c)Horizontal cut of the same flame.

These phenomena are caused by the complicated turbulent mixing and chemical reactions, and we measured the scales of turbulence across the flame region. Firstly the distributions of spatial macroscale of temperature fluctuation for different burner Reynolds numbers determined by the two point cross correlation of gas temperature measured by the two point Rayleigh scattering method with changing the distance between the measuring points. The typical results are shown in Fig. 1.9 [10]. The macro scale of temperature at higher Reynolds number shows a little smaller. The macroscale in Region I shows a slight increase with the radial distance. In Region II it increases gradually, and in Region III after showing the maximum value it decreases steeply. The macroscale in Region IV decreases down to almost zero.

On the other hand to this macroscale we tried to measure the microscopic temperature distribution within the measuring area of single measuring point for Rayleigh scattering method. The incident light was sheeted under 0.1 mm thick. The time–resolved video images were converted into RGB planes of 511 x 511 pixels, in which one pixel corresponds to 1.79 x 1.79 μm², The LRS signal stored within the region 125 x 250 pixels in a G–image plain was converted into the temperature signal. This area corresponds to 0.22 x 0.44 mm² in the real dimension. The inhomogeneity of incident laser intensity was corrected before the conversion. The results are shown in Fig. 1.10 for Regions I and III. The temperature in Region II was almost flat and higher than 1500 K, and the observable distribution of temperature could not be found out. The cluster analysis was applied to the converted pictures. The spatial filters larger than 7 x 7 pixels in which the same weight was applied to every pixel within the area gave almost the same results. The area in which the temperature is higher than 1500 K was colored dark. The typical time–resolved clear images obtained for the Regions I and III are shown in Fig. 1.11 [11]. Region II gives monotonically dark picture, and it shows that the combustion takes place there and temperature there is always higher than 1500 K. The figure shows that in Region I the high temperature clusters diffuse into the fuel flow to rise up the gas temperature there. On the contrary to this in Region III the finer clusters of high temperature gas disperse in the air or reversely the finer clusters of low temperature gas or air disperse in the high temperature gas regions.

As the reference of cluster scale, Taylor's dissipation length was calculated based on the data obtained by a two point cross correlation of the gas velocity measured by a two point LDV system. This dissipation length was calculated as 200 μm. The cluster scales shown in Fig. 1.11 ranges from 40 to 200 μm, and the dimension of temperature inhomogeneities can be considered as lower than that of momentum ones.

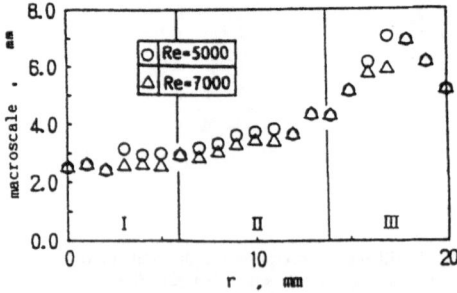

Fig. 1.9   Radial distribution of macroscale of temperature fluctuation

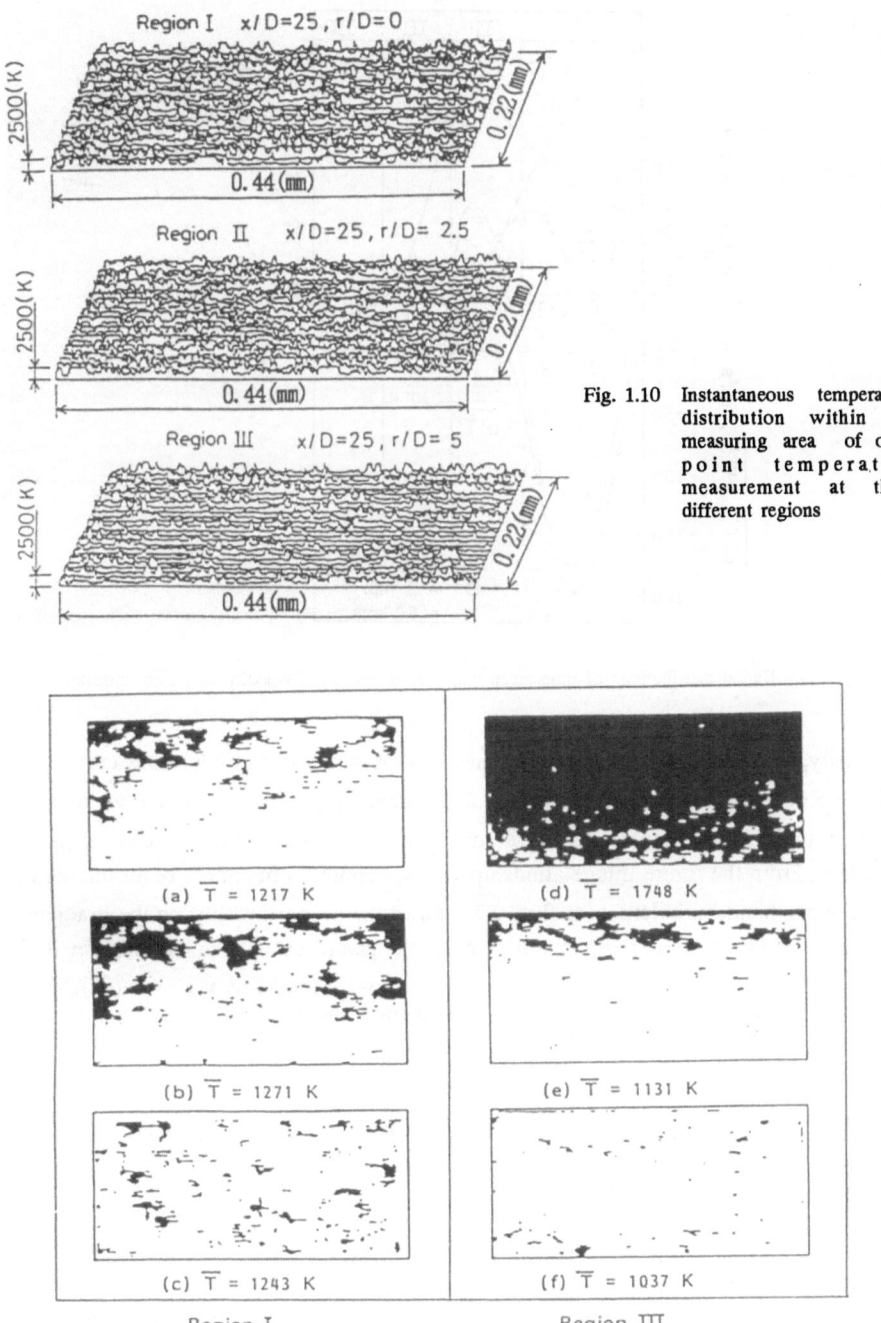

Fig. 1.10 Instantaneous temperature distribution within the measuring area of one-point temperature measurement at three different regions

Fig. 1.11 Typical cluster analysis within the same region in Fig. 1.12, T means the spatial mean temperature

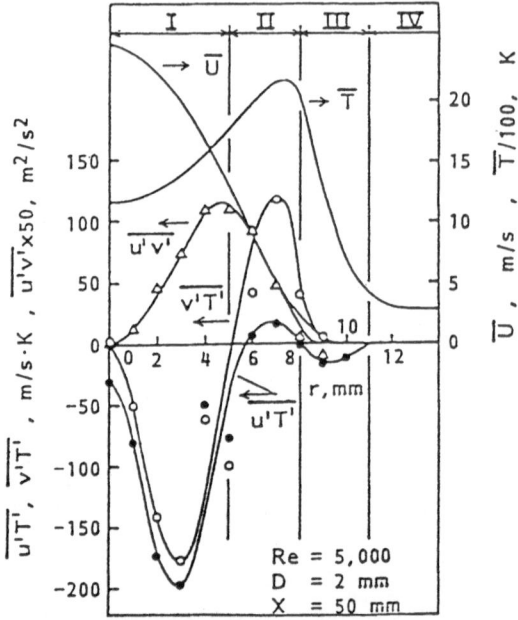

Fig. 1.12   Radial distributions of time mean temperature and axial velocity and their correlations

Finally, we would like to show the radial distributions of various cross correlations at Re = 5,000 and x/D = 25 in Fig. 1.12 The turbulent Reynolds stress $\overline{u'v'}$ is usually assumed to be proportionate to the gradient of time mean axial velocity as $\overline{u'v'} \propto -(\partial \overline{U}/\partial r)$. From the figure this relationship can be supported in almost of all the regions. On the other hand, turbulent heat flux $\overline{v'T'}$ shows the proportionality to the gradient of time mean temperature $(\partial \overline{T}/\partial r)$ at almost of all the places except the combustion region (Region II). This means that the simple turbulent closure model as $\overline{v'T'} \propto -(\partial \overline{T}/\partial r)$ may not be applied at the combustion zone especially near the place at maximum time mean temperature.   (Kazutomo Ohtake)

1.3 Effect of Surrounding Gas Motion on Turbulent Diffusion Flames

1.3.1 Introduction

A turbulent jet diffusion flame penetrating and spreading transversely into a uniform stream of fluid is of great relevance to a wide variety of engineering applications including the industrial furnaces and the internal combustion engines. To understand these kinds of combustion conditions including the flame structures would lead to the improvement of thermal efficiency and the reduction of pollutants emissions from these industrial facilities.

Although several experimental work[12–15] have been carried out on this kind of subject, not so much is known yet about the internal flame structures which are inherently more complicated than such cases as the jet flames flowing into the stagnant or co–axially flowing air regimes. Few attempts have been performed for the simultaneous measurements of velocity, temperature and concentration profiles in these kinds of flames. The detailed data have been needed for verifying the validity of the predicted results on the flame structures [13,16–19].

The primary objective of this section is to obtain and discuss the detailed structure of a steady turbulent jet diffusion flame penetrating into the free air stream at uniform velocity normal to the fuel jet. The velocity ratio of fuel jet to the cross air flow affects the flame structures so much. Fine compensated fine thermocouple was used to measure the gas temperature and LDV was used to obtain the gas velocity. The concentration profiles of major species in the flame was obtained by a gas chromatograph. The flame configurations were determined by taking the photos. The measured results were compared with the results obtained by three dimensional numerical analysis.

1.3.2 Experimental apparatus and procedures

The schematic diagram of the adopted apparatus is shown in Fig. 1.13 A steady horizontal free stream of air with a uniform velocity profile was established in the test section behind the outlet of an open circuit wind tunnel. Dimensions of the outlet of the wind tunnel are 200 mm wide and 640 mm high, which are proved as the enough dimensions for this kind of experiment by the study of a drop–laden air jet structure [20]. A horizontal flat plate with a sharp leading edge was installed 50 mm above the bottom wall of the tunnel, and a fuel nozzle was placed 100 mm downstream the leading edge. The fuel nozzle consisted of coaxial stainless steel pipes. Propane was fed vertically upward normal to the uniform stream of cross air flow from the inside pipe

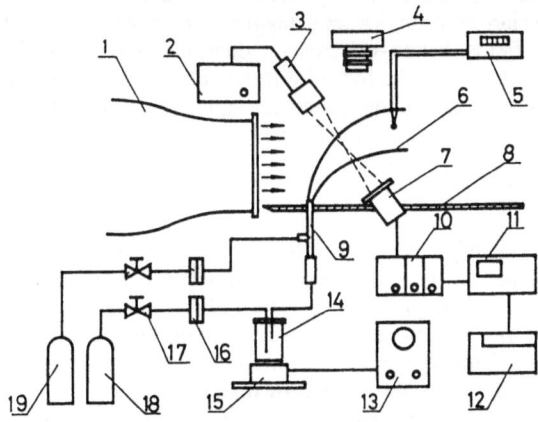

1.wind tunnel 2.power supply 3.He–Ne laser 4.camera 5.thermometer 6.flame t.photomultiplier 8.flat plate 9.fuel nozzle 10. frequency tracker 11.oscilloscope 12.recorder 13.speed controller 14.particle seeder 15.vibrator 16.flowmeter 17.control valve 18.propane 19.hydrogen

Fig. 1.13 Experimental apparatus

of inner diameter of 2 mm after controlled and measured its flow rate and seeded with talc fine particles (nominal diameter of 2 μm). The particle seeder was vibrated by an adjustable speed motor to keep the constant and stable seeding.

From the annular space between the coaxial double tube of the fuel nozzle, hydrogen was fed at the low velocity to stabilize the high velocity propane diffusion flame. Surplus unburned hydrogen was carried away by the cross air flow.

A forward scattering LDV (DISA LDA03 system) was used to measure the profile of gas velocity in the flame. A 10 mW He–Ne laser was used as a light source. The beam intersecting angle was set at 11.42 deg. and the probed volume was estimated as 0.31 mm in diameter and 3.2 mm in length with a fringe spacing of 3.2 μm. The axis of LDV optical system was set in a horizontal plane and perpendicular to the direction of cross air flow. Velocity components were measure by rotating the beam plane. The output electric signal from a photomultiplier (RCA 4526) was processed with a Doppler frequency tracker (DISA 55N20). Flame temperature was measured by a silica–coated Pt–Pt/Rh 13% thermo–couple of 0.1 mm in diameter. A 35 mm film camera was also used to take the flame photos to determine the flame shapes.

The flame gas was sampled with a water–cooled probe of the tip inner diameter of 2 mm. Condensed water along the sample gas line was eliminated by a water trap and the sampled gas was analyzed by a gas chromatograph. Molecular Sieve 13X–S or Gaskuropack 54 was packed and a TCD was used to analyze C_3H_8, CO, O_2 and CO_2. The following set of experimental conditions were chosen as: propane jet velocity $U_0 =$ 10 – 40 m/s, velocity of cross air flow $U_\infty = 1.1 – 15.3$ m/s.

1.3.3 Numerical analysis

The mathematical model employed was a $k - \varepsilon$ two equation turbulent model[16] to calculate the distribution of Reynolds stress, coupled with a combustion model based on the approach of Magnussen where the local rate of combustion is assumed to be controlled by the rate of combustible gaseous species and air. Solution was obtained applying a finite–difference procedure applicable to three–dimensional elliptic equation. The grid was in Cartesian coordinates and staggered to prevent uncoupling between the pressure and velocity fields. Finite–difference equations were derived using the control volume technique together with the hybrid differencing of connective terms and central differencing for other terms in the modeled equations. The obtained quasi–linear algebraic equations were solved by a line Gauss–Seidel method in an iterative procedure.

1.3.4 Results and discussions

Figure 1.14 shows the sketch of flame and the coordinate system as has been used most commonly in the studies on the isothermal jet in a cross flow.

A system of axes, which expresses the nature of the deflected flame, is determined from the experimental measurements and pertinent flame quantities are described along these principal axes. X, Y and Z axis are defined as the longitudinal, lateral and vertical Cartesian coordinates relative to the cross flow direction setting their origin at the center of the fuel nozzle. The ξ–axis, called as a jet axis herein, is defined by the locus of a maximum velocity in the flame. The ζ–axis is in the transverse cross section of the flame normal to the ξ–axis intersected by the plane of symmetry in the $X–Z$ plane. The η–axis is perpendicular to the ξ– and ζ– axes.

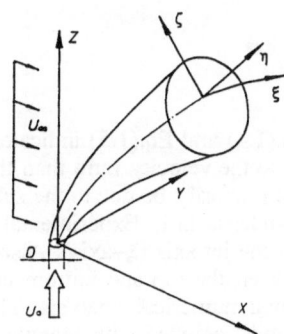

Fig. 1.14   Coordinate system

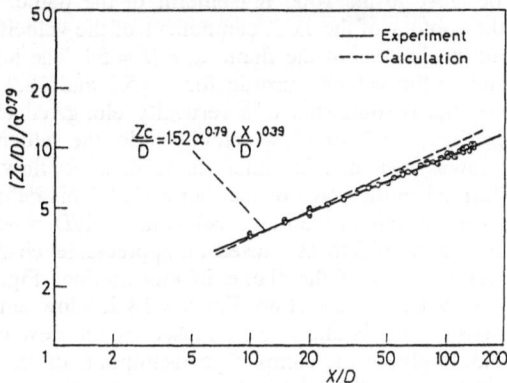

Fig. 1.15   Location of the central axis of the flame, solid line: locus of a middle point between the top and bottom edge of the visible flame determined from photographic observation, dashed line: locus of a middle point between the top and bottom edges of the 1000 K isotherm predicted, U_0=10 – 40 m/s, U_∞=1.1 – 15.3 m/s

Figure 1.15 shows the location of the central axis of flame in the X–Z plane. The dots represent the middle point between the top and bottom edges of the visible flame determined from the photographic observation. The least square curve fitting to these data points results the solid line which is expressed as;

$$\frac{Z_c}{D} = 1.52 \; \alpha^{0.79} \; (\frac{X}{D})^{0.39} \tag{1.4}$$

where Z_c is the Z coordinate of the central axis and D is the inner diameter of the fuel nozzle.

The α represents the ratio of the fuel jet velocity at the nozzle exit to the cross flow velocity, called simply as the velocity ratio hereafter. The location of the central axis of an isothermal air jet in a cross flow is given as ;

$$\frac{Z_c}{D} = 2.05 \ \alpha^{0.72} \ (\frac{X}{D})^{0.28}$$ (1.5)

The comparison between Eq.(1.4) and Eq.(1.5) indicates that the jet diffusion flame in a cross flow is more sensitive to the velocity ratio than the isothermal air jet. The less deflection of the jet flame would primarily be due to the substantial increase in the axial velocity resultant from the heat release in it. Experimental evidence also indicates that the central axis is located below the jet axis (ξ–axis). Also shown with a broken line is the locus of a middle point between the top and bottom edges of the 1000 K isotherm predicted by the three dimensional numerical analysis. The turbulent combustion was modeled via the $k-\varepsilon$ turbulent model coupled with Magnussen's turbulent rate of mixing approach. The measured results are in agreement with the predicted results. The above comparison might be rather qualitative, but significant since the edge of visible flame region was found to be close to the 1000 K isotherm of the flame.

Figure 1.16 gives the contour of the axial component of the velocity in the transverse cross section normal to the jet axis of the flame at $\xi/D = 54$. The left and right halves of the figure correspond to the velocity profile for α =5.3 and 18.2, respectively. The velocity profile for α = 18.2 is somewhat of a vertically elongated elliptic shape while it is of the kidney shape for α = 5.3. At a low velocity ratio, the velocity in the periphery of the flame is much lower than that in either flame or cross flow region. A pair of attached vortices are formed in the wake of the flame [18]. This deformed the flame in the kidney shape. The central axis of the flame is located at $\zeta/D \approx -5$. The variation of velocity ratio within the range of 5 to 19 caused no appreciable change in the distance between the jet and central axis of the flame in this section. Figure 1.17 gives the isotherm in the same transverse cross section. For α = 18.2, a low temperature region is observed near the jet axis, which is likely to be caused by the flow of the unburnt fuel. It disappears for α = 5.3, implying the complete consumption of the fuel. The isotherm is in kidney shape again for α = 5.3, while it does not shape for α = 18.2 . The high

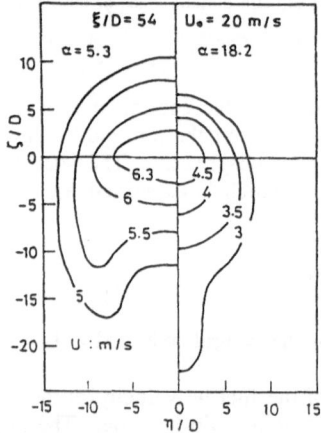

Fig. 1.16   Contour of constant axial component of velocity in the transverse cross section of the flame at
ξ/D=54,left:α=5.3, right:α=18.2

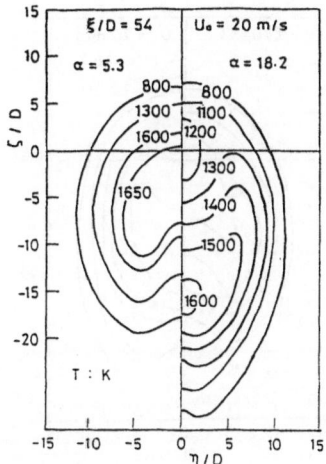

Fig. 1.17 Isotherm in the transverse cross section of the flame at ξ/D=54, left:α=5.3, right:α=18.2

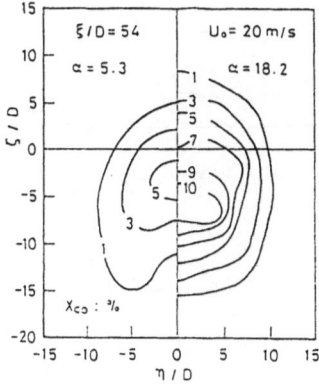

Fig. 1.18 Contours of constant concentration of carbon monoxide in the transverse cross section of the flame at ξ/D = 54, left:α=5.3, right:α=18.2

temperature region is located below the jet axis. The formation of attached vortices is likely to result in the well–mixed region of the kidney shape in the wake of the flame where the chemical reaction proceeds very actively and consequently the flame temperature is rather high. It is also evident that the decrease in a velocity ratio makes the high temperature region move up to approach the axis of jet flame.

Figures 1.17, 1.18, 1.19 and 1.20 present the contours of isotherm, equi–concentration of CO, O_2 and CO_2 in terms of mole fraction in the transverse cross section at ξ/D = 54. The CO concentration profile is nearly of concentric circle centered at the central axis of the flame for α = 18.2. It becomes kidney shape and high concentration region moves up with decreasing the velocity ratio. The O_2 concentration profile is also of kidney

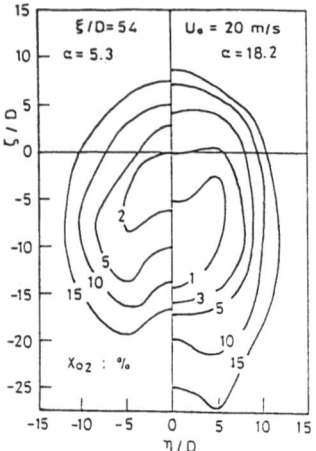

Fig. 1.19   Contours of constant concentration of oxygen in the transverse cross section of the flame at
ξ/D = 54, left:α=5.3, right:α=18.2

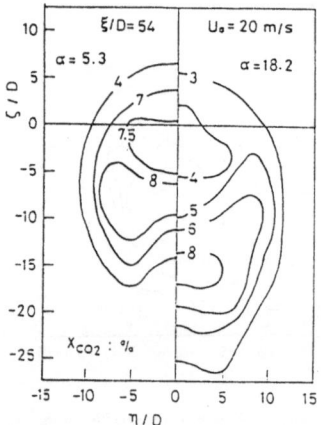

Fig. 1.20   Contours of constant concentration of carbon dioxide in the transverse cross section of the flame
at ξ/D = 54, left:α=5.3, right:α=18.2

shape with a minimum between the jet and central axes of the flame for $\alpha = 5.3$, while
it has a minimum below the central axis of the flame for $\alpha =18.2$. The CO_2 concentration
profile shows a peak below the central axis and a valley between the both axes.

Figure 1.21 indicates that the isotherm predicted is of kidney shape which bears
resemblance to the empirical results. Fig. 1.22 shows the flow field predicted. it is
evident that a pair of attached vortices are formed. This is likely to results in the well-
mixed region of the kidney shape in the make of the flame where chemical reaction
proceeds very actively and consequently the flame temperature is rather high.

                                                                    (Toshikazu Kadota)

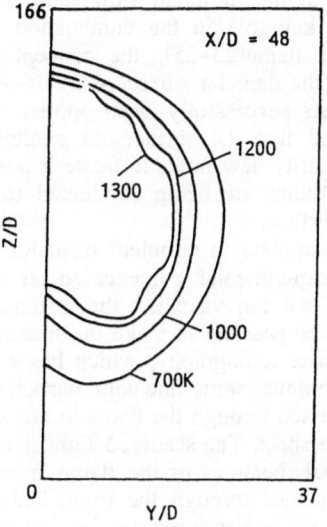

Fig. 1.21 Isotherm (Calculation)

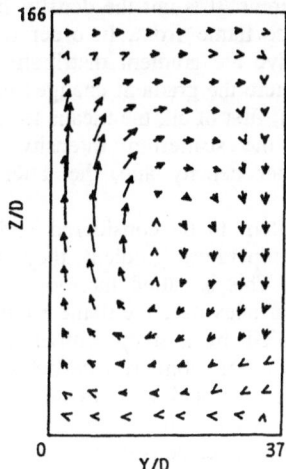

Fig. 1.22 Velocity profile (Calculated)

1.4 Visualization of Turbulent Diffusion Flame Surface

1.4.1 Introduction

The laminar flamelet concept is becoming very popular as a mean of describing turbulent combustion in the regime where chemical time is short compared to turbulent time [21,22]. In this regime, chemical reactions take place within asymptotically thin layers, called flamelets, which are embedded in the turbulent flow, and the instantaneous flamelet surface configuration plays a key role in the combustion process. in our previous studies on turbulent premixed flame[23–25], the concept of fractals was introduced to examine experimentally if the flamelet surface is a self–similar fractal or not. The laser tomography technique has successfully been applied to visualize the flamelet surface, and it has been found that the surface do exhibit a fractal–like character, in the sense that the self–similarity law holds to make it possible to specify a value of fractal dimension. Further studies are being conducted to study dynamic behavior of turbulent premixed flame surface.

The laminar flamelet concept can be applied to turbulent diffusion flames as well. However, there have been reported no experimental evidence so far that the flamelet surface exhibit the fractal character. If we can visualize the instantaneous flamelet surface by some simple means, it would be possible to make the fractal analysis of the surface. One possible technique is the laser tomography, which has successfully been applied to the premixed flame. In the technique, some fine solid particles are introduced into a mixture, and a thin laser sheet is passed through the flame to illustrate the seeding particles present in the section cut by the sheet. The scattered light is collected at right angles to the sheet, and two–dimensional imaging of the flame front in the section becomes possible, if the gas density changes through the front. This is because the scattering intensity is proportional to the particle number density and the latter is again proportional to the gas density. In the premixed flame, the gas density decreases very rapidly through the front due to the rapid temperature increase. In the diffusion flame, however, the situation is somewhat different. It is not the density itself but the density gradient that changes very rapidly at the flame front. In order to identify the flame surface, in the beginning we have to drive the gradient distribution of Mie scattering intensity, and then to find the location where the gradient changes very rapidly. In order to make this possible, we have to develop, first of all, the means to measure the accurate scattering intensity distribution. And the scattering intensity should exactly be proportional to the local particle number density and, the latter should be exactly proportional to the local gas density.

In the connection, there is one problem to be considered seriously. When small particles are suspended in a gas with a temperature gradient, they experience a force in the direction down to the gradient [26]. This is called thermophoresis, and there is a possibility that the behavior of seeding particles near the flame front may be affected by this effect, since the front is accompanied by a steep temperature gradient. In our previous paper the thermophoretic behavior of submicron particles in premixed flames was studied theoretically. A simplified inert model was developed and the model was applied to the plane one–dimensional premixed flame of methane–air mixture to study the thermophoretic effects on the particle number density distribution in the flame zone. The derived position of tomography image was found to be affected by the thermophoretic effects to shift it downstream approaching the luminous reaction zone. However, the shift was too small to distinguish clearly the thermophoretic effects. The objective of the present study is to study experimentally and theoretically the thermophoretic behavior of submicron particles in the diffusion flame to asses if we can apply the laser tomography technique to the visualization of diffusion flames. In the study the laminar counterflow diffusion flame was adopted since the flame structure is eventually one–dimensional, which makes it very easy to make the detailed comparison of theoretical prediction with experimental observation. In addition, the temperature distribution has a sharp peak and we can control this location relative to the stagnation plane so that we have the pronounced effects of thermophoresis.

1.4.2 Experimental apparatus and procedures

Figure 1.23 shows the cross section of the axisymmetric vertical burner used in this experiment. The burner has two identical, upper and lower, nozzles to eject oxidizer and fuel, respectively, and the diffusion flame is established in the vicinity of the stagnation plane of these opposing flows. The nozzles have porous plates of 58 mm diameter at the exit to make the flow uniform, and the distance between the upper and lower nozzle was 15 mm. The fuel used was methane and the oxidizer was air. Methane was diluted by N_2 (CH_4 25% and N_2 75% by volume), while the oxidizer was O_2 diluted by N_2 (O_2 50% and N_2 50% by volume).The fuel injection velocity u_F was kept equal to the oxidizer injection velocity u_O. The flat one–dimensional flame was established in the vicinity of the stagnation plane of the opposing two jets.The same laser tomography technique as the previous ones [23–25] was adopted to study the Mie scattering intensity distribution of the flame.

Figures 1.24 and 1.25 show the laser tomography images obtained for the case when the injection velocities were 5.24 cm/s. Figure 1.24 shoes the image when the flow was not ignited, whereas Fig. 1.25 shows that when the flame was established. The streamlines are visualized by the streaks of the particles, size of which seems much

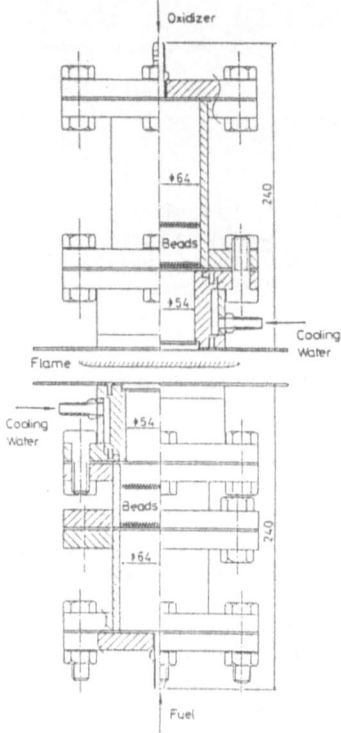

Fig. 1.23   Counter flow diffusion flow burner

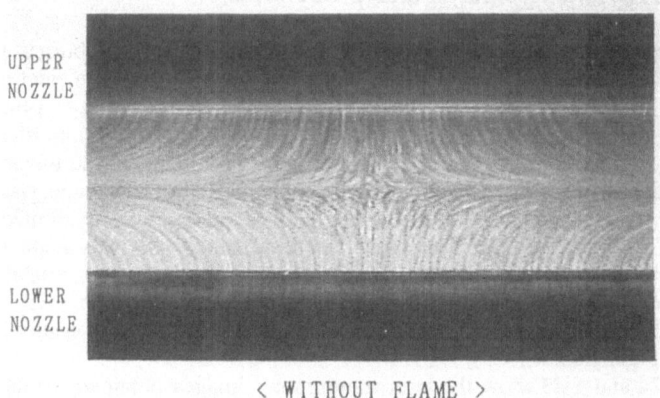

< WITHOUT FLAME >

Fig. 1.24   Laser tomography image for flow without flame

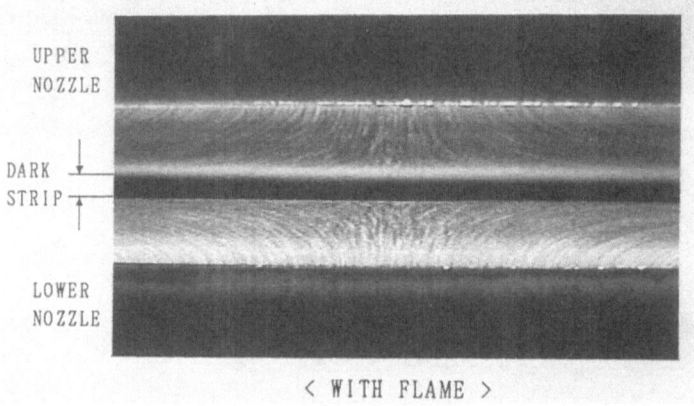

< WITH FLAME >

Fig. 1.25   Laser tomography image for flow with flame

larger than that of the original fine particles. It can be seen clearly that when there is no flame the both particles contained in the oxidizer and fuel flows can reach the stagnation plane. These apparent large particles may be formed by some agglomeration processes when the flows pass through the respective pores of porous plates at the nozzle exits. When the flame was established, There appeared a dark strip in the center parallel to the nozzle exit planes. There are no scattering lights in this strip, suggesting no particles exist in this strip. The particles cannot be convected into this zone because of the opposing velocity induced by the large temperature gradients. This is a direct evidence of the thermophoretic effect, and the similar observations was made in the past [8].

Figure 1.26 shows the intensity distribution of the scattering light along the axis, corresponding to the image shown in Fig. 1.25. The intensity in the ordinate is normalized by the respective values at the both nozzle exits. This kind of distribution was obtained fir 135 horizontal positions in one video frame, parallel to the flame and

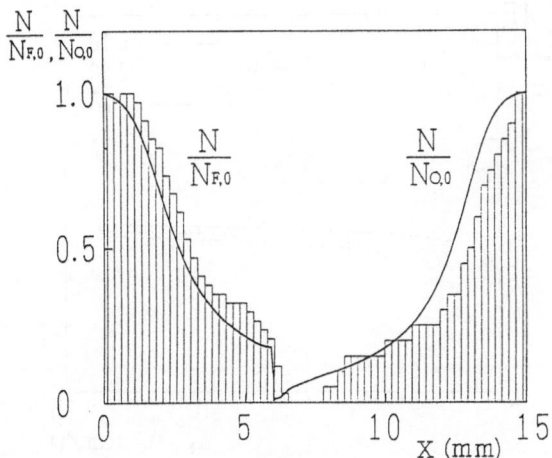

Fig. 1.26   Comparison of measured and calculated particle number density distribution for $\alpha_T L_p$=0.20

the nozzle exit. Then 30 video frames were used to obtain 135 x 30 = 4,050 data and they were averaged to give the distribution shown in the figure. Each distribution was usually accompanied by some sharp indents due to the streaks observed in the original images, which are presumably produced by the coagulated large particle. In the process of averaging, these indents were carefully removed to obtain the smooth distribution. In addition, there existed intensity distribution along the axis due to the nonuniformity of the light source. The distribution shown in Fig. 1.26 was corrected also for this nonuniformity. Now, it may be considered that the distribution represents the number density distribution of the seeding particles. As is seen in the distribution there exists the region around the center where no particles are present. This corresponds to the dark strip observed in Fig.1.25, and is seen in the figure the fuel side edge of the strip is more clearly identified than the oxidizer side edge. Figure 1.27 shows how the both side edges changed with the injection velocities. The solid circles represent the averaged values for the 30 video frames, while the vertical lines indicate the scattered ranges of the data. As the injection velocities are increased, the width of the strip became smaller.

1.4.3 Numerical calculation

In order to study the correlation between the theoretical prediction and the experiment, the numerical calculation was performed for the particle number density distribution in the flame. The theoretical model and the assumptions for the counter flow flame are familiar [28] and hence are not described here. The reaction scheme adopted was so called C_2 chemistry recommended by Kee et al.[28] and involves 29 species and 134 elementary reactions. The necessary thermochemical and transport properties were obtained from CHEMKIN data base. The solution procedure was similar to that developed for the calculation of normal one–dimensional flame. The central diffusion formula was used for the convective terms. The adaptive placement of the mesh points to form the finer meshes done in such a way that the total number of mesh points needed to represent the solution accurately is minimized.

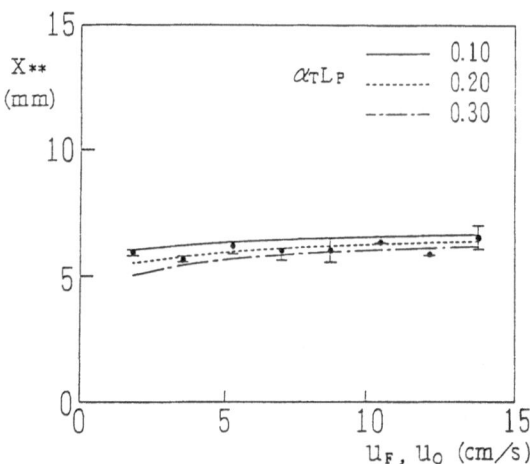

Fig. 1.27   Comparison of measured and calculated change with injection velocities of edges of dark strip for three different values of $\alpha_T L_p$

In the previous experiment [23–25], the number density of seeding particles required to visualize the flame was of the order of 1000 particles per cubic millimeter. The size of each particle was about 1 μm. Then the volume fraction occupied by the particles in the unburned mixture X_p is about 0.5×10^{-6}, while the mass fraction Y_p becomes about 1.4×10^{-3}. The particle is small enough to follow accurately the gaseous flow, and we may think that there is no slip between the gas and particle velocities. The thermal effect is not important either, and we may assume that the particle temperature is always equal to the local gas temperature. The most crucial problem is the chemical effect. Some heterogeneous reactions should proceed on the heated particle surface, and this may affect reactions in the gas stream. In view of our meager knowledge of these surface reaction at present, and of the small volume fraction of the particle, we will neglect this effect. The resulting model is a complete inert model, in which the seeding particles merely move passively through the given flow, temperature and concentration fields of the counter flow diffusion flame. Then the equation governing the particle number density in the flame could be derived very easily. The equation contains only one thermophoretic parameter $\alpha_T L_p$, where α_T and L_p are thermal diffusion factor and particle Lewis number, respectively.

The numerical calculation was performed for the conditions corresponding to the flame shown in Fig. 1.25, and the calculated particle number density distribution for $\alpha_T L_p$ = 0.20 is shown in Fig. 1.26. The solid curves represent the calculated distribution as compared to the experimental distribution. The particle number density is made nondimensional by those at the nozzle exits. As is seen in the figure, the calculation predicts that there exists a point where the number density becomes zero. It is interesting to note that at this point the density decreases to zero almost discontinuously as we approach from the fuel side. On the other hand, the decrease is very gradual from the oxidizer side. The calculation predicts fairly well the general trend of the observation except one thing. It cannot predict the no particle region with a finite width as observed in the experiment. In Fig. 1.27 the predicted position of discontinuous decrease in the fuel side is plotted against the injection velocities as compared to the experimental observation. The results for three values of $\alpha_T L_p$ are compared, and 0.20 seems to give the best fit with the experimental data.

1.4.4 Visualization of turbulent diffusion flame surface

The above results have shown that the visualization of diffusion flames by using Mie scattering is accompanied by the serious difficulty. In the premixed flame the temperature gradient in the flame is in one direction and the effects of thermophoresis is just to shift slightly the particle number density distribution. Then the effect is not serious when we make the fractal analysis of the obtained tomography image. In the diffusion flame, however, the effect is serious since in certain flow conditions, such as encountered in the counterflow diffusion flame, the particles cannot reach the flame front itself. This is because there exist two steep temperature gradients in the opposite directions. In the turbulent diffusion flame structure, such local flow conditions can easily be realized and we should have several locations where the particles are absent in the concentrated reaction zone. These interruptions should be fatal to the fractal analysis, and Mie scattering technique cannot be adopted for that purpose. We should adopt some other means, such as laser induced fluorescence method, which do not use the submicron particles of condensed phases.

Acknowledgement

The authors would like to express their sincere thanks to Dr. Kee RJ of Sandia National Laboratories for his kindness in making CHEMKIN code available, and to Professor Weinberg FJ for his comments. They also would like to express their thanks to Uchida MN and Tanaka H for their help in the course of the present study. (Tadao Takeno)

1.5 Colorimetric Analysis of Turbulent Flames

1.5.1 Introduction

Optical measurement of combustion is one of the most advantageous measurement methods because of its minimal interference with the characteristics to be measured. Optical measurement is divided into two categories: The first method relates light emission from the flame with the characteristics to be measured, while the second uses another one light source such as laser.

The most popular method using light emission is spectroscopic analysis which is used to obtain information on the chemical reactions occurring inside the flame [29]. In the present, there are many researchers focusing on a particular radical emission whose distribution is used to characterize the flame structure and determine the extent of the different reactions involved [30–32]. The ratio of light emission intensities of more than two kinds of radicals are used by several researchers to know the combustion characteristics (e.g. equivalence ratio, exhaust emission and so on)[33,34]. Another method measured the total intensity of light emission to clarify the flame structure [35]. Infrared emission is also used to determine the temperature distribution of a flame [36]. Researches to reconstruct the three–dimensional structure of a flame have been tried by synthesizing the radical emission images taken from more than two or three different directions [37,38].

On the other hand, many laser diagnostics have been developed throughout the world and detailed description of these methods are found in Ref.[1,39].

The present work deals with a new combustion diagnostics method using flame color information of the objective flame. The flame color changes with changes in combustion conditions. This is caused by changes in relative light emission intensity corresponding to the changing conditions and reflects well the combustion situation. The flame color determined by the combustion conditions is discernible by the human eye and have been utilized for combustion control based on the accumulated experience or knowledge of an expert. In the present work the quantitative expression of flame color without relying on human visual sensation and its utilization in combustion diagnostics and combustion control have been investigated. The method that is most similar to this that has been tried is the use of flame color to estimate the flame temperature of boiler flames. However, that method is based on the relation between the ratio of soot radiation and temperature and is essentially different from this research.

With recent progress in image processing techniques or colorimetric equipment, this research leads to the development of two dimensional simultaneous combustion diagnostics.

1.5.2 Principle of flame color utilization

(a) CIE standard colorimetric system

Since color is one of the human senses, it is necessary, first of all, to express flame color numerically to discuss the relations between the flame color and burning conditions. Although there are several ways to characterize colors, we adopted chromaticity coordinates (x,y) defined by CIE (Commission International de l'Eclairage). The specifications of colors according to the coordinates will be outlined below.

In general, visible light L is specified by the color equation,

$$L = X[X] + Y[Y] + Z[Z]$$

where $[X]$, $[Y]$ and $[Z]$ are fundamental color stimuli called reference stimuli. The X, Y and Z are tristimulus values which reveal the amount of each reference stimulus. According to JIS Z 8701–1982, the tristimulus values are calculated by the following equations:

$$X = k \cdot \int_{380nm}^{780nm} P(\lambda) \bar{x}(\lambda) \cdot d\lambda \tag{1.6}$$

$$Y = k \cdot \int_{380nm}^{780nm} P(\lambda) \bar{y}(\lambda) \cdot d\lambda \tag{1.7}$$

$$Z = k \cdot \int_{380nm}^{780nm} P(\lambda) \bar{z}(\lambda) \cdot d\lambda \tag{1.8}$$

where $p(\lambda)$ is a special distribution of the light L, λ is wavelength, $x(\lambda)$, $y(\lambda)$ and $z(\lambda)$ are color matching functions, and κ is a proportional coefficient which makes Y agree with the luminous quantity of the light. In the present studies, the value of κ is regarded as unity because only the relative tristimulus values as,

$$(x,y,z) = \frac{1}{(X+Y+Z)}(X,Y,Z) = \frac{1}{S}(X,Y,Z) \tag{1.9}$$

where S is the sum of the tristimulus values are used and $x + y + z = 1$. The colors of visible light are usually specified by the coordinates (x, y) alone.

(b) Quantitative expression of flame color

Propane premixed flame stabilized on a circular burner (i.d. 8 mm) was investigated for various air ratios and its color quantitatively expressed by chromaticity coordinates. Figure 1.28 shows the chromaticity coordinates of flames at $m = 0.7 \sim 1.2$ on a chromaticity diagram. General classification of color according to JIS Z 8110–1984 (names of light source colors) is also shown in the figure.

The results show that the coordinate x rarely changes but the coordinate y decreases by about 0.15 for an increase of 0.5 in m. The color corresponding to the chromaticity coordinates changes from blue–green to blue–purple with the increase in m. This agrees well with results of visible observation. The colorimeter (e.g. Minolta CS–100) can measure the chromaticity coordinates with accuracy higher than ±0.004. Accordingly, the differences in flame color which correspond to the change of 0.02 in m can be distinguished by the measurement of the coordinate y. Thus, it is found that the chromaticity coordinates introduced here are extremely effective as a method by which to determine flame color.

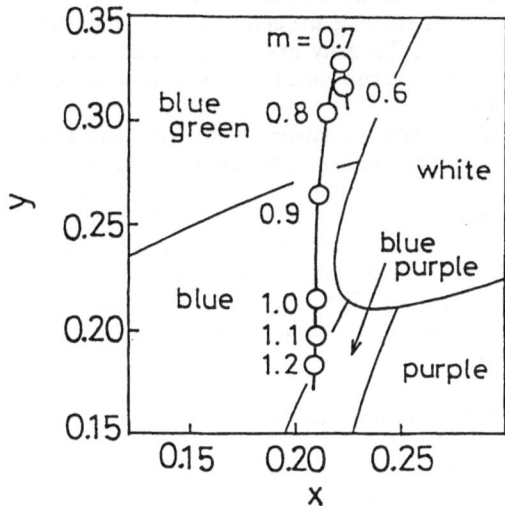

Fig. 1.28 Chromaticity coordinates of propane premixed flame

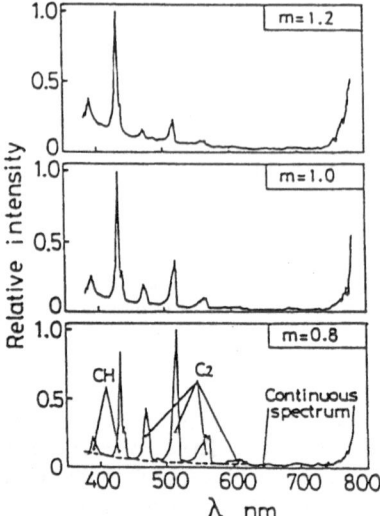

Fig. 1.29   Flame spectra (m=0.8,1.0,1.2)

1.5.3 Factors affecting the flame color

To apply the present method to practical combustion systems, it is important to know
the effects of various parameters on the flame color. Figure 1.29 shows the flame spectra
corresponding to three kinds of air ratio of the propane premixed flame. These spectra
are also the original data for calculating the chromaticity coordinate plotted in Fig. 1.28.
The flame spectra, especially the intensity of C_2 Swan band system, change with changes
in air ratio. The change of the spectra makes air ratio to be a most important factor to
determine the flame color.

Besides air ratio, there are many parameters which have an effect on flame color. The
effect of mixture velocity, mixture temperature, kind of fuel, location of measured point,
and those effects on flame color are reported.

Considering only hydrocarbon premixed flames, the flame color is determined by the
additive mixture of a few radical light emissions. Because the color of each radical light
emission is independent of air ratio, flame color is determined by relative intensity of
each radical light emissions [40]. Then, reversing the process of the calculation to
determine the chromaticity coordinates makes it possible to estimate the flame spectrum
from the coordinates [40].

1.5.4 Application to practical flames

Some methods to quantify the flame color are available. Figure 1.30 shows the approach
in formulating a quantitative expression of flame color. The lower path in the figure
represents the method using flame spectra to calculate chromaticity coordinate. The value
plotted in Fig. 1.28 is obtained through this way. On the other hand, the upper path is
a method using some equipments to read flame color directly, e.g. color VTR camera
with color image processing board or colorimeter which are commercially available. The

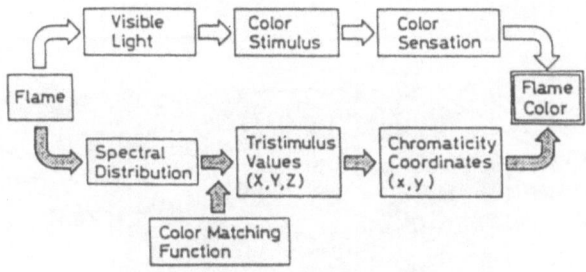

Fig. 1.30   The concepts of flame color discernment and expression

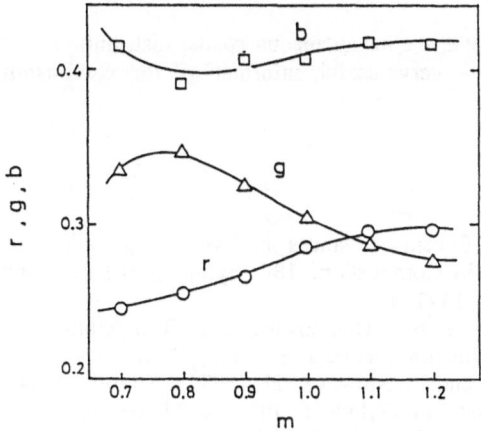

Fig. 1.31   The relationship between air ratio m and coordinates (r,g,b)

numerical expression of flame color depends on the equipment used here. For TV cameras, the flame color is expresses by RGB chromaticity coordinates and for colorimeters, the *XYZ* system is usually used. To apply this method to practical Flames, the most common way is to use the relation between some combustion characteristics and flame color information which are obtained from standard flame. Figure 1.31 shows the relation between chromaticity coordinates expressed by RGB colorimetric system and air ratio, m, of propane premixed flame. According to this figure, the g coordinate basically decreases with an increase in air ratio.

If the air ratio is larger than 0.8, the air ratio can be estimated from flame color using this relation. Figure 1.32 shows the distribution of g coordinate values of a vortex flame. This flame is stabilized behind a bluff–body with propane injection opposite to the air flow directions[41]. Combustion occurs at the shearing layer between air flow and recirculation zone. A region of high g value inside the vortex and downstream of the flow can be seen. If the relationships as contained in Fig. 1.31 is applicable to this flame, then it can be said that the air ratio is higher at the inside and downstream side of vortex.

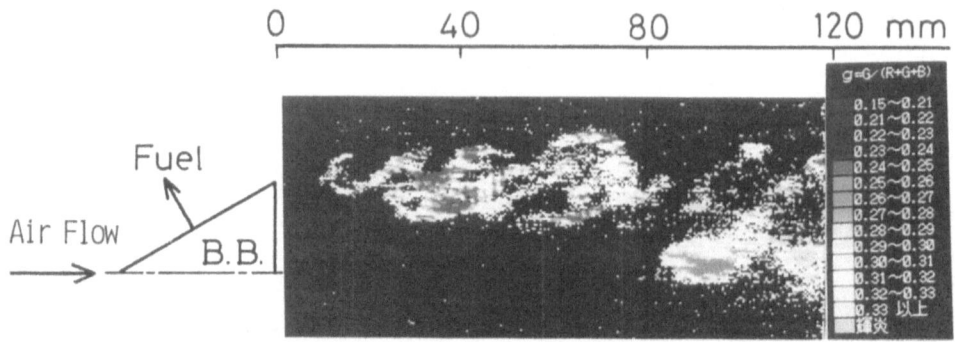

Fig. 1.32   The value of g coordinate distribution in a vortex flame stabilized behind a bluff–body

As demonstrated above, the instantaneous spatial distribution of flame color is easily obtained and can be a very useful information for combustion diagnostics and combustion control.   (Kenichi Ito)

References

[1]Eckbreth A (1981) Recent Advances in Laser Diagnostics for Temperature and Species Concentration in Combustion. 18th Symp. (Intl.) on Comb.,The combustion Institute, Pittsburgh, pp 1471–1488
[2]Ohtake K, (1989) Laser Diagnostics for Temperature and Concentration Measurements in Combustion Fields, J. Soc. High Temp.,15: pp 7–14
[3]Ohtake K, Ida T and Yoshikawa N (1986) Simultaneous Measurements of Temperature and Velocity in Turbulent Diffusion Flames by Rayleigh Scattering and LDV System, Trans. JSME,52: pp 3571–3576
[4]Ohtake K (1987) Simultaneous Measurement of Temperature and Velocity in Turbulent Diffusion Flames by Rayleigh Scattering and LDV,Laser Diagnostics and Modeling of Combustion, Springer–Verlag, Tokyo, pp 29–34
[5]Ida T and Ohtake K (1990) Experimental Study of Turbulent Diffusion Flame Structures and Their Similarities, Trans. JSME, 56: pp 3514–3521

[6]Driscoll JF, Schefer RW and Dibble RW (1982) Mass Fluxes $\overline{\rho' u'}$ and $\overline{\rho' v'}$ Measured in a Turbulent Nonpremixed Flame, 19th Symp.(Intl.) on Comb., The Combustion Institute, Pittsburgh, pp 477–485
[7]Ohshima H and Yanagi T (1981) Time Constant of Fine Thermocouple and Temperature Fluctuation Measurement, 19th Japan Symp. Comb., pp 320–232
[8]Suzuki T, Vetura JMP, Yule AJ and Chigier NA (1981) Measurement of Turbulent Diffusion Flame Structure by Ion Probe, ibid, pp 221–223
[9]Ohtake K, Naruse I, Horiuchi K and Tsuji H (1991) Structure of Turbulent Diffusion Flame by High–Speed TV Image Processing and Rayleigh Scattering, Trans. JSME, 57: pp 1135–1140
[10]Ida T, Tsuji H and Ohtake K (1991) Microscopic Structure of Turbulent Diffusion Flame, 29th Japan Comb. Symp., pp 61–63
[11]Ida T, Ohtake K (1992) Microscopic Structure of Turbulent Diffusion Flames, Trans. JSME, 58: pp 1918–1924

[12]Brzustowski TA, Gollahalli SR and Sullivan HF (1975) The Turbulent Hydrogen Diffusion Flame in a Cross-Wind, Comb. Sci. Tech.,11: pp 29-33

[13]Bros PE and Brzustowski TA (1978) An Experimental and Theoretical Study of the Turbulent Diffusion Flame in Cross-Flow, 17th Symp. (Intl.) on Comb., The Combustion Institute, Pittsburgh, pp 389-398

[14]Becker HA, Liang D and Downey CI (1981) Effect of Burner Orientation and Ambient Airflow on Geometry of Turbulent Free Diffusion Flames,18th Symp.(Intl.) on Comb.,The Combustion Institute, Pittsburgh, pp 1061-1072

[15]Rao VK and Brzustowski TA (1982) Tracer Studies of Jets and Diffusion Flames in Cross-Flow, Comb. Sci. Tech.,27: pp 229-239

[16]Escudier MP (1972) Aerodynamics of a Burning Turbulent Gas Jet in a Crossflow,Comb. Sci. Tech.,4: pp 293-301

[17]Allemand JB (1984) Three-Dimensional Steady Parabolic Calculations of Large Scale Methane Turbulent Diffusion Flames to Predict Flare Radiation Under Cross-Wind Conditions, 20th Symp. (Intl.) on Comb., The Combustion Institute, Pittsburgh, pp 531-540

[18]Tsue M and Kadota T (1992) Numerical Analysis on the Jet Diffusion Flame in a Cross Flow, Proc. of JSME Annual Meeting, pp 319-321

[19] Fairweather M Jones WP, Lindstedt RP and Marquis AJ (1991) Predictions of Turbulent Reacting Jet in a Cross-Flow, Comb. and Flame, 84: pp 361-375

[20]Kadota T and Mibe T (1988) Structure of Two-phase Jet in a Cross Flow, Trans. JSME, 54: pp 1337-1342

[21]Bray,KNC (1987) In Complex Chemical Reaction Systems (Warnatz J and Jager W eds.) Springer-Verlag, pp 356-375

[22]Peters N (1986) Laminar Flamelet Concepts in Turbulent Combustion, 21st Symp. (Intl.) on Comb., The Combustion Institute, Pittsburgh, pp 1231-1250

[23] Murayama M and Takeno T (1988) Fractal-like Character of Flamelets in Turbulent Premixed Combustion, 22nd Symp. (Intl.) on Comb., The Combustion Institute, Pittsburgh, pp 551-559

[24]Takeno T, Murayama M and Tanida Y (1990) Fractal Analysis of Turbulent Premixed Flame Surface, Experiments in Fluids, 10: pp 61-70

[25]Takeno T, Baba N, Kushida G and Murayama M (1989) Dynamics of Turbulent Premixed Flame Surface, Joint Meeting of the Australia/New Zealand and Japanese Sections of The Combustion Institute, pp 145-147

[26]Talbot L (1981) Thermophoresis-A Review, Progress in Astronautics and Aeronautics, 74: pp 467-488

[27]Pandya TP and Weinberg FJ (1964) The Structure of Flat, Counter-Flow Diffusion Flames, Proc. of Roy.Soc.London, A 279: pp 544-561

[28] Kee RJ, Miller JA, Evans GH and Dixon-Lewis G (1988) A Computational Model of the Structure and Extinction of Strained, Opposed Flow, Premixed Methane-Air Flames, 22nd Symp. (Intl.) on Comb., The Combustion Institute, Pittsburgh, pp 1479-1494

[29]Gaydon AG, The Spectroscopy of Flames 2nd Ed.(1974) Chapman and Hall

[30]Kato F and Hashimoto T (1991) Estimation of Flame Equivalence Ratio by the Simultaneous Measurements of Three Components of Radical Emissions, Japan Symp on Comb, pp 487-489

[31]Nakabe K, Mizutani Y, Hirao T and Tanimura S (1988) Burning Characteristics of Premixed Sprays and Gas-Liquid Coburning Mixture, Comb. and Flame, 74: pp.39-51

[32]Mizutani Y, Matsumoto Y and Matsui T (1986) Visualization of the Reaction Zone of a Flame by Image Processing, Trans. JSME,52: pp 1931-1937

[33]Ito H, Hommo Y, Song JI and Gomi T (1986) An Instantaneous Measuring Method of Air-fuel Ratio by Luminous Intensity of Radicals (Application to Practical Burner Flames), Trans. JSME, pp 3362-3371

[34]Ito F, Fujimoto K, Ikebe H, Tagami I Shimomugi S and Miyamae S (1985) The Evaluation of Flame Behavior by Spectrum Analysis of Flame (1st Rep. Basic Investigation of Small Oil Flame), Trans. JSME, 51: pp 1731-1735

[35]Suzuki T,Oba M, Hirano T and Tsuji H (1978) An Experimental Study of Premixed Turbulent Flames (Flame-Structure Study by Measuring Light-Intensity Distributions), Trans. JSME, 44: pp 3534-3542

[36]Uchida H, Nakajima M and Yuta S (1985) Measurement of flame temperature distribution by IR emission computed tomography, Appl. Opt., 24: pp 4111-4116

[37]Doi J, Sato and Miyake T (1987) Three-Dimensional Measurement of the Shape of Combustion Flames, Laser Diagnostics and Modeling of Combustion, Springer-Verlag, Tokyo, pp 195-202

[38]Nakayama M and Ogiwara G (1990) Image Processing of Methanol Spray Flame, 28th Japan Symp. on Comb., pp 428-430

[39]JSME, Laser Diagnostics and Modeling of Combustion (1987) Maruzen

[40]Fujita O, Ihara S, Tatsuta S and Ito K (1990) Prediction of Visible Emission Spectrum by Flame Colorimetry (Laminar Propane Flame), 28th Japan Symp. on Comb., pp 395-397

[41]Ito K and Fujita O (1990) Structure of Diffusion Wake Flame behind a Bluff-Body, Proc. of Conference on Mechanism of Non-Uniform Combustion, pp 55-64

Chapter 2
Modeling of Turbulent Diffusion Flames

2.1 Introduction

In turbulent combustion fields, local heat release takes place and intense fluctuation exists in temperature, concentration and density as well as in flow velocity. These condition makes it very difficult to measure and understand phenomena in actual combustors precisely. Therefore, the numerical prediction of turbulent combustion is of great technological use. Considerable progress has been made in recent years in the modeling technique of combustion fields with the development of computers, and then, the present group research selected the modeling of non–premixed combustion flows as one of the main subjects.

Short reports will be made in Sections 2.2~2.7 on the individual studies relating to modeling. In this section, the present status of the modeling technique will be described briefly to clarify the background of the each study.

There are two principal problems in the modeling of turbulent diffusion flames. The first is the way how to represent flow fields, and the second is how to obtain the local reaction rate in turbulent flows. Relating to the former, the large eddy simulation or direct simulation method in which the Navier–Stokes equations are solved directly without a turbulence model has been lately used to obtain instantaneous flow condition for non–combustion cases. Such a study will be reported in Section 2.5. Though these methods will become a more useful technology in the future, they have hardly been applied to the simulation of combustion fields at present and the numerical prediction depends on turbulence models, where only time–averaged properties are dealt with.

Since another model is required to represent the reaction rate in turbulent fields, two models, turbulence and combustion models, are necessary in the numerical simulation of turbulent diffusion flames.

Conservation equations of mass, momentum, concentration and energy are solved on time–averaged values in the simulation of combustion flows. If the equations are written in terms of conventional time–average (Reynolds average), density–related correlation terms, such as $\overline{\rho' u_i'}$ or $\overline{\rho' \phi_\alpha'}$, are contained in them because intense density fluctuation exists in combustion fields, and another model is required to obtain these new unknown terms. To circumvent this difficulty, density–weighted averaging (Favre averaging) defined in Eqs.(2.1) and (2.2) is often used for the prediction of flames.

$$s = \tilde{s} + s''$$
(2.1)

$$\tilde{s} = \overline{\rho s}/\overline{\rho} \ , \quad \overline{\rho s''} = 0 \ (\overline{s''} \neq 0)$$
(2.2)

The notations, "~" and "–", denote density–weighted and unweighted quantities, respectively. s'' is fluctuating component of a quantity s in density–weighted average. With this averaging, the conservation equations of mass, momentum and a scalar are written as follows:

$$\frac{\partial \bar{\rho}}{\partial t} + \frac{\partial}{\partial x_i}(\bar{\rho}\tilde{u}_i) = 0 \tag{2.3}$$

$$\frac{\partial}{\partial t}(\bar{\rho}\tilde{u}_i) + \frac{\partial}{\partial x_j}(\bar{\rho}\tilde{u}_i\tilde{u}_j) = -\frac{\partial \bar{p}}{\partial x_i} - \frac{\partial}{\partial x_j}(\bar{\tau}_{ij} + \bar{\rho}\,\widetilde{u_i''u_j''}) + \bar{\rho}g_i \tag{2.4}$$

$$\frac{\partial}{\partial t}(\bar{\rho}\tilde{\phi}_\alpha) + \frac{\partial}{\partial x_i}(\bar{\rho}\tilde{u}_i\tilde{\phi}_\alpha) = -\frac{\partial}{\partial x_i}(\bar{j}_{\alpha i} + \bar{\rho}\,\widetilde{u_i''\phi_\alpha''}) + \overline{\rho S(\phi_\alpha)} \tag{2.5}$$

where τ_{ij} is viscous stress, $j_{\alpha i}$ is flux of a scalar ϕ_α, p is pressure, S is a source term and g_i is gravitational acceleration. Density–related correlation terms do not appear in the above conservation equations, and the forms are the same as conventional time–averaged equations for constant density flows only if "~" is replaced by "–".

A turbulence model is required to obtain the Reynolds stress $\bar{\rho}\,\widetilde{u_i''u_j''}$ and the scalar flux $\bar{\rho}\,\widetilde{u_i''\phi_\alpha''}$. The k–$\varepsilon$ two–equation model has been widely used as the turbulence model for the simulation of turbulent flames. However, it is known that the k–ε model is difficult to provide good prediction for elliptic flows such as containing recirculating flow regions. The Reynolds stress closure model, in which the Reynolds stress and the scalar flux are obtained from solving their transport equations, has been proposed as an alternative. Though the Reynolds stress model can remove some of the deficiencies existing in the k–ε model, it does not appear to be developed to a satisfactory stage. The comparison of the above two models has been conducted in the numerical prediction of recirculating flows [1]~[4]. Such a study will be reported in Section 2.7. Generally speaking, the Reynolds stress model does not always provide better results for recirculating flows than the k–ε model in the present stage of development [1][4]. In connection with this fact, it may be considered that turbulence models do not have large effect on the simulated results because recirculating flows are mainly controlled by convection [5]. The causes for the discrepancy between measured and simulated results may be attributed to numerical errors for the calculation of these complex flows. Moreover, the intense fluctuation of large structure is observed in these flows [6], and then, it may be doubtful if time–average modeling methods themselves can represent the structure of such flows.

The increase of kinematic viscosity due to temperature rise suppresses turbulence in relatively low turbulence regions of flames. This phenomenon is called the laminarization due to combustion, and it has been observed in coaxial jet diffusion flames for a long time[7][8]. Recently, it was reported that the laminarization has a large effect on the structure of bluff body diffusion flames with recirculating flow regions[9]. Since this phenomenon cannot be represented by ordinary turbulence models, the models need to be modified for the prediction of such condition in flames. A study relating this problem will be reported in Section 2.6.

The conserved scalar approach is often used as a combustion model for turbulent diffusion flames. In this model, it is assumed that all species and heat have equal effective diffusivities. And the fast chemistry assumption is often invoked for the situation where the chemical reaction is sufficiently faster than the mixing. As a result, the thermochemical state of gas mixture, such as mass fractions of species and temperature, can be determined in terms of a strictly conserved scalar variable, which has no source term in its conservation equation. The mixture fraction can be typically used as the conserved scalar. For combustion flows formed through mixing of fuel stream 1 and oxidizer stream 2, the mixture fraction f is defined as follows:

$$f = \frac{M_1}{M_1 + M_2} \tag{2.6}$$

where M_1 and M_2 are the masses of elements originated in streams 1 and 2.

The mixture fraction f fluctuates in turbulent diffusion flames. The simplest method to take account of the fluctuation is the introduction of density weighted probability density function(p.d.f.) $P(f, x_i)$ for f. Various forms, such as the clipped Gaussian distribution or β-probability density function, are chosen as p.d.f.. Local p.d.f. is most often specified in terms of the mean and variance of f, obtained from the following transport equations:

$$\bar{\rho}\tilde{u}_j \frac{\partial \tilde{f}}{\partial x_j} = \frac{\partial}{\partial x_j}\{\frac{\mu_t}{\sigma_t}\frac{\partial \tilde{f}}{\partial x_j}\} \tag{2.7}$$

$$\bar{\rho}\tilde{u}_j \frac{\partial \tilde{f'^2}}{\partial x_j} = \frac{\partial}{\partial x_j}\{\frac{\mu_t}{\sigma_t}\frac{\partial \tilde{f'^2}}{\partial x_j}\} + 2\frac{\mu_t}{\sigma_t}(\frac{\partial \tilde{f}}{\partial x_j})^2 - C_D\frac{\bar{\rho}\varepsilon}{k}\tilde{f'^2} \tag{2.8}$$

where k and ε are turbulent kinetic energy and its dissipation rate respectively, μ_t is turbulent viscosity and σ_t is turbulent Plandtl number.

As mentioned above, instantaneous values of chemical species concentrations, temperature and density in gas mixture is determined in terms of an instantaneous value of the conserved scalar f. If these dependent variables are denoted $\phi(f)$, the density-weighted and unweighted average of $\phi(f)$ can be obtained from Eqs.(2.9) and (2.10), respectively.

$$\tilde{\phi}(x_i) = \int_0^1 \phi(f)P(f, x_i)df \tag{2.9}$$

$$\bar{\phi}(x_i) = \bar{\rho}\int_0^1 \frac{\phi(f)}{\rho(f)}P(f, x_i)df \tag{2.10}$$

A number of methods have been proposed to relate composition, temperature and density of mixture to f. The simplest method is the so-called flame sheet model, in which the fast single irreversible step is assumed in the reaction process;

fuel + oxidant → *products*.

This model has the disadvantage that dissociation and intermediate products, such as CO and H_2 in hydrocarbon flames, can not be represented. An alternative method is based on the assumption of full chemical equilibrium. Though dissociation and intermediate products are taken into consideration in this model, concentrations of CO and H_2 are overpredicted in fuel rich mixture. The assumption of the fast chemistry must be removed to avoid this overprediction. However, inclusion of chemical kinetics is very difficult in the conserved scalar approach representing the fluctuation of f with p.d.f.. Meanwhile, there exist a number of combustion phenomena, such as ignition and extinction and NO_x formation, in which a chemical reaction with finite rate plays an important role, and then, the development of the model of finite reaction rate appears to be the most important subject in the study of combustion models.

One of such models is the laminar flamelet approach [10][11], in which experimental results or simulated ones considering chemical kinetics for laminar diffusion flames are used to relate the thermochemical state to f. This method can also give good prediction for CO concentration. The laminar flamelet approach has been improved to consider the effect of turbulent flow condition on the laminar flamelet structure[12]. This model will be presented in Section 2.2. As the other promising models capable to represent the chemistry of finite reaction rate like NO_x formation, there are the two–variable formalism adopted by Janicka and Kollmann[13], the perturbation method formulated by Bilger[14] and the direct calculation of p.d.f.s[15]. A Lagrangian approach may be better to represent the finite rate chemistry correctly than Eulerian, and such a model will be proposed in Section 2.3.

The modeling of reacting flows in actual combustor will be presented in Section 2.8. Turbulence and combustion models available at present are incomplete as mentioned above. In the modeling of actual combustors, there are the other error sources in connection with the complex geometry and the numerical calculation technique, and distinguishing each error is very difficult. However, the separate estimation of various models and calculation techniques is very important in the development of the modeling method, and well–programmed experiments are required for that estimation. An example of these experiments will be shown in Section 2.4.   (Yoshiaki Onuma)

2.2 Modelling of Turbulent Diffusion Flames

Diffusion Flames owe their name to the rate controlling step, namely diffusion, in the process of macroscopic and microscopic mixing and subsequent reaction in non-premixed combustion. The interaction between these processes may be understood in terms of the respective time scale of convection, diffusion and reaction. The convective and diffusive time scales are in general of the same order of magnitude but the chemical time scale is very much smaller. Therefore, the assumption of local chemical equilibrium has been used quite successfully for diffusion flames, in particular those of hydrogen or hydrogen/carbon–monoxide mixtures.

However, whenever a more detailed analysis is required departure from equilibrium must be taken into account. Non–equilibrium effects are not only important for the prediction of CO and H_2 levels in hydrocarbon flames, they also provide the basic mechanism that leads to local quenching and eventually to lift–off and blow–off of diffusion flames. A more detailed review of non–premixed combustion models was given in [16]. Here, only the basic features and additional considerations will be presented.

2.2.1 The mixture fraction variable

If the reactants—fuel and oxidizer—are fed separately into a combustion system, they diffuse towards each other and burn in a flame structure called a diffusion flame. Combustion occurs preferentially at those locations in the flow field where mixing is stoichiometric. The global reaction equation for complete combustion of a hydrocarbon fuel F, written as

$$v'_F F + v'_{O_2} O_2 \rightarrow v''_{CO_2} CO_2 + v''_{H_2O} H_2O \tag{2.11}$$

defines the stoichiometric coefficients v'_{O_2} and v'_F. The reaction equation relates the change of mass fractions of oxygen dY_{O2} and fuel dY_F that are consumed, by

$$\frac{dY_{O_2}}{v'_{O_2} M_{O_2}} = \frac{dY_F}{v'_F M_F} \tag{2.12}$$

where M_i is the molecular weight. This equation may be integrated from the unburnt state to any later state as

$$v Y_F - Y_{O_2} = v Y_{F,u} - Y_{O_2,u} \tag{2.13}$$

where $v = v'_{O_2} M_{O_2} / v'_F M_F$ is the stoichiometric mass ratio. Away from the stoichiometric location, the mixture is either fuel lean or fuel rich and therefore leaves either some oxygen or the fuel (which may partially be oxidized to CO and H_2) unreacted. Therefore it cannot produce a high enough temperature to maintain a reaction rate, since combustion chemistry is very temperature sensitive.

In order to describe the mixture field and to identify the location of the stoichiometric mixture, it is useful to introduce the mixture fraction Z as a dependent variable. In a system of only two mass streams, where 1 denotes the fuel stream, $\dot{m}_1$, and 2 the oxidizer stream, $\dot{m}_2$, Z represents the local mass fraction of the fuel stream in the unburnt mixture,

$$Z = \frac{\dot{m}_1}{\dot{m}_1 + \dot{m}_2} \tag{2.14}$$

Since both the fuel and oxidizer streams may contain inerts such as nitrogen, the local mass fraction $Y_{F,u}$ of the fuel is the same fraction as in the original fuel stream(if effects of differential diffusion are neglected),

$$Y_{F,u} = Y_{F,1} Z \tag{2.15}$$

where $Y_{F,1}$ denotes the mass fraction of fuel in the fuel stream. Similarly, since $1-Z$ represents the local mass fraction of the oxidizer stream in the unburnt mixture, one obtains for the local mass fraction of oxygen,

$$Y_{O_2,u} = Y_{O_2,2}(1-Z) \tag{2.16}$$

where $Y_{O2,2}$ represents the mass fraction of oxygen in the oxidizer stream ($Y_{O2,2}$ =0.232 for air). Introducing Eqs.(2.15) and (2.16) into Eq.(2.13) , the mixture fraction is

$$Z = \frac{\nu Y_F - Y_{O_2} + Y_{O_2,2}}{\nu Y_{F,1} + Y_{O_2,2}} \tag{2.17}$$

For a stoichiometric mixture with $\nu Y_F = Y_{O2}$,

$$Z_{st} = [1 + \frac{\nu Y_{F,1}}{Y_{O_2,2}}]^{-1} \tag{2.18}$$

where Z_{st} is the stoichiometric mixture fraction.

2.2.2 The flamelet concept of diffusion flames

The mixture fraction appears naturally as an independent variable for diffusion flames. Under the condition that equal diffusivities of chemical species and temperature can be assumed(an assumption that is good for hydrocarbon flames but much less realistic for hydrogen flames) all Lewis numbers

$$L_{e_i} = \frac{\lambda}{c_p \rho D_i} \qquad (i = 1, 2, ..., n) \tag{2.19}$$

are unity, such that a common diffusivity coefficient D can be introduced. The balance equations for Z and the temperature T are

$$\rho \frac{\partial Z}{\partial t} + \rho v_\alpha \frac{\partial Z}{\partial x_\alpha} - \frac{\partial}{\partial x_\alpha}(\rho D \frac{\partial Z}{\partial x_\alpha}) = 0 \tag{2.20}$$

$$\rho \frac{\partial T}{\partial t} + \rho v_\alpha \frac{\partial T}{\partial x_\alpha} - \frac{\partial}{\partial x_\alpha}(\rho D \frac{\partial T}{\partial x_\alpha}) = -\frac{1}{c_p} \sum_{i=1}^{n} h_i \dot{m}_i \tag{2.21}$$

Here h_i are the specific heats and m_i the chemical production rates of the reacting species($i=1,2,...,n$). The specific heat capacities c_{pi} are all assumed constant and equal to c_p for simplicity. Eq.(2.20) does not contain a chemical source term, since Z represents the chemical elements originally contained in the fuel, and elements are conserved during combustion. We assume the mixture fraction Z to be given in the flow field as a function of space and time by solution of Eq.(2.20). Then the surface of stoichiometric mixture can be determined from

$$Z (x_\alpha , t) = Z_{st} \tag{2.22}$$

Combustion takes place in a thin layer in the vicinity of this surface if the local mixture fraction gradient is sufficiently high. Let us locally introduce a coordinate system attached to the surface of stoichiometric mixture. We replace the coordinate x_1 by the mixture fraction Z and define the original coordinate system such that the coordinate x_1

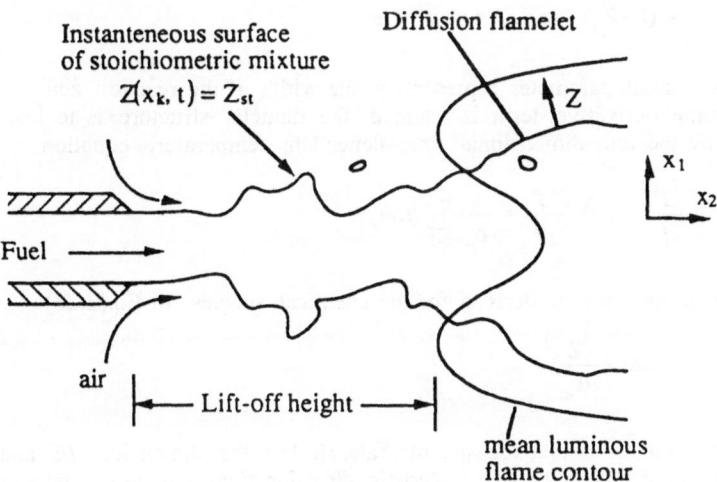

Fig. 2.1 Schematic illustration of diffusion flamelets attached to the surface of stoichiometric mixture.

does not lie within this surface. By definition, the new coordinate Z is locally normal to the surface of stoichiometric mixture.This is schematically shown in Fig. 2.1 for a lifted jet diffusion flame. Using $Z_2=x_2$, $Z_3=x_3$, $t^*=t$ as the other independent variables, we obtain with the transformation rules

$$\frac{\partial}{\partial t} = \frac{\partial}{\partial t^*} + \frac{\partial Z}{\partial t}\frac{\partial}{\partial Z}, \quad \frac{\partial}{\partial x_1} = \frac{\partial Z}{\partial x_1}\frac{\partial}{\partial Z},$$

$$\frac{\partial}{\partial x_k} = \frac{\partial}{\partial Z_k} + \frac{\partial Z}{\partial x_k}\frac{\partial}{\partial Z}, \quad (k = 2, 3) \tag{2.23}$$

the temperature equation in the form

$$\rho(\frac{\partial T}{\partial t^*} + v_2\frac{\partial T}{\partial Z_2} + v_3\frac{\partial T}{\partial Z_3}) - \frac{\partial(\rho D)}{\partial x_2}\frac{\partial T}{\partial Z_2} - \frac{\partial(\rho D)}{\partial x_3}\frac{\partial T}{\partial Z_3} - \rho D[(\frac{\partial Z}{\partial x_\alpha})^2\frac{\partial^2 T}{\partial Z^2}$$

$$+ 2\frac{\partial Z}{\partial x_2}\frac{\partial^2 T}{\partial Z\partial Z_2} + 2\frac{\partial Z}{\partial x_3}\frac{\partial^2 T}{\partial Z\partial Z_3} + \frac{\partial^2 T}{\partial Z_2^2} + \frac{\partial^2 T}{\partial Z_3^2}] = -\frac{1}{c_p}\sum_{i=1}^{n} h_i\dot{m}_i \tag{2.24}$$

If the flamelet is thin in the Z–direction,an order of magnitude analysis similar to that for a boundary layer shows that the second derivative with respect to Z is the dominating term on the left hand side of Eq.(2.24). This term must balance the reaction term on the right hand side. The term containing the time derivative is only important if very rapid changes,such as extinction, occur. Formally this can be shown by introducing the stretched coordinate ζ and the fast time scale τ

$$\zeta = (Z-Z_{st}) / \varepsilon , \quad \tau = t^* / \varepsilon^2 \tag{2.25}$$

where ε is a small parameter representing the width of the reaction zone.

If the time derivative term is retained, the flamelet structure is to leading order described by the one–dimensional time–dependent temperature equation

$$\rho \frac{\partial T}{\partial t} - \rho \frac{\chi}{2} \frac{\partial^2 T}{\partial Z^2} = \frac{1}{c_p} \sum_{i=1}^{n} h_i \dot{m}_i \tag{2.26}$$

Similar equations may be derived for the chemical species. In Eq.(2.26)

$$\chi = 2D \left(\frac{\partial Z}{\partial x_\alpha}\right)^2 \tag{2.27}$$

is the instantaneous scalar dissipation rate. It has the dimension $1/s$ and may be interpreted as the inverse of a characteristic diffusion time. Due to the transformation it implicitly incorporates the influence of convection and diffusion normal to the surface of stoichiometric mixture. In the limit $\chi \rightarrow 0$ the local equilibrium model and the flame-sheet model are obtained. Local quenching of the flamelet occurs, if χ exceeds a critical value χ_q [16]. For the counter–flow geometry, the scalar dissipation rate at the location where the mixture is stoichiometric may be approximated, assuming constant density and diffusivity, by

$$\chi_{st} = \frac{a}{\pi} \exp\{-2[erfc^{-1}(2Z_{st})]^2\} \tag{2.28}$$

where a is the velocity gradient and $erfc^{-1}$ the inverse of the complementary error function. For example, $(erfc^{-1}(2Z_{st})$ is 1.13 for methane–air flames with $Z_{st}=0.055$ and it is 1.34 for H_2–air flames with $Z_{st}=0.0284$.

The basis for the universal coordinate transformation is the assumption that the reaction zone around Z_{st} is asymptotically thin, of $O(\varepsilon)$, in the mixture fraction coordinate. This may be verified by asymptotic analysis for real flames or simply by evaluating the reaction rate. Such an asymptotic analysis for methane–air diffusion flames has been performed in [17] where it is shown that the thickness ε is that of the broadest reaction layer, the H_2–CO–oxidation layer, on the lean side of the flame structure.

2.2.3 Comparison of time scales

Chemistry introduces a characteristic time scale. Since chemistry occurs close to $Z=Z_{st}$, χ is to be replaced by χ_{st} in eq.(2.26) in the limit $\varepsilon \rightarrow 0$. Then the scalar dissipation rate χ_{st} represents the inverse of a characteristic diffusion time which is to be compared to the chemical time represented by the inverse of the reaction rate. As noted above the extinction of the diffusion flamelet occurs at a particular value of the imposed scalar dissipation rate, namely χ_q, which therefore represents the inverse of a chemical time scale for extinction. We will define the extinction time scale as

$$t_c = \frac{Z_{st}^2(1-Z_{st})^2}{\chi_q} \tag{2.29}$$

This time scale is representative for the oxidation chemistry. Its definition is motivated by the expression for χ_q using one step large activation energy asymptotics[16] with equal temperatures T_1 and T_2 for the fuel and oxidizer stream

$$\frac{\chi_q}{Z_{st}^2(1-Z_{st})^2} = \frac{4B\rho_{st}\nu'_{O_2}e}{M_F(T_{st}-T_u)^3}(\frac{RT_{st}}{E})^3\exp(\frac{-E}{RT_{st}}) \tag{2.30}$$

where the inverse of the right hand side is proportional to the equivalent chemical time for a stoichiometric premixed flame[17].

For methane flames as an example the scalar dissipation rate at extinction was estimated in [16] as χ_q=8/s based on a_q=320/s for methane flames which leads to t_c=0.34×10^{-3}s.

Other chemical time scales to be considered are those for pollutant formation. The formation of thermal NO_x in turbulent diffusion flames may be assumed to proceed essentially through the Zeldovich mechanism [18].

$$O + N_2 \rightarrow N + NO$$

$$N + O_2 \rightarrow O + NO$$

where the first reaction is rate determining. Comparing the rate of this reaction to that of the most important chain branching reaction

$$H + O_2 \rightarrow OH + O$$

which is rate determining for the oxidation process, one obtains

$$\frac{k_{N1}}{k_1} = \frac{7\cdot10^{13}\exp(-37750 / T)}{2\cdot10^{14}\exp(-8400 / T)} = 1.5\cdot10^{-6} \qquad at \quad T=2000K$$

This indicates that the time scale for NO_x formation is between 5 and 6 orders of magnitude slower than the time scale of oxidation in the characteristic range of combustion temperatures.

These chemical time scales must be compared to the time scales of turbulence. Many numerical calculations of turbulent reacting flows are nowadays being based on k-ε-type models, and Favre-averaging is being used to account for strong density changes due to combustion. In terms of the Favre averaged turbulence kinetic energy $\tilde{k}$ and its dissipation $\tilde{\varepsilon}$, the turbulent time scale may be defined

$$t_t = \frac{\tilde{k}}{\tilde{\varepsilon}} \tag{2.31}$$

At the other end of the turbulent spectrum, the smallest time scale is the Kolmogorov time

$$t_K = (\nu \, / \, \bar{\varepsilon})^{1/2} \tag{2.32}$$

where ν is the kinematic viscosity. The Kolmogorov time represents the turnover time of the smallest eddies. For the example of a round jet one may estimate these turbulent time scale approximately as [19]

$$t_t = \frac{0.16 \, d}{\bar{Z}_{cL} \, u_0} \quad , \quad t_K = \frac{1.33}{\bar{Z}_{cL}^2} Re^{\frac{1}{2}} \frac{d}{u_0}$$

where d is the diameter and u_0 is the exit velocity and Re the Reynolds number based on these quantities. The mixture fraction on the centerline may be estimated as $Z_{cL}=5.3d/x$. This leads,for instance,at $x/d=30$, $d=10$mm and $u_0=10$m/s and a Reynolds number of 5000 to $t_t=5.1\times10^{-3}$s and $t_K=0.60\times10^{-3}$s. The extinction time scale for methane estimated before is at this position an order of magnitude smaller than the turbulent time and also still smaller than the Kolmogorov time. Therefore methane jets typically fall into the flamelet regime.

2.2.4 Applications of the flamelet concept

The flamelet concept postulates that a turbulent diffusion flame consists of an ensemble of thin diffusion flamelets where reaction takes place. In [16] five different states of a diffusion flamelet have been identified.

1. the steady unreacted initial mixture

2. the unsteady transition after ignition

3. the quasi–steady burning state

4. the unsteady transition after quenching

5. the unsteady transition after reignition

If one assumes that the unsteady transitions are not very frequent, only the two steady states 1 and 3 contribute to the overall statistical description of a turbulent diffusion flame. The unreacted state 1 is independent of χ_{st} but the burning state 3 depends on two parameters, Z and χ_{st}. In a turbulent flow field these parameters are statistically distributed. To predict non–equilibrium effects in turbulent diffusion flames, it is therefore necessary to predict the joint distribution function of Z and χ_{st}. In [16] the properties of the joint probability density function of Z and χ_{st} have been discussed in detail and the relation to semi–empirical turbulence models of the k–ε type have been pointed out.

   Local quenching effects which lead to a disruption of the flame surface may have important consequences for turbulent diffusion flame stability. Starner and Bilger[20] have measured the electrical conductivity between the nozzle and the main flame brush in a specially designed piloted diffusion flame. They found intermittency in the electrical conductivity which points towards an interruption of the reacting(and therefore electrically conducting) flame surface and therefore towards local flame quenching.

Likewise, Dibble et al.[21], using C_2-fluorescence as well as Rayleigh scattering, observed increasing local flamelet extinction in a turbulent methane jet diffusion flame as they increased the jet exit velocity.

In turbulent jet flames the mean scalar dissipation rate decreases with distance from the nozzle. Therefore,if a flame is burning far downstream, the probability of quenching of a flamelet increases with decreasing distance from the nozzle. But also there may be flamelets which were not reached by an ignition source and therefore stay unignited. Even within the turbulent flame brush there may be burnable yet unignited clusters of flamelets that are not connected to burning flamelets. A theory that is able to account for such a situation is percolation theory [16]. Percolation theory describes the conduction in randomly distributed networks. For example, if holes are punched randomly into carbon paper, there will be a threshold, beyond which the probability that 0an electric current can pass from one side of the paper to the other decreases to zero. There is an analogy to lifted flames, where local quenching of diffusion flamelets corresponds to the holes in the carbon paper and the lift–off height to the percolation threshold. In a first approximation, assuming zero variance of the probability distribution of χ_{st} and statistical independence between Z and χ_{st}, the lift–off height should correspond to the downstream location where the mean scalar dissipation rate is equal to the laminar quenching value χ_q. This prediction provides a basis for a verification of the flamelet concept.

In [22] measurements of stabilization heights in round methane–air jet flames diluted with nitrogen were performed. The stoichiometric mixture fraction Z_{st} was kept constant by also diluting the fuel. The residence times d/u_0 for each dilution,were scaled with the corresponding value of χ_q obtained from an evaluation of the laminar counterflow flame results of Ishizuka and Tsuji[23]. Fig. 2.2 shows χ_q for the different dilutions, multiplied with the residence time d/u_0, plotted over the lift–off height H, divided by d. It is seen

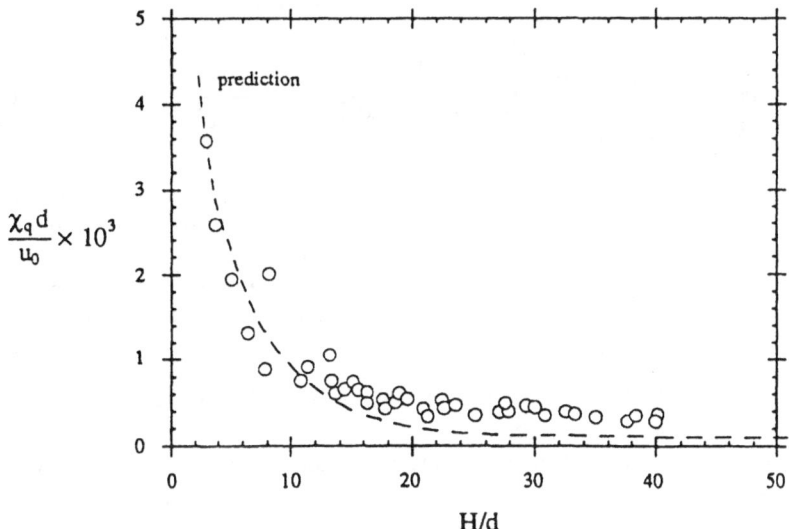

Fig. 2.2 The laminar quenching value χ_q,non–dimensionalized with the turbulent flame residence time d/u_0 plotted over the non–dimensional lift–off height H/d for different mole fractions X_{O2} of oxygen in air.

that this scaling of turbulent flame data with the laminar flamelet quenching parameter reduces the lift–off data to a single curve. The prediction is based on a k–ε–type turbulence model using statistical independence of Z and χ_{st}.

2.2.5 Conclusions

In summary, the flamelet concept has proven to be very useful to describe non–equilibrium effects in non–premixed combustion. It is a straight–forward extension of the local equilibrium model and results in a two–variable statistical formation, which was derived from the flame–attached coordinate transformation. It appears to be a promising tool for the investigation of important questions like flame stability, but also of NO_x– and soot–formation, which are yet to be explored.   (Norbert Peters)

2.3 A Lagrangian Stochastic Model for Non–Premixed Reacting Flows

In most reacting flows, the different species are introduced separately into the flow and so have to be mixed by turbulence and molecular motion. This mixing and reaction determine the distribution of chemical species in the reacting flow and the effectiveness of turbulent combustion or chemical reactions in many industrial processes. Therefore, a better understanding of reacting flows should improve the engineering design and control of many combustion and chemical processes.

When considering the one–step irreversible, second–order chemical reaction between chemical species A and B in an isothermal flow; $A+B{\rightarrow}C+D$, a general turbulent diffusion (mass–conservation) equation can be written for the mean concentration C of A or B:

$$\frac{\partial \overline{C}}{\partial t} + \overline{U_i}\frac{\partial \overline{C}}{\partial x_i} = \frac{\partial[\kappa(\frac{\partial \overline{C}}{\partial x_i}) - \overline{u_i c}]}{\partial x_i} - k[\overline{C_A}\ \overline{C_B} + \overline{c_A c_B}] \tag{2.33}$$

where C and c are the instantaneous and fluctuating concentrations, U_i and u_i the instantaneous and fluctuating velocities in the x_i–direction, respectively, κ the molecular diffusivity, k the chemical reaction rate constant and the overbars indicate mean(averaged) values. The second term on the right is the mean chemical reaction rate term, and it can be split into the mean concentration product $\overline{C_A}\ \overline{C_B}$ and the concentration fluctuation product $\overline{c_A c_B}$. Clearly, the latter correlation between the species concentration fluctuations, $\overline{c_A c_B}$, is of great importance, if the segregation parameter α ($=\overline{c_A c_B}\ /\ \overline{C_A C_B}$) is not close to zero.

In some studies of chemical processes, the correlation of the concentration fluctuation $\overline{c_A c_B}$ has been given by a constant value, or plausible assumptions such as Toor's hypothesis [24] have been used even for a moderately fast reaction. Thus, the concentration correlation has not been discussed explicitly in any detail. Recent measurements of Komori & Ueda[25], Mudford & Bilger[26], Bennani et al.[27] and others have showed that $\overline{c_A c_B}$ is not negligible compared with the mean concentration product $\overline{C_A}\,\overline{C_B}$, and it is significant in estimating the mean reaction rate. However, the previous experiments differed considerably in their measurements of $\overline{c_A c_B}$ and the large difference have not been physically explained. The experimental studies have suggested that theoretical or numerical discussion should be done to explain all the measurements.

Direct numerical simulations of the full equation is the only complete computational approach to reacting flows. However, the direct simulations are limited to very low–Reynolds–number flows of order of 10^2 and they have not been able to explain the eddy motions with small scales comparable to the Kolmogorov scale, which play an important role in reacting flows. Therefore, two types of turbulence models have been developed for turbulent reacting flows. Most of the models are based on the conventional methods of turbulence closure but they cannot predict the correlation $\overline{c_A c_B}$ without further closure assumptions, such as taking α equal to a constant. At present, it is very difficult to derive the concentration correlation from the conventional models, because of the lack of conclusive measurements of the concentration fluctuations.

Another type of model is based on computing the evolution of the probability density function(p.d.f.) for the concentrations of A and B. Most of these (p.d.f.) methods involve some assumptions about diffusion and mixing processes, and their scales for these processes are essential inputs to the transport equations for the joint p.d.f. for the species concentrations. It is possible with the p.d.f. model to predict $\overline{c_A c_B}$ exactly, since it does not need a closure assumption if the molecular mixing terms can be neglected.However, the connection between fluid motions and mixing processes is far from clear, so such models give little insight into the physical processes controlling mixing and reactions although they are proving to be rather successful in modelling many reacting flows. On the other hand, a p.d.f.model based on a Lagrangian formulation can be postulated by using a comparatively simple stochastic equation, such as a Langevin equation, for modelling the fluid particle motions. Such a model avoids solving a complicated transport equation for the p.d.f.. Moreover, the model affords a clearer interpretation than the Eulerian model.

This section therefore aims to introduce a Lagrangian stochastic model developed by Komori et al.[28] which can explain the previous measurements, and which helps to define the essential parameters for non–premixed reacting flows. The model computations concentrate on discussing the segregation parameters α ($=\overline{c_A c_B} / \overline{C_A C_B}$) or the intensity of segregation, defined by the ratio of the concentration fluctuation $\overline{c_A c_B}$ to the mean concentration product $\overline{C_A}\,\overline{C_B}$, because of its special importance, as mentioned above.

2.3.1 A Lagrangian stochastic model

According to the statistical theory of marked particles, (hereafter we consider very small marked particles which may be thought of as molecules), the ensemble mean concentration is

$$\overline{C(z,\ t)} = \int_{-\infty}^{\infty} P_1(z'_1,\ 0;\ z,\ t)C_0^{(1)}(z'_1)dz'_1 \tag{2.34}$$

and the ensemble mean–square concentration at a single time and position is

$$\overline{C^2(z,\ t)} = \overline{C^{(1)}(z,\ t)C^{(2)}(z,\ t)}$$

$$= \int\!\!\int_{-\infty}^{\infty} \ddot{P}_2(z_1',\ z_2',\ 0;\ z,\ z,\ t)C_0^{(1)}(z_1')C_0^{(2)}(z_2')dz_1'dz_2' \tag{2.35}$$

if the separation of the two particles vanishes at a time t. Here, P_1 and P_2 are one– and two–point displacement p.d.f.'s and $C_0^{(i)}(z_i')$ is the source distribution of particle i at a time $t=0$ and position $z=z_i'$. For statistically stationary motions of the particles, the concept of reversed dispersion can be applied. If the initial concentrations of A and B are C_{A0} and C_{B0}, and if they are non–premixed $\overline{C_{A0}(z')C_{B0}(z')} = 0$ at $t=0$, then from Eq.(2.35) with $C_A C_B$ replacing C^2

$$\overline{C_A C_B(z,\ t)} = \frac{1}{2}[\overline{C_A^{(1)}(z,\ t)C_B^{(2)}(z,\ t)} + \overline{C_A^{(2)}(z,\ t)C_B^{(1)}(z,\ t)}] \tag{2.36}$$

Here, a marked particle is assumed to have one chemical component. If an exact specification of P_1 and P_2 can be found, Eqs.(2.34)–(2.36) give exact concentration statistics. However, there is no analytically or computationally efficient method for obtaining exact solutions of these p.d.f.'s.

Sawford & Hunt[29] partitioned the particle displacement into a turbulent part (due to the motion of turbulent fluid particle elements containing very small marked particles at that time), and a random Brownian component(due to molecular motion), as

$$dz = w_p dt + (2\kappa)^{1/2}dW_d \tag{2.37}$$

where w_p is the velocity of the fluid element containing the small marked particle at the given instant, and dW_d is a Gaussian white noise process. Here, the assumption that the Brownian(molecular) motion is independent of the turbulent motion is, of course, used.

Model equations for the rate of separation of pairs of particles were given by

$$d\Delta = R^{1/2}(\Delta)U^{(2)}dt + \kappa^{1/2}dW_d^{(2)} \tag{2.38}$$

and for the displacement of the center of mass of the two particles

$$d\Sigma = [2\cdot R(\Delta)]^{1/2}U^{(1)}dt + \kappa^{1/2}dW_d^{(1)} \tag{2.39}$$

where Δ is the separation of two small particles, $\Delta = (z_1-z_2) / \sqrt{2}$, $\Sigma = (z_1+z_2) / \sqrt{2}$, $R(\Delta)$ the Eulerian structure function, and $U^{(i)}$ is the independent random velocity of particle i which is the solution of the Uhlenbeck–Ornstein process:

$$dU^{(i)} = -[U^{(i)} / T_L]dt + \sigma_w[2 / T_L]^{1/2}dW_t\ ;\quad U_{t=0}^{(i)} = \sigma_w N \tag{2.40}$$

Here, T_L is the Lagrangian integral time–scale, which is assumed to be proportional to the ratio of the turbulent integral space scale L to the r.m.s. fluctuating velocity σ_W, dW_t is a Gaussian white noise process,and N is a zero–mean, standard Gaussian random variable. This equation is equivalent to the Langevin equation when dt approaches zero. It is well established that the equation can describe the random velocity of a single particle in homogeneous turbulence.

The structure function $R(\Delta)$ in Eqs.(2.38) and (2.39) is an important function that determines the separation of two particles and is given by Sawford & Hunt[29] based on Kolmogorov's results as

$$R(\Delta) = [\frac{\Delta^2}{L^2+\Delta^2}]^{1/3}[\frac{\Delta^2}{\eta^2/\phi^3+\Delta^2}]^{2/3} \tag{2.41}$$

where η is the Kolmogorov scale,$(\nu^3/\varepsilon)^{1/4}$, determined by the viscous dissipation ε and kinematic viscosity ν, and ϕ is a constant which is equal to 0.358 [29]. In the presence of mean shear, the relative streamwise displacement of a particle pair is given by $d(x_2-x_1)/dt = T_L(d\overline{U}/dz)(z_2-z_1)$, where $\overline{U}$ is the mean streamwise velocity. Then, the structure function is given by Durbin's [30] formula:

$$R(\Delta) = [\frac{\Omega^2}{L^2+\Omega^2}]^{1/3}[\frac{\Omega^2}{\eta^2/\phi^3+\Omega^2}]^{2/3}[1+\frac{L^2\Delta_x^2}{\Omega^4+\Omega^2L^2}] \tag{2.42}$$

where $\Delta_x = \dfrac{x_2-x_1}{\sqrt{2}}$ and $\Omega^2 = \Delta^2+\Delta_x^2$

In the case of reacting flows, however, we have to consider the effect of the concentration change(reaction) of a marked particle, because a marked particle reacts with another marked particle if two turbulent fluid elements with marked particles start from different concentration–species sources. The reaction is assumed to begin within the smallest turbulent eddy scale η at the interface. That is, we assume that when the distance between two marked particles of different species becomes less than the Kolmogorov scale, η, i.e. $|z_1-z_2|\leq\eta$, the reaction begins between the fluid elements A and B at the same rates as under well–mixed conditions. The reaction is assumed to continue until the particular meeting of two marked particles at (z,t). This represents the effect on the reaction of the micro–mixing at scales less than η. Then, the concentration change of a marked particle consisting of many molecules is given by the chemical law

$$\frac{dC_{i(n)}^{(1)}(z,\ t)}{dt} = \frac{dC_{j(n)}^{(2)}(z,\ t)}{dt} = -kC_{i(n)}^{(1)}(z,\ t)C_{j(n)}^{(2)}(z,\ t)[1-\sigma_{ij}] \tag{2.43}$$

where $C_{j(n)}^{(1)}(z,\ t)$ is the concentration of a marked particle 1 of the chemical species i for the n–th particle pair, $C_{j(n)}^{(2)}(z,\ t)$ the concentration of another marked particle 2 of the species j which meets the marked particle 1 at z and t,and δ_{ij} is the Kronecker delta. The problems of the assumptions used in this model are in details discussed by Komori et al.[28].

2.3.2. Model computations

Here, the simplest non-premixed reacting flow with an initial concentration profile of a mixing-layer type and a uniform mean shear is considered and the segregation parameter α only on the averaged interface between fluids A and B, i.e. on the plane $z=0$, is discussed. The effect of inhomogeneity on α was small, and initial concentration profiles did not affect α much on the averaged interface ($z=0$) except in the region near the source.

(a)Dimensionless groups in turbulent reacting flows

By considering the straining motion on the smallest scales, if follows that in a turbulent flow the smallest length-scale for the concentration field is $(v^3/\varepsilon)^{1/4}Sc^{-1/2}$, where the Schmidt number $Sc=v/\kappa$. For estimating the effect of reactions we need to estimate the smallest time-scale of the concentration field, which is of the same order as that of the smallest fluid motions,namely $(v/\varepsilon)^{1/2} \approx T_L Re_t^{-1/2}$, where $Re_t = \sqrt{\overline{u^2}}L / v$ is the turbulent Reynolds number based on the integral scale. Therefore a measure of the effect of the rate of reaction on the smallest concentration gradients of C_A and C_B across an interface is the ratio of this time-scale to that of the reaction rate $[k(C_A C_B)^{1/2}]^{-1}$, i.e. the Damköhler number for the micro-scale motions

$$Da_{Kol} = (\frac{v}{\varepsilon})^{1/2}k(C_A C_B)^{1/2} \tag{2.44}$$

If $C_A C_B$ is represented by the initial concentrations,Da_{Kol} may be given by $Da_{Kol} = (\frac{v}{\varepsilon})^{1/2}k(C_{A0}C_{B0})^{1/2}$. If $Da_{Kol}{>}1$, then the reaction occurs faster than the time for the smallest eddies to feed the interface layer with the species A and B. This results in C_A tending to zero on B's side of the interface and vice versa. This kind of reaction-diffusion layer occurs in flames but not in more slowly reacting chemical engineering problems where the time-scale of the reaction is more comparable with the integral scale of the turbulent motions T_L. For a moderately fast or slow reaction, an integral-scale Damköhler number is defined by

$$Da_I = T_L k(C_{A0}C_{B0})^{1/2} \tag{2.45}$$

and it is related to Da_{Kol} by

$$Da_I = Da_{Kol} Re_t^{1/2} \tag{2.46}$$

Significant changes in the development of the reaction are found when Da_I is of order 1. If $Da_{Kol} {<}1$, the micro-mixing is the same as for a non-reacting scalar, but wherever this mixing has occurred there is slow decay of C_A and C_B. Further, to promote the chemical reaction, the turbulent motion has to distort or stretch the interface layer occupied by the chemical product and to promote the turbulent and molecular mixing between A and B. High shear is expected to aid this process.

Thus the key dimensionless parameters for a turbulent mixing process with reaction are Re_t, Sc, $\tau[=T_L(d\bar{U}/dy)]$ and Da_I or Da_{Kol}. Several computations showed that when Re_t, Sc and τ are given, the segregation parameter α is uniquely determined by Da_I. Also it was found that the ratio of the initial concentrations $\beta(=C_{A0}/C_{B0})$ also significantly affects α. This suggested that when we estimate α in a reacting flow we have to consider the effect of β in addition to the effects of Da_I, Re_t, Sc and τ.

(b) Comparisons with the measurements of α

Figure 2.3 shows the comparisons of the predictions of the segregation parameter α with the measurements for six reacting and one non–reacting flows. In the seven flows, the segregation parameter, i.e., the concentration fluctuations were directly measured. In a non–reacting grid–generated turbulence($Da_I=0$) Komori et al.[31] measured the concentration fluctuations by using a combined laser–induced fluorescence and Mie–scattering technique. For the non–reacting flow, α grows from a negative value to zero with the relative time of mixing t/T_L. The measurements[31] shown by solid circles rather well agree with the predictions. In a grid–generated turbulence with a rapid reaction ($Da_I=1.7\times10^8$), Komori et al.[32] measured the concentration fluctuations by using a combined laser–induced fluorescence and electrode–conductivity technique. The measured value of α approach -1. This means that the chemical reaction is so fast that the interfacial part where A coexists with B becomes narrow. The predictions also agree with the measurements. For the reacting plume of Bennani et al.[27] in grid–generated liquid turbulence, the predicted value of α is a little larger than the measurements, but it tends to settle to a constant value of about 0.55 in the developed mixing region of $t/T_L>10$.

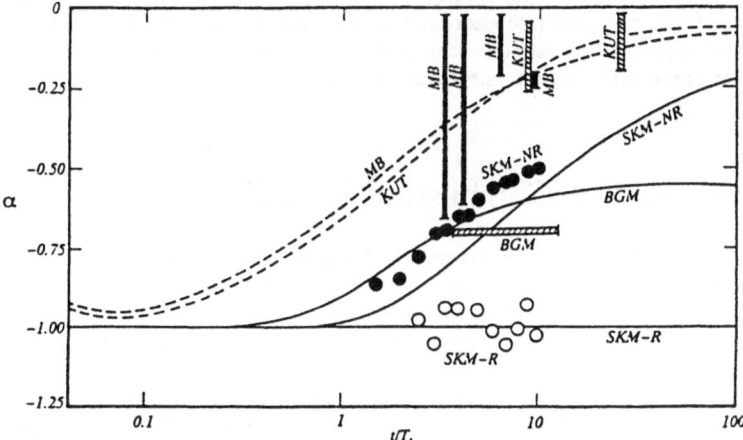

Fig. 2.3 Comparisons of the segregation parameter α between the predictions (curves) and the measurements(data bands and open and solid circles):MB,counter jets[26];BGM,grid–generated liquid turbulence[27];KUT,atmospheric surface layer[28];SKM–NR,non–reacting grid–generated liquid turbulence[31];SKM–R,grid–generated liquid turbulence[32,33], from [28] with permission.

For the reacting counter jets of Mudford & Bilger[26] with a low Schmidt number of $Sc=1$, the values of α measured at four locations plotted against t/T_L have long data bands in the figure. The prediction curve passes through their data bands except for the relative time of $t/T_L=6.3$. In particular, the measured α at $t/T_L=9.5$ is in good agreement with the prediction.

For the two-dimensional reacting plume in the atmospheric surface layer[28], the value of α measured at two positions are scattered, like those of Mudford & Bilger[26], but they seem to show a slight increase of α with the relative time of mixing t/T_L, again like Mudford & Bilger[26]. The predicted α also shows a similar increase, passing through the average of the measured α. Thus, the stochastic model can well explain the measurements, though there exists quantitative difference.

(c)Summary of the simulations

The main results from the numerical simulations by the above Lagrangian stochastic model can be summarized as follows.

(1)The relative mixing time t/T_L, the turbulent Reynolds number Re_ρ, the Schmidt number Sc, the Damköhler number based on the integral time-scale Da_I and the ratio of the initial concentrations β are the important parameters which determine the segregation parameter α, and therefore the mean chemical reaction rate in a non-premixed reacting flow. By estimating these five parameters, the present model enables us to indicate an approximate value for the segregation parameter α. Also, it can qualitatively explain previous measurements of the segregation parameter, about which there has been substantial disagreement among the experimentalists.

(2)In the initial region near the source, the segregation parameter is negative and it grows towards zero with increasing mixing time. The growing rate for $Sc{\approx}1$ is larger than for $Sc{\approx}1000$. For a high Damköhler number, the segregation parameter remains at a large negative value, even for a large mixing time, and therefore for a large mixing time the chemical reaction rate decreases in the downstream region.

(3)Shear can rapidly increase the segregation parameter towards zero, and therefore promote the chemical reaction rate even for a high Damköhler number.

(Satoru Komori)

2.4 Estimation of Combustion Models

A combustion model is estimated frequently through the comparison of simulated result with experimental one. Usually profiles of concentration and temperature are used in that comparison, because the local reaction rate cannot be directly measured. However, since their profiles are also influenced largely by a turbulence model, it is difficult to estimate the combustion model properly with this method. Therefore, it is desirable to obtain local reaction rate from experimental results anyway and compare it with simulated one. Such situation is the same also in the estimation of diffusion coefficient. In the present study[34], from this point of view, local reaction rate and effective diffusion coefficient were tried to obtain by numerical calculation using measured values for flow velocity, concentration and temperature. Then, by comparing these results with simulated ones, estimation of models for reaction rate and diffusion coefficient was conducted. A coaxial jet diffusion flame of hydrogen/nitrogen mixture was used as an object of the study.

2.4.1 Local reaction rate

Under the assumption that effective diffusion coefficients D_{eff} are the same for all chemical species, a conservation equation of parabolic type is written for hydrogen element which is strictly conserved matter without a source term in the equation. Equation(2.47) was obtained by integrating that equation from central axis to any stream line[35].

$$\frac{\partial}{\partial x}\int_0^{r_s} \bar{\rho}\tilde{U}r\tilde{m}_H \, dr = [\bar{\rho}D_{eff} \, r\frac{\partial \tilde{m}_H}{\partial r}]_{r=r_s} \qquad (2.47)$$

where r_s is radial distance of the stream line, U is flow velocity in x–direction, m_H is mass fraction of hydrogen element, and "~" and "–" indicate Favre average and conventional time–average, respectively. Using hydrogen element as a tracer, effective diffusion coefficient D_{eff} can be calculated by substituting measured values to Eq.(2.47).

Equation(2.48) is obtained from a conservation equation of chemical species i through the same operation as in Eq.(2.47).

$$\frac{\partial}{\partial x}\int_0^{r_s} \bar{\rho}\tilde{U}r\tilde{m}_i dr - [\bar{\rho}D_{eff} \, r\frac{\partial \tilde{m}_i}{\partial r}]_{r=r_s} = \int_0^{r_s}\bar{R}_i \, rdr \qquad (2.48)$$

Local reaction rate R_i can be calculated by substituting D_{eff} obtained with Eq.(2.47) and measured values to Eq.(2.48).

2.4.2 Modeling

The basic equations consist of Favre averaged conservation equations of mass, momentum and scalar, to which boundary layer approximation was applied. A modified k–ε model was used as a turbulence model, considering the so–called laminarization phenomenon due to combustion. The model will be explained later in Section 2.6.

A combustion model used was the conserved scalar approach, where an one–step irreversible reaction with fast chemistry was assumed and the probability density function (p.d.f.) was introduced for mixture fraction f to take account of turbulent mixing. A clipped Gaussian distribution was used as p.d.f., and a form of p.d.f. at local position was specified in terms of the mean and the variance of f, obtained from the solution of their transport equations.

2.4.3 Results and discussion

Figures 2.4 and 2.5 show radial profiles of effective diffusion coefficient and integrated reaction rate $\int R_i \, rdr$ for three cross sections. In each figure, (a) is the result calculated with measured values, which will be called experimental result hereafter, and (b) is the simulated result. The experimental results of diffusion coefficient and reaction rate proved to be reasonable, because their profiles were correspond well to the measured radial profiles of fluctuating velocity, concentration and temperature. Meanwhile, comparing (a) and (b) in each figure, the simulation is found to represent the experimental result fairly well. It means that the combustion model used was proper in this simulation. It may be concluded in this study that the technique developed here, to obtain local reaction rate with experiment and numerical calculation, can provide a good method for the estimation of combustion models.   (Yoshiaki Onuma)

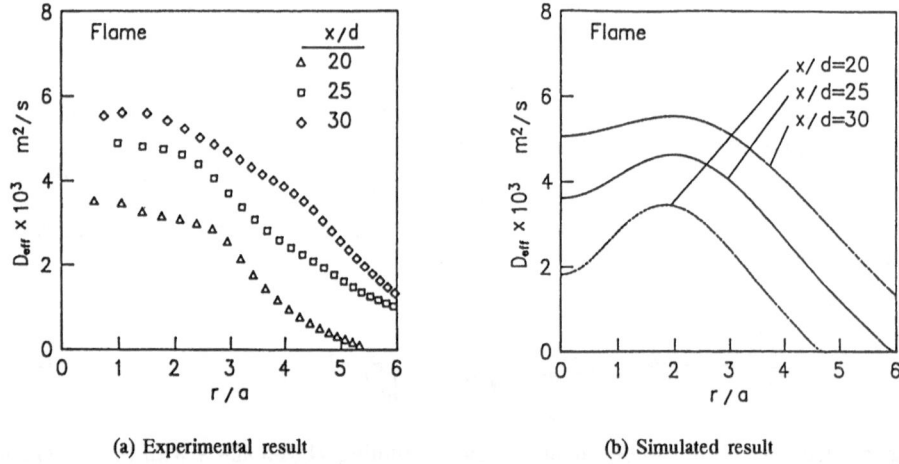

(a) Experimental result   (b) Simulated result

Fig. 2.4   Radial profiles of effective diffusion

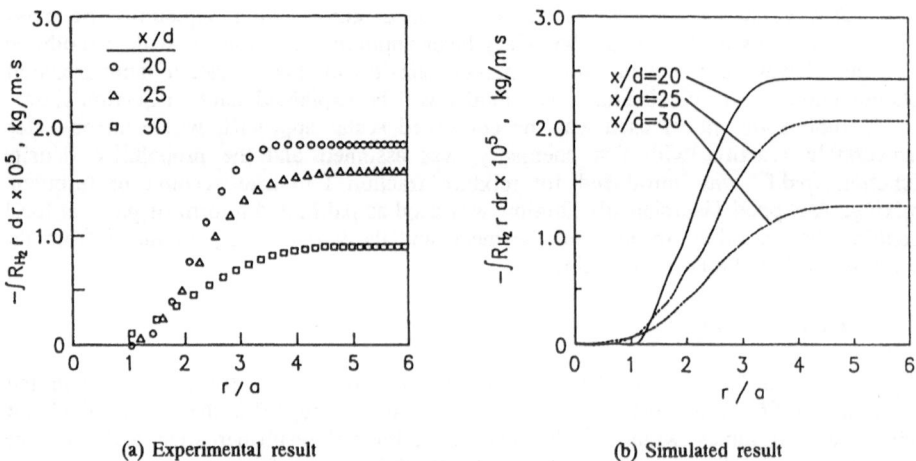

(a) Experimental result   (b) Simulated result

Fig. 2.5   Radial profiles of integrated reaction

2.5 Simulation of the Vortex Generation and Mixing Process in Gas Jets

In various combustors where gas fuel jets are used,the combustion process is dominated by the mixing of fuel and ambient air due to turbulence or vortices generated by the jets. In this study, the vortex generation and mixing process in gas jets are studied fundamentally by numerically simulating the behavior of two–dimensional(plane) gas jets using a higher–order finite difference method.

2.5.1 Method of numerical analysis

For simulating the vortex generation in jets, it is necessary to solve the Navier-Stokes(N–S) equations either directly or by using the LES (Large Eddy Simulation) method [36]. These methods, however, need a lot of computation time. Alternatively, as a practical method, the third–order upwind differencing scheme may be applicable for this purpose. In this study, the UTOPIA(Uniformly Third–Order Polynomial Interpolation Algorithm) scheme [37], whose numerical viscosity is small, is used for the convection terms in the N–S equations without any turbulence models. Gas is assumed to be incompressible fluid and a predictor–corrector algorithm [38] is used for pressure–velocity calculation. For solving the transport equation of fuel concentration accurately,the CIP(Cubic–Interpolated Pseudo–Particle) method [39] is applied. This method can calculate the convection of a substance in a Lagrangian manner. Therefore, its numerical diffusivity is sufficiently small and furthermore over/under–shooting is much smaller than ordinary higher–order schemes. In this study, the concentration c denotes the mass fraction of the injected gas (fuel) which is assumed to have the same physical properties as those of the ambient air.

2.5.2 Numerical simulation

(a)A Steady free jet

A two–dimensional free jet issuing at constant velocity(5m/s,20m/s) from a slit(10mm width) on a wall into quiescent air is simulated. A calculation grid of 111×71 is used. Figs.2.6(a) and (b) show the calculated steam lines and concentration c at an instant. It is observed that pairs of symmetric vortices are succeedingly generated in the upstream region and they undergo a transition to Kármán vortices in the downstream region and consequently the concentration distribution meanders according to the motion of these vortices. Obviously, the jet is not under the steady state although the issuing velocity is kept constant. This calculated result agrees qualitatively with experimental observations. Figs.2.6(c) and (d) show the mean stream lines and the mean concentration distribution averaged during 240ms (about 12000 time steps), respectively. The time–averaged velocity u and concentration c distributions in y(cross–stream) direction at a distance of x=150mm from the jet outlet are shown in Figs.2.6(e) and (f), where b_u and b_c are half the half–value widths of velocity and concentration distributions in the section,and also $\bar{u}_m$ and c_m are the maximum values of velocity and concentration. In Figs.2.7(a)–(d), the variations in the calculated values of b_u, b_c, $\bar{u}_m$ and c_m with distance x are compared with empirical formulae [40]. (d=width of the jet outlet. Subscript 0 denotes the jet outlet.)

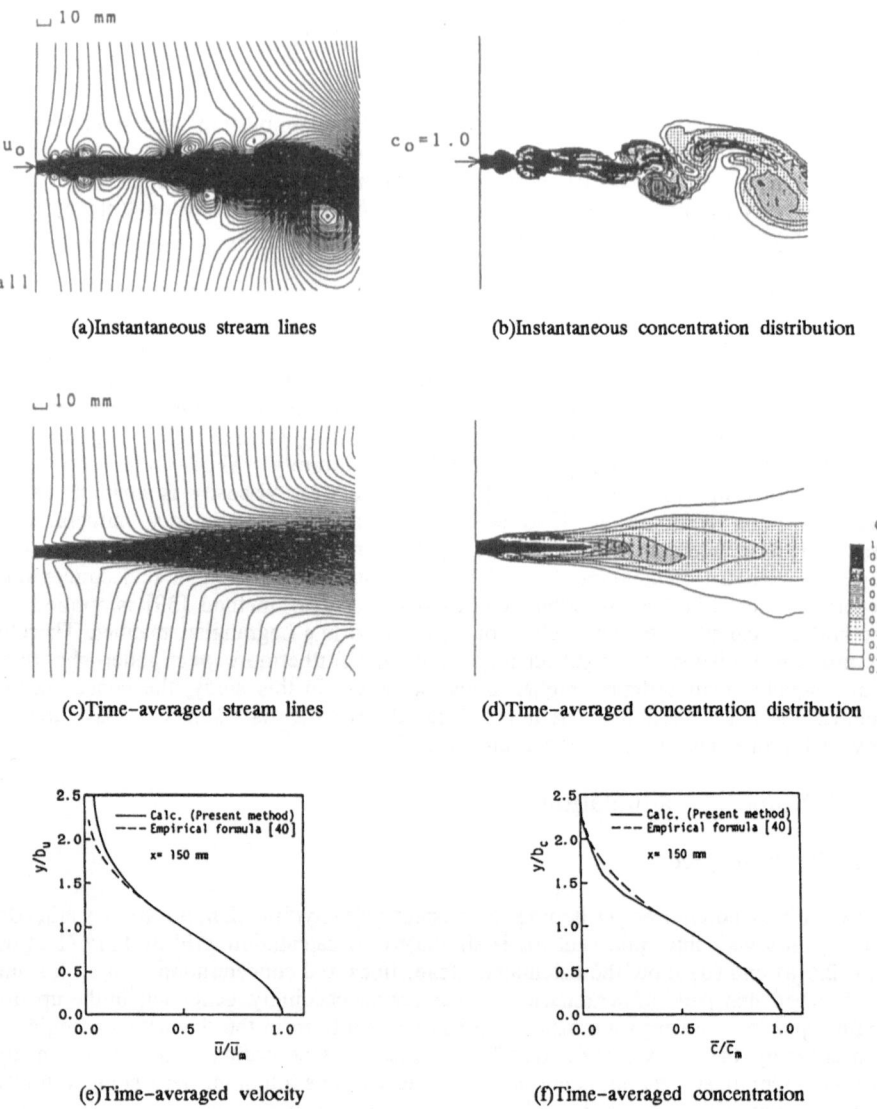

(a)Instantaneous stream lines   (b)Instantaneous concentration distribution

(c)Time-averaged stream lines   (d)Time-averaged concentration distribution

(e)Time-averaged velocity   (f)Time-averaged concentration

Fig. 2.6 Stream lines and concentration distribution in a 2-dim,free jet issuing at a constant velocity
         (u_0=20m/s)

It is shown that the time-averaged values calculated by the present method agree
relatively well with the empirical curves. As a result, it is found that this calculation
method can reproduce well both instantaneous and mean fields of the velocity and
concentration in the jet. For reference, the calculated values of $b_u, b_c, \bar{u}_m$ and c_m by a
conventional method(the control volume method with the k-ε turbulence model)are
shown in Figs.2.7(a)-(d).

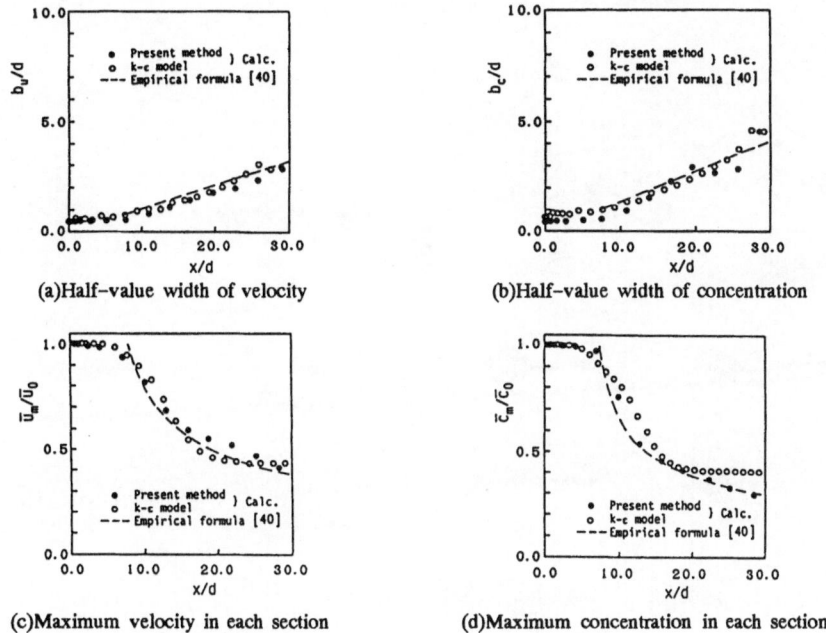

(a)Half-value width of velocity

(b)Half-value width of concentration

(c)Maximum velocity in each section

(d)Maximum concentration in each section

Fig. 2.7 Comparison of the calculated results with the empirical formulae [40] for a 2-dim. free jet issuing at a constant velocity (u_0=5m/s)

(b)An impulsively started jet

The behavior of a two-dimensional gas jet which starts impulsively from a slit(5mm width) on a wall into equiscent air is simulated. The issuing velocity is increased linearly from 0 at t=0 to u_0=50m/s at t=1 ms and then kept constant. A calculation grid of 100×85 is used. Fig.2.8 shows the temporal variations of velocity , concentration and pressure distributions. It is found that initially a pair of large vortices are generated at the tip of the jet, and afterwards they separate from the succeeding second pair of large vortices which grow by merging the following small vortices"2", "3" and "4"(see numerals on vortices). This second pair of vortices "a" and "b"(see alphabets on vortices) with a vortex "c" and the other vortices "d" and "e" are diverted from the jet center-line in opposite directions. According to such vortex motion, the concentration distribution is folded complicatedly.

(c) A jet impinging on a wall

The behavior of a two-dimensional gas jet which impinges on a wall is simulated. As shown in Fig.2.9, the jet starts impulsively from a slit (5mm width) into quiescent air between two parallel walls under the same issuing condition as that in Fig.2.8. The temporal variations of velocity and concentration distributions are demonstrated in Fig.2.9. It is found that a pair of vortices generated initially at the tip of the jet develop along the wall just after impingement,and afterwards they roll up so that the ambient air is engulfed.

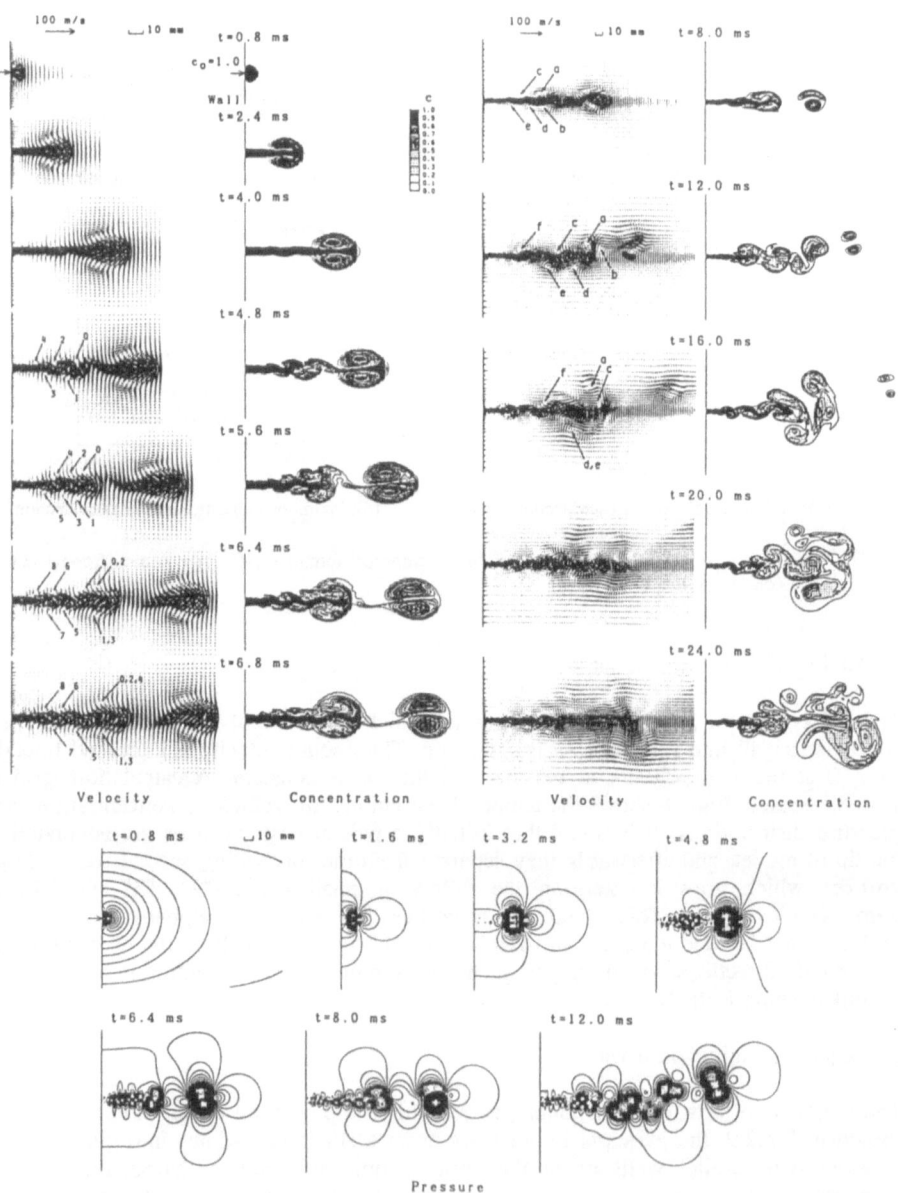

Fig. 2.8 Velocity,concentration and pressure distributions in an impulsively started 2-dim. jet (u_0=50m/s)

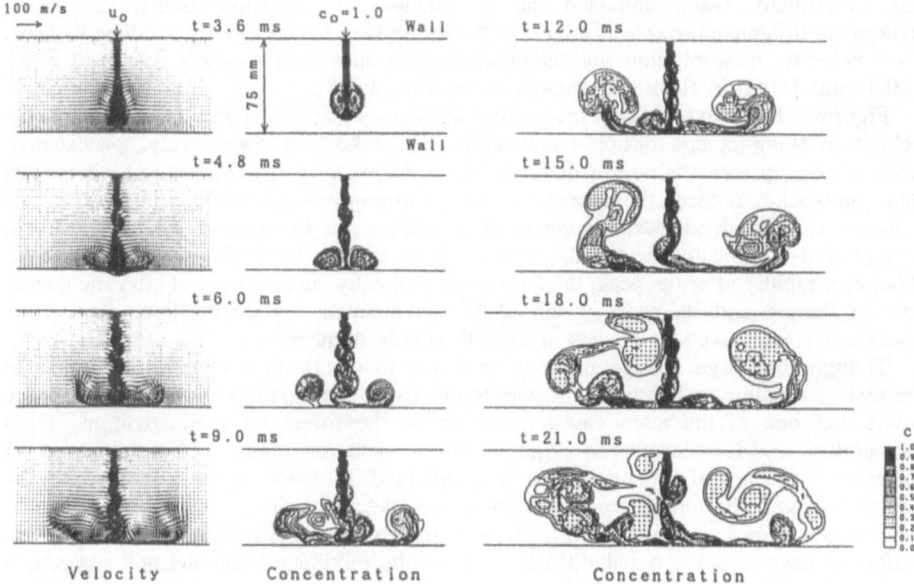

Fig. 2.9 Velocity and concentration distributions in a 2–dim. jet impinging on a wall (u_0=50m/s)

2.5.3 Conclusions

By simulating both steady and transient gas jets, it has been revealed that pairs of
vortices are succeedingly generated in the upstream region, and some of them grow by
combination and meander due to mutual interaction. Such behavior of vortices plays an
important roll on the mixing process of the injected gas and ambient air.

(Yuzuru Shimamoto and Tomoyuki Wakisaka)

2.6 Laminarization Due to Combustion and Its Modeling

The phenomenon that the development of turbulence is suppressed around nozzle exit
in jet diffusion flames has been known for a long time as the laminarization due to
combustion [8]. This is supposed to be caused by following mechanism. The increase
of kinematic viscosity due to high temperature decreases local turbulence Reynolds
number in flames. The balance between generation and dissipation of turbulence existing
in high Reynolds number regions is not preserved in low Reynolds number regions and
the dissipation rate is superior to the generation rate, which results in the suppression of
turbulence. In the present study, the effect of this laminarization on the flame structure
and its importance were experimentally examined, and a model was proposed to
represent the laminarization phenomenon.

2.6.1 Laminarization due to combustion

An experiment was conducted on a coaxial jet diffusion flame, in which hydrogen/nitrogen mixture was issued vertically upward parallel to sorrounding air flow. Flow velocity, concentration and temperature were measured on three flames of 5000, 10000 and 15000 in Reynolds number of nozzle jets[41].

Figures 2.10(a) and (b) show the profiles of time–averaged velocity U and fluctuating velocity u' along jet axis for non–combustion and combustion, respectively. x is distance from nozzle tip and d is nozzle diameter 6mm. Comparing (a) and (b) in U, it is seen that combustion reduces the decreasing rate of time–averaged velocity; hence, a high velocity is maintained even downstream in flames. In fluctuating velocity profiles, though non–combustion jets make a sharp peak near nozzle exit and the turbulence dissipates rapidly after the peak, the turbulence is greatly suppressed just after the nozzle exit in flames and its peak is shifted far downstream by combustion. And these phenomena exist even in the jets of high Reynolds number.

Though this large change of flow field due to combustion may be attributed to thermal expansion or buoyancy, it was found from examination of the experimental result that one of the main cause seems to be the foregoing laminarization. High temperature and low–turbulence regions exist in upstream region and periphery of jet diffusion flames, and the turbulence is suppressed in those regions because of the laminarization. The turbulence suppression may delay the development of turbulence field in jet flames comparing with non–combustion jets. The reduced turbulence in periphery decreases the radial diffusion rate or the mixing rate of fuel and oxygen. It means that the laminarization gives a great influence to the whole structure of jet diffusion flames.

2.6.2 A triple jet diffusion flames[42]

The applicability of various models to turbulent diffusion flames is usually examined by comparing simulated result with experimental one. The jet diffusion flame is often used for that test to minimize the numerical error, because it forms two–dimensional boundary

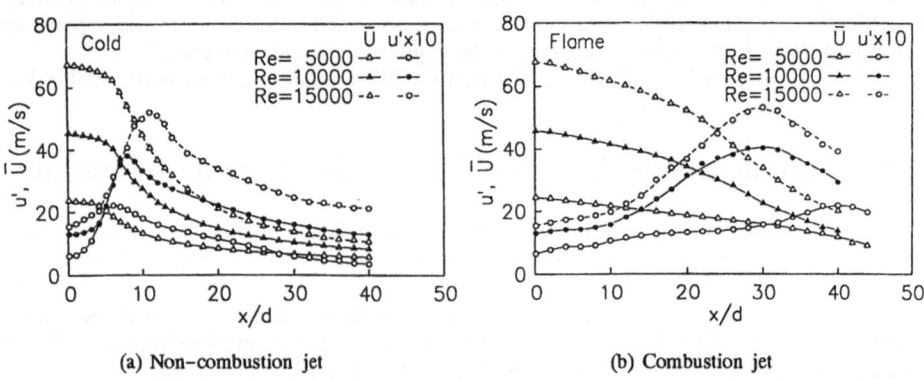

(a) Non–combustion jet   (b) Combustion jet

Fig. 2.10 Axial profiles of time–averaged and fluctuating velocities in a jet diffusion flame, from [41] with permission.

layer. The agreement of simulated result with experimental one is much worse in jet diffusion flames than in cold jets. It may be considered to be one major reason for the discrepancy that, though turbulence suppression caused by the laminarization is significant in jet diffusion flames, an ordinary turbulence model is applicable only to fully turbulent region, but not to low–turbulence laminarized by combustion.

As shown before, the laminarization happens mostly in the periphery, because that region of relatively small turbulence becomes hot due to combustion. Then, a triple coaxial jet diffusion flame was formed using a combustion apparatus shown in Fig.2.11. In this flame, since a broad turbulent region is formed in the shear layer between the high and low velocity air flows, all the combustion takes place within this turbulent region and all the high–temperature regions exist in the high–turbulence flow field. As the local turbulence Reynolds number is kept high even in high–temperature region, the laminarization effect is considered to be weak in this flame. After the experiment, numerical simulation was conducted on this combustion flow field using the ordinary k–ε model as a turbulence model and the calculated result was compared with the experimental one.

The basic equations used in the modeling consist of Favre averaged conservation equations of mass, momentum and scalar, to which boundary layer approximation was applied. The combustion model used is the conserved scalar approach shown in Section 2.4.

Figure 2.12 shows the profiles of time–averaged and fluctuating velocities along jet axis in combustion and non–combustion. Figures (a) and (b) are the results of experiment and simulation, respectively. Comparing (a) and (b), it is seen that the calculated result agrees with the experimental one in the flame as well as in the cold jet. Such agreement was obtained in the radial profiles as well. As mentioned before, numerical simulation using the k–ε model does not present good prediction for usual coaxial jet diffusion flames. In contrast to this, in the triple jet diffusion flame the calculated result agreed well with the experimental one. This fact may suggest that turbulent combustion flow fields of parabolic type can be predicted well using the ordinary k–ε model and the Favre averaging, if the region of low turbulence Reynolds number is not formed by combustion or the influence of the laminarization is weak.

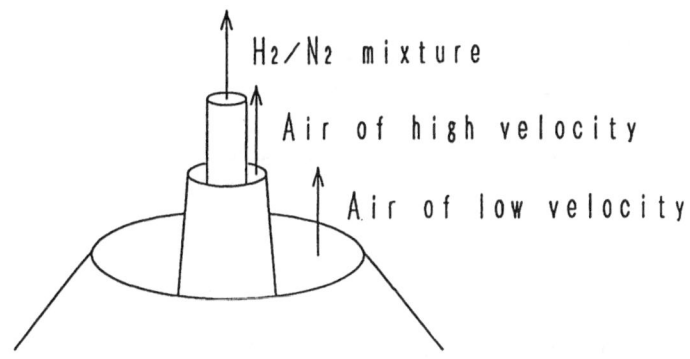

Fig. 2.11   A combustor of triple jet diffusion flames

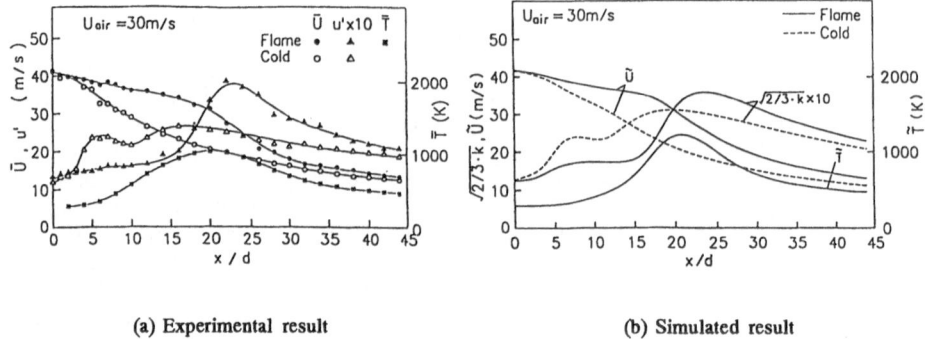

(a) Experimental result   (b) Simulated result

Fig. 2.12 Axial profiles of time–averaged and fluctuating velocities and temperature in a triple jet diffusion
flame

2.6.3 Modeling of the laminarization

The above study suggests that originating a turbulence model to represent the laminarization is an important subject in the modeling of turbulent combustion flow field. A turbulence model applicable to turbulent flames with the laminarization must represent the turbulence behavior in low Reynolds number flow field as well as in high Reynolds number. Such models have been proposed to represent a wall boundary layer flow with low–turbulence region[43]. Then, modification of the k–ε model was attempted based on those models.

The basic equations and combustion model used here are the same as mentioned above. A modified k–ε model represented by Eqs.(2.49)–(2.53) was used as a turbulence model.

$$\tilde{U}\frac{\partial k}{\partial x} + \tilde{V}\frac{\partial k}{\partial r} = \frac{1}{r}\frac{\partial}{\partial r}\{r(\nu+\frac{\nu_t}{\sigma_k})\frac{\partial k}{\partial r}\}+\nu_t(\frac{\partial \tilde{U}}{\partial r})^2-\varepsilon \qquad (2.49)$$

$$\tilde{U}\frac{\partial \varepsilon}{\partial x} + \tilde{V}\frac{\partial \varepsilon}{\partial r} = \frac{1}{r}\frac{\partial}{\partial r}\{r(\nu+\frac{\nu_t}{\sigma_\varepsilon})\frac{\partial \varepsilon}{\partial r}\}+C_{\varepsilon 1}\frac{\varepsilon}{k}\nu_t(\frac{\partial \tilde{U}}{\partial r})^2-C_\varepsilon f_2\frac{\varepsilon^2}{k} \qquad (2.50)$$

$$\nu_t = C_\mu f_\mu \frac{k^2}{\varepsilon} \qquad (2.51)$$

f_μ and f_2 are empirical constants and $f_\mu=f_2=1$ in the ordinary k–ε model. The form of these functions was decided to be Eqs.(2.52) and (2.53) through computer optimization in numerical simulation of hot air jets[44].

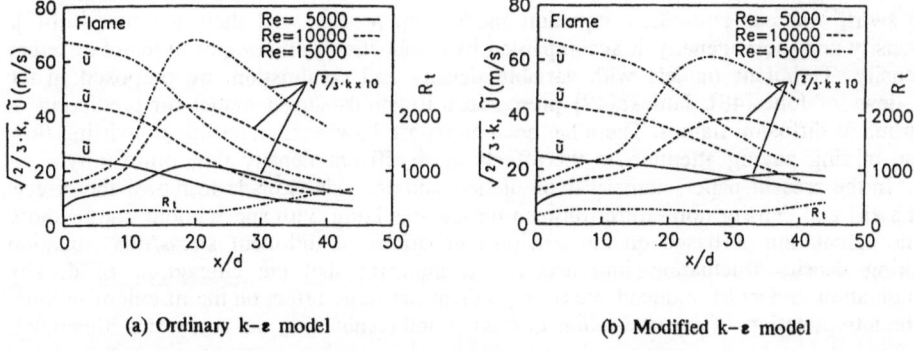

(a) Ordinary k-ε model   (b) Modified k-ε model

Fig. 2.13   Simulated axial profiles of time–averaged velocity and turbulence kinetic energy in jet diffusion flames, from [44] with permission.

$$f_\mu = 1-0.6\exp\{-(R_t / 250)^2\} \tag{2.52}$$

$$f_2 = 1-0.05\exp\{-(R_t / 1000)^2\} \tag{2.53}$$

R_t is local turbulence Reynolds number, which was calculated from $R_t = k^2/\nu\varepsilon$.

Figure 2.13 shows simulated results corresponding to the experimental ones indicated in Fig.2.10(b). Figures (a) and (b) were calculated using the ordinary k–ε model and the modified k–ε model, respectively. Comparing the two simulations and the experiment, it is found that the prediction was greatly improved by taking account of the laminarization with the modification functions. The same effect was noticed in the radial profiles. These results may suggest that the consideration to the laminarization due to combustion is important in the modeling of turbulent combustion flow fields.

(Yoshiaki Onuma)

2.7 Simulation of Mixing and Combustion in Swirling Flames

Turbulent swirling flames are often encountered in furnaces and combustors. Swirling flows have been used to stabilize the flame and to obtain rapid combustion. Counter effects of swirl, however, have been reported in the study of a turbulent swirling flame formed in a circular tube[45]. The turbulent mixing is retarded and consequently, the flame is elongated when swirl is introduced to the surrounding air in a confined tube. Understanding the effects of swirl on turbulent transport induced by interactions of turbulence, swirl–induced pressure gradient and density non–homogeneity is essential to control mixing and combustion in swirling flows.

Previous numerical computations of non-reacting turbulent swirling flows have shown that k-ε two equation model fails to predict the characteristics of swirling flow and mixing, which are those of laminarization phenomena and retardation of mixing due to swirl, whereas stress/flux equation model can predict well their features[46][47]. Density non-homogeneity is accompanied by combustion or mixing of fluids of different density. Turbulent models with variable density and combustion are proposed in the reviews of Jones[48]. Janicka[49] presented a Reynolds-stress model for prediction of turbulent diffusion flames. There has been no report, however, on turbulent swirling flow and mixing paying attention to the effects of significant density non-uniformity.

In the present paper, numerical prediction and analysis is made to reveal the effects of swirl and density non-uniformity on turbulent mixing with and without combustion. The calculation is based on the transport equations of turbulent stress/flux equation taking density fluctuation into account. It indicates that the interaction of density fluctuation and swirl-induced pressure gradient has large effect on the turbulent mixing. The interpretation of the retardation of mixing and combustion due to swirl is illustrated.

2.7.1 Formulation

Local velocity and concentration of gas species are evaluated from conservation equations of mass, momentum and gas species and transport equations of turbulent fluxes in density-weighted Favre(mass weighted)-averaged form in an axisymmetric cylindrical coordinate system. By considering that there is no recirculation zone due to swirl in the present condition where swirling flow goes through a straight round tube without expansion and swirl strength is not so large, boundary layer approximation is valid. The conservation equations are:

$$\frac{\partial}{\partial x}(r\bar{\rho}\tilde{U}) + \frac{\partial}{\partial r}(r\bar{\rho}\tilde{V}) = 0 \tag{2.54}$$

$$\bar{\rho}(\tilde{U}\frac{\partial\tilde{U}}{\partial x} + \tilde{V}\frac{\partial\tilde{U}}{\partial r}) = -\frac{\partial\bar{P}}{\partial x} - \frac{1}{r}\frac{\partial}{\partial r}(r\bar{\rho}\ \widetilde{u''v''}) + \bar{\rho}\ \bar{g} \tag{2.55}$$

$$\bar{\rho}(\tilde{U}\frac{\partial\tilde{W}}{\partial x} + \tilde{V}\frac{\partial\tilde{W}}{\partial r} + \frac{\tilde{V}\tilde{W}}{r}) = -\frac{1}{r^2}\frac{\partial}{\partial r}(r^2\bar{\rho}\ \widetilde{v''w''}) \tag{2.56}$$

$$\frac{\partial\bar{P}}{\partial r} = -\frac{\partial}{\partial r}\bar{\rho}\tilde{v}'^2 + \frac{\bar{\rho}}{r}(\tilde{W}^2 + \tilde{w}'^2 - \tilde{v}'^2) \tag{2.57}$$

$$\bar{\rho}(\tilde{U}\frac{\partial\tilde{M}_j}{\partial x} + \tilde{V}\frac{\partial\tilde{M}_j}{\partial r}) = -\frac{1}{r}\frac{\partial}{\partial r}(r\bar{\rho}\ \widetilde{v''m_j''}) + \bar{R}_j \tag{2.58}$$

where x and r are coordinates in axial and radial directions, respectively. Double prime["] designates fluctuation with respect to Favre-averaged quantity. u'', v'' and w'' are fluctuations of velocity components with respect to Favre-averaged axial, radial and tangential velocity U, V, and W, respectively. m_j'' is the fluctuation of gas species concentration with respect to Favre-averaged gas species concentration M_j. $^-$ and $^\sim$ indicate conventional and Favre average, respectively. ρ is fluid density; P, pressure; R_j, reaction rate of j species. As for energy conservation equation, M_j and m_j are replaced by enthalpy h, and R_j is set to zero. As for conservation equation for mixture fraction F which indicates the mass fraction of nozzle fluid reduced before reactions, M_j and m_j are replaced by F and f, respectively and R_j is equal to zero. Transport of momentum or scalar by turbulent motion, represented by correlation of two fluctuating velocities or fluctuating velocity and gas species concentration in Eq.(2.55)–(2.58), are evaluated from stress/flux equation model in Favre–averaged form. In the present computations, transport equations of turbulent stress and flux are modeled following Launder, Reece and Rodi[50] for pressure strain–rate term and Daly and Harlow[51] for diffusion term in the stress transport equation and Launder[52] for pressure scrambling term and diffusion term in the scalar–flux transport equation. The modeled terms in density–unweighted averaged form are re–written in terms of density–weighted ones.

The transport equation of $v''f''$ is shown below in an axisymmetric coordinate system with boundary layer approximation.

$$\tilde{U} \, \frac{\partial \overline{v''f''}}{\partial r} + \tilde{V} \, \frac{\partial \overline{v''f''}}{\partial r} - \frac{\overline{w''f''}}{r} \tilde{W}$$

$$= - v'^2 \frac{\partial \tilde{F}}{\partial r} + \frac{\overline{w''f''}}{r} \tilde{W} - \overline{f''} \frac{\partial \overline{P}}{\partial r} \frac{1}{\bar{\rho}} - C_{1c} \frac{\bar{\varepsilon}}{k} \overline{v''f''} - C_{2c} \frac{\tilde{W}}{r} \overline{w''f''}$$

$$- C'_{1c} \frac{\bar{\varepsilon}}{k} \{ \frac{1}{k} (\tilde{v}'^2 \, \overline{v''f''} + \overline{v''w''} \, \overline{w''f''} + \overline{u''v''} \, \overline{u''f''}) - \frac{2}{3} \, \overline{v''f''} \}$$

$$+ 2 \frac{\partial}{\partial r} C_c \frac{k}{\bar{\varepsilon}} (\tilde{v}^2 \frac{\partial \overline{v''f''}}{\partial r} - \frac{1}{r} \overline{v''w''} \, \overline{w''f''}) - \frac{2}{r^2} C_c \frac{k}{\bar{\varepsilon}} (\tilde{w}^2 \, \overline{v''f''} + \overline{v''w''} \, \overline{w''f''})$$

$$+ \frac{2}{r} C_c \frac{k}{\bar{\varepsilon}} (\tilde{v}^2 \frac{\partial \overline{v''f''}}{\partial r} - \overline{v''w''} \frac{\partial \overline{w''f''}}{\partial r}) \qquad\qquad (2.59)$$

The first–third terms in Eq.(2.59) which are denoted by $Pv''f''1$, $Pv''f''2$ and $Pv''f''3$ are the production terms in transport equation of $\overline{v''f''}$. f''^2 is evaluated from its transport equation[48]. The empirical constants in Eq.(2.59) is adopted from Launder[52]. For computations in a flame, equilibrium assumption is used where reactions go to equilibrium as soon as the fuel and the oxidizer are mixed. Thermochemical state of the mixture was determined in terms of the mixture fraction F.

2.7.2 Analysis and discussions

The authors studied experimentally the effects of swirl on turbulent flow, mixing and combustion in a tube[45]. Turbulent swirling or non–swirling air flows are supplied in a tube of 60mm inner diameter and gaseous fuel (mixture of C_3H_3 and H_2 whose volume rate is 1:1) issues from a round tube nozzle installed coaxially at the central axis. The computed axial velocity U and tangential velocity W, in the swirling flame by using above methods were proved to predict well the experimental ones.

In Fig.2.14 experimental and computed mixture fraction F along the central axis are compared for swirling and non–swirling flames. X denotes distance from the nozzle tip. Experimental mixture fraction in the swirling flow is higher as compared with that in the non–swirling flow in the downstream region. It is noted that turbulent mixing is retarded due to swirl. The calculation can predict the retardation of mixing due to swirl.

Figure 2.15 illustrates the processes of promotion and suppression of turbulent mixing in the swirling flame based on the evaluation of production terms in the transport Eq.(2.59) for $\widetilde{v''f'}$. There are three production terms as indicated in Eq.(2.59). The term $Pv''f''1(=-v''^{2}\partial\tilde{F} / \partial r)$ is concerned with radial gradient of $\tilde{F}$ and contributes significantly to production of $\widetilde{v''f'}$. It has an effect to drive gradient diffusion in radial direction. The term $Pv''f''2(= \widetilde{w''f'}\tilde{W} / r$) including swirl velocity component $\tilde{W}$ becomes a negative production term which reduces the magnitude of $\widetilde{v''f'}$ and has an effect to suppress the turbulent transport. The term $Pv''f''3(=-\overline{f''}(\partial\overline{P} / \partial r)\times 1 / \overline{\rho})$ includes radial pressure gradient $\partial\overline{P} / \partial r$ which becomes large in a swirling flow due to centrifugal force. It is a negative production term and has an effect of suppression on turbulent mixing because lighter(burnt) gas prevails in the central part of the swirling flame. $Pv''f''3$ consists of the term including density fluctuation and swirl–induced radial pressure gradient $\partial\overline{P} / \partial r$.

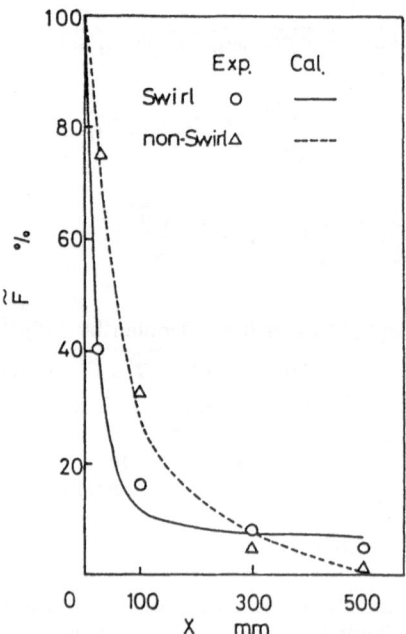

Fig. 2.14 Comparison of computed and experimental mixture fraction along the central axis.

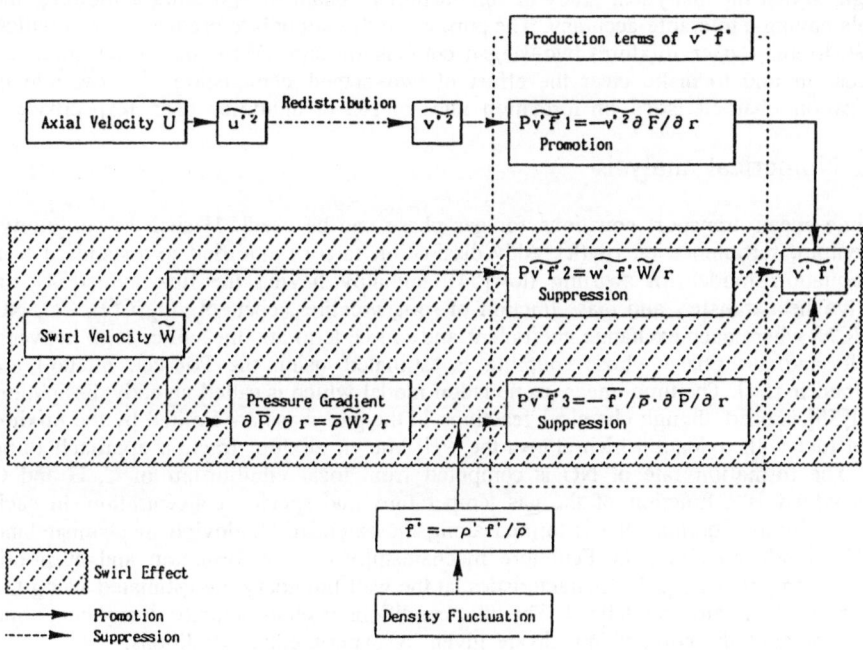

Fig. 2.15 Process of promotion and retardation of mixing in swirling flows.

2.7.3 Conclusions

Factors affecting turbulent mixing were investigated in variable density swirling flames based on the computations employing transport equations of stress and fluxes.

The computations can predict the retardation of turbulent mixing due to swirl in the flow with combustion. Retardation of turbulent mixing is induced by two production terms which arise in the transport equation of radial flux of mixture fraction. One of the production terms includes swirl velocity and the other involves density fluctuation and pressure gradient. Both of them become negative production terms and have effect of suppression on turbulent mixing in swirling flame.

(Toshimi Takagi and Shuichirou Hirai)

2.8 Modeling of Spray Combustion of Slurry Fuels

In the spray combustion simulation the following items are solved simultaneously; atomization characteristics of liquid fuel, turbulent flow pattern accompanied with swirling flow, temperature field transferred by convection and radiation heats, and reaction rates by combustion. However details of each item have not been elucidated enough so that the analytical study of combustion mechanism is desired combining the models having a tolerable accuracy. The purpose of this study is to predict characteristics of PWM(pitch–water mixture) two–staged combustion and CWM(coal–water mixture) combustion and to make clear the effect of two–staged combustion air flow rate or atomization characteristics on formation rates of NO or unburned char, respectively.

2.8.1 Numerical analysis

All time–mean transport equations for gas phase can be used[53] such as continuity, momentum, turbulence energy(k), eddy dissipation rate(ε) modified to a standard k–ε two–equation model for swirling flow[54], enthalpy having the flux model[54] for radiative heat transfer, and mass fraction of soot, volatiles, char, N_2, O_2, CO_2, CO, NO and HCN, where each exchange coefficient and source term and the source term between particle and gas phase whose coupling is estimated using the PSI cell method are described in [54]. The time–mean combustion model which is called an eddy dissipation model[55] is used, though chemical reactions in the gas phase are limited by the mixing rate of fuel with oxidant which allows the well known mixture fraction approach to be used. The formation rate of NO is computed from local equilibrium of C, H and O atoms which is a function of the gas temperature and species concentration in each control volume. Thermal NO is formed along the extended Zeldovich mechanism and fuel NO produced along the Fenimore mechanism[56]. Soot formation and burn–out rates are computed by [57]. Characteristics at the wall boundary are estimated using the ordinary wall function model[58]. The inlet conditions such as velocity components and temperatures of the combustion air are given by experimental conditions.

(a) Dispersion model of the droplet

The atomization process is modeled as shown in Fig.2.16 under the following assumptions as the liquid atomization mechanism is not yet resolved; 1) the position of

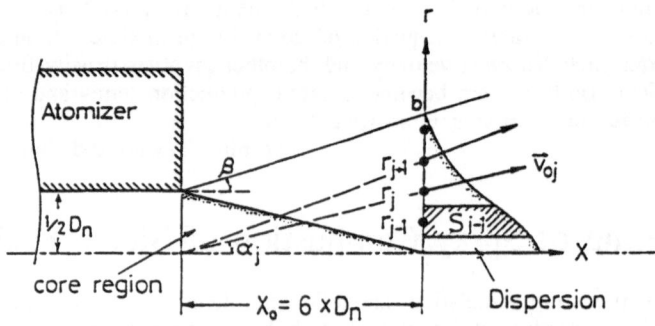

Fig. 2.16 Droplet dispersion model

ejection is at a burner exit, 2) droplets are already spread out toward the radial direction at the end of a core region, $x=x_0(x_0=6Dn,$ Dn=nozzle diameter), beyond which there is no break-up or coalescence of droplets. For convenience of computing trajectories, let the location (x_0,r_j) be the initial point of the computation. The number fraction P_j ejected at the initial point to the total droplet numbers is expressed as;

$$P_j = 2S_j r_j \Delta r / M_l, \quad S_j = S_{max}\exp[-0.65(r_j/b)^2\{1+0.027(r_j/b)^4\}] \quad (2.60)$$

where S_j is the dispersion in the annular region between radius $(r_j-\Delta r/2)$ and $(r_j+\Delta r/2)$. S_j is based on the correlation for the concentration distribution in an axisymmetrical co-axial jet[59]. S_{max} is a normalizing constant, such that

$$M_l = 2\int_0^b S_j r_j dr_j, \quad b = (12\tan\alpha+1)D_n / 2 \qquad (2.61)$$

where M_l is the liquid mass flow rate, b is the jet radius and α is the ejection angle. The droplet number rate, n_{ij}, having the diameter D_{pi} ejected from the core end is P_iP_jN, where N is a total ejected droplet number per unit time and P_i is the droplet number fraction with D_{pi} whose frequency is followed by the logarithmic normal distribution.

First of all, the atomization experiment is carried out to measure the droplet size distribution and dispersion at 150[mm]below the burner tip, while the dispersion is estimated based on the Lagrangian treatment under the same condition as the experiment. The initial condition such as P_j is then given to adjust the simulated dispersion to the experimental result at the same location.

(b) Droplet reaction model

The slurry droplet reaction model is shown in Fig.2.17. The model of a reaction process is divided into four steps; 1) heating up to 373[K] by the radiative and convective heat transfer, 2) evaporation of water at 373[K] including the blowing effect[53], 3) devolatilization of pitch or coal with swelling whose index is determined by the experiment and 4) char combustion (the fragmentation cannot be correlated quantitatively). The convective and radiative heat transfer rates, the combustion heat release of particle and the latent heat of evaporation are included in the droplet energy conservation equation.The devolatilization process is modeled by the single overall reaction [54]. The rate coefficients are given by Arrhenius type expressions whose constants are quoted from an another experiment[54] as shown in Table 2.1 with the

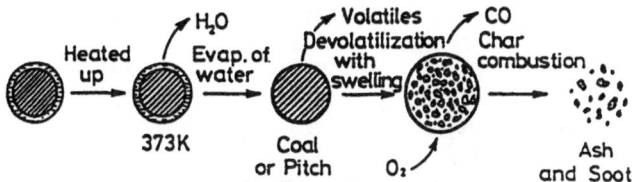

Fig. 2.17 Slurry droplet reaction model

proximate and ultimate analyses of pitch and coal. Following devolatilization the char burnout is also given based on the overall reaction rate coefficient that is limited by both rates of chemical reaction k_c and oxygen diffusion k_d[54]. Carbon in the char is once converted to CO with a heterogeneous reaction and then CO converts to CO_2 with a homogeneous reaction whose rate is estimated by the eddy dissipation model[55].

(c) Numerical solution

The prediction procedure is based on a conventional axial–symmetrical solution of the governing time–averaged gas phase conservation equations on to which are superimposed the computations of the particle flights and the particle/gas interactions being handled by appropriate source terms appended to the gas phase equations. The SIMPLE method is applied to compute the derivative of pressure. The Lagrangian droplet equations are simultaneously integrated with RKF method for each droplet having any droplet number rate which is divided into 7 classes in the initial ejection angle and 15 classes in the initial droplet diameter distribution. The number of grid nodes for the furnace is 45 for axial nodes and 30 for radial ones.

Table 2.1 Proximate and ultimate analyses of pitch and coal and operating conditions for PWM and CWM combustion

Species		PWM	CWM
Water content [wt%]		30	2
Proximate analysis [wt%]			
Volatiles		38.9	28.37
Char		60.88	53.6
Fly ash		0.22	18.03
Ultimate analysis [wt%]			
C		86.22	58.51
H		5.69	4.36
N		1.19	1.26
S		5.69	0.42
O		–	7.42
Devolatilization parameters			
Frequency factor	$[s^{-1}]$	$3.69*10^2$	$5.5*10^2$
Activation energy	$[kJmol^{-1}]$	27.4	78.7
Operating condition			
Fuel flow rate	$[kghr^{-1}]$	150	80
Primary air flow rate	$[Nm^3hr^{-1}]$	102	20
temperature	[K]	283	283
Secondary air flow rate	$[Nm^3hr^{-1}]$	1140	362
temparature	[K]	423	523
Excess air ratio	[–]	1.32	1.2

2.8.2 Experimental apparatus and conditions

The experiments are carried out in a 1.4[m] I.D. and 6.5[m] length horizontal furnace[53] surrounded with water–cooled stainless steel jackets for PWM. The two–staged combustion air is introduced through one tube located at 1.07[m] away from the burner tip. On the other hand, the furnace for CWM is 0.6[m] I.D. and 5[m] length[54]. Both operating conditions are given in Table 2.1.

2.8.3 Simulation of combustion characteristics for PWM and CWM

(a) Analysis of two–staged combustion for PWM

Figure 2.18 shows an example of effect of two–staged combustion air rate fraction to all air rates on NO and soot concentrations in the exhaust gas. The estimated NO has the maximum value at about 30[%] of the two–staged air fraction and furthermore the unburned matter increases with it[53].

(b) Analysis of CWM combustion

Figure 2.19 shows a comparison of estimated unburned ratio with results measured on the centerline of the furnace. The maximum unburned ratio is obtained at the largest M_a/M_l in both results[53], though the average droplet diameter as shown in the figure is decreased with an increase of M_a/M_l. Fig.2.20 shows the effect of swirl number on the unburned ratio and NO concentration in the exhaust gas with variations of effective swirl number S_{eff} which is computed by correcting the tangential momentum from inlet to exit of swirler through a straight tube[54]. The optimum condition to decrease the unburned ratio is found out, while there are some differences between experimental and estimated

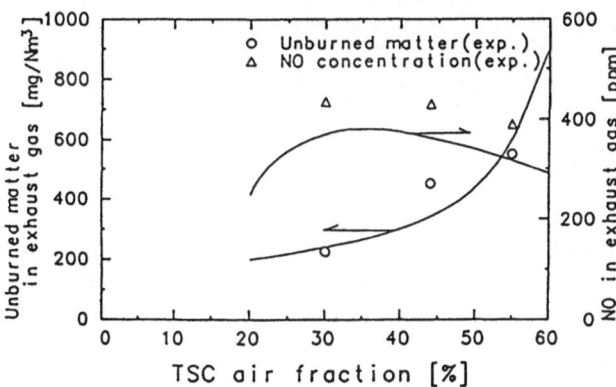

Fig. 2.18 Effect of two–staged combustion air rate fraction to all air rates on NO and unburned matter concentrations in the exhaust gas.

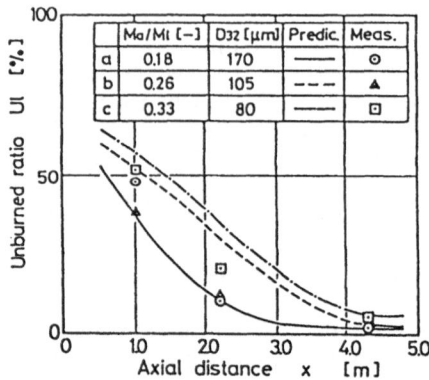

Fig. 2.19 Comparison of estimated unburned ratio with results measured on the center-line of the furnace

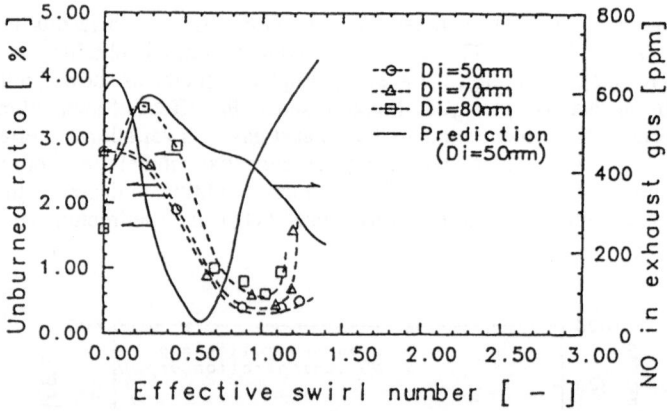

Fig. 2.20 Effect of swirl number on the unburned ratio and NO concentration in the exhaust gas with variations of S_{eff}

results. On the other hand, the highest NO concentration is estimated at S_{eff} =0.3. Both results suggest that the optimum operating condition in this furnace exists at a swirl number of S_{eff} =0.5.

2.8.4 Conclusions

A simulation for the pitch–water and coal–water slurry combustion has been performed and the simulated results for the unburned ratio and NO concentrations are in good agreement with measurements. This simulation method would be applied to study the optimum operation and design of slurry combustion furnace with the two–staged or swirling flow system.   (Takatoshi Miura and Hideyuki Aoki)

References

[1] Pope SB and Whitelaw JH (1976) The calculation of near–wake flows. J. Fluid Mech. 73:9–32

[2] Hirai S, Takagi T and Higashiya T (1989) Numerical prediction of flow characteristic and retardation of mixing in a turbulent swirling flow. Intl. J. Heat Mass Transfer 32:121–130

[3] Nikjooy M, Karki KC, Mongia HC, McDonell VG and Samuelsen GS (1989) A numerical and experimental study of coaxial jets. Intl. J. Heat Fluid Flow 10:253–261

[4] Senda M, Nishimura M, Hayama K and Taira T (1991) Numerical analysis of an axisymmetrical confined jet with a bluff body. Trans. Jpn. Soc. Mech. Eng., Ser. B, 57:360–365

[5] Jones WP and Whitelaw JH (1984) Modeling and measurements in turbulent combustion. 20'th Symp.(Intl.) on Combust., The Combustion Institute, pp233–249

[6] Roquemore WM, Bradley RP, Stutrud JS, Reeves CM and Britton RL (1983) Influence of the vortex shedding process on a bluff body diffusion flame. AIAA Paper 83–0335

[7] Scholefield DA and Garside JE (1953) The structure and stability of diffusion flames. 3'rd Symp.(Intl.) on Combust., The Williams and Wilkins Company, pp102–110

[8] Takagi T, Shin HD and Ishio A (1980) Local laminarization in turbulent diffusion flames. Combust. Flame 37:163–170

[9] Lee CE and Onuma Y (1991) Experimental study of turbulent diffusion flames stabilized on a bluff body. Trans. Jpn. Soc. Mech. Eng., Ser. B, 57:276–281

[10] Bilger RW (1977) Reaction rates in diffusion flames. Combust. Flame 30:277–284

[11] Liew SK, Bray KNC and Moss JB (1981) A flameret model of turbulent non–premixed combustion. Combust. Sci. Tech. 27:69–73

[12] Peters N (1983) Local quenching due to flame stretch and non–premixed turbulent combustion. Combust. Sci. Tech. 30:1–17

[13] Janicka J and Kollmann W (1979) A two–variables formalism for the treatment of chemical reactions in turbulent H_2–air diffusion. 17'th Symp.(Intl.) on Combust., The Combustion Institute, pp421–430

[14] Bilger RW (1980) Perturbation analysis of turbulent non–premixed combustion. Combust. Sci. Tech. 22:251–261

[15] Pope SB (1981) A Monte Carlo method for PDF equations of turbulent reactive flow. Combust. Sci. Tech. 25:159–174

[16] Peters N (1984) Laminar diffusion flameret models in non–premixed turbulent combustion. Progr. Energy Combust. Sci. 10:319–339

[17] Seshadri K, Peters N (1988) Asymptotic structure and extinction of methane–air diffusion flames. Combust.Flame 73:23–44

[18] Peters N, Donnerhack S (1981) Structure and similarity of nitric oxide production in turbulent diffusion flames. 18'th Symp.(Intl.) on Combust., The Combustion Institute, p33–42

[19] Peters N (1990) Length scales in laminar and turbulent flames, to appear in:"Numerical Approaches to Combustion Modeling " Oran ES, Boris JP, Eds.

[20] Starner SH, Bilger RW (1985) Characteristics of a piloted diffusion flame designed for study of combustion turbulence interactions. Combust. Flame 61:29–38

[21] Dibble RW, Long MB, Masri A (1985) Two–dimensional imaging of C_2 in turbulent non–premixed jet flames, 10'th Intl. Colloquium on Dynamics of Explosions and Reactive Systems, Berkeley

[22] Donnerhack S, Peters N (1984) Stabilization heights in lifted methane–air jet

diffusion flames diluted with nitrogen. Combust. Sci. Technol. 41:101–108

[23] Ishizuka S, Tsuji H (1981) An experimental study of effect of inert gases on extinction of laminar diffusion flames. 18'th Symp.(Intl.) on Combust., The Combustion Institute, pp.695–703

[24] Toor HL (1969) Turbulent mixing of two species with and without chemical reaction. Indust. Engng Chem. Fundam., 8:655–659

[25] Komori S and Ueda H (1984) Turbulent effects on the chemical reaction for a jet in a nonturbulent stream and for a plume in a grid–generated turbulence. Phys. Fluids, 27:77–86

[26] Mudford NR and Bilger RW (1984) Examination of closure models for mean chemical reaction rate using experimental results for an isothermal turbulent reacting flow. 20'th Symp.(Intl.) on Combust., The Combustion Institute, pp387–394

[27] Bennani A, Gence JN and Mathieu J (1985) The influence of a grid–generated turbulence on the development of chemical reaction. AIChE J., 31:1157–1166

[28] Komori S, Hunt JCR, Kanzaki T and Murakami Y(1991) The effects of turbulent mixing on the correlation between two species and on concentration fluctuations in non-premixed reacting flows. J.Fluid Mech.,228:629–659

[29] Sawford BL and Hunt JCR (1986) Effects of turbulence structure, molecular diffusion and source size on scalar fluctuations in homogeneous turbulence. J. Fluid Mech., 165:373–400

[30] Durbin PA (1980) A stochastic model of two–particle dispersion and concentration fluctuations in homogeneous turbulence. J. Fluid Mech., 100:279–302

[31] Komori S, Kanzaki T and Murakami Y (1989) Simultaneous measurements of instantaneous concentrations of two species being mixed in a turbulent flow by using a combined laser–induced fluorescence and laser–scattering technique. Phys. Fluids A, 1:349–351

[32] Komori S, Kanzaki T and Murakami Y (1991) Simultaneous measurements of instantaneous concentrations of two reacting species in a turbulent flow with a rapid reaction. Phys. Fluids A, 3:507–510

[33] Komori S, Kanzaki T and Murakami Y (1991) Concentration statistics in shear–free grid–generated turbulence with a second–order rapid reaction. in Advances in Turbulence 3 , ed. Johansson A and Alfredsson H, Springer, pp.271–278

[34] Furushima K, Aoyama K and Onuma Y (1991) Local reaction rates in a jet diffusion flames. 29'th Symp.(Jpn.) on Combust., pp640–642

[35] Senecal JA and Shipman CW (1979) Mass transfer and reaction rates in a ducted propane–air diffusion flame. 17'th Symp.(Intl.) on Combust., The Combustion Institute, pp355–362

[36] Moin P, Mansour NN, Reynolds WC and Ferziger JH (1979) Large eddy simulation of turbulent shear flows. Proc. 6'th Intl. Conf. on Numer. Meth. in Fluid Dynamics, Springer–Verlag, pp 400–409

[37] Leonard BP (1981) Computational Techniques in Transient and Turbulent Flow, Vol.2, Pineridge Press

[38] Wakisaka T, Shimamoto Y and Isshiki Y (1989) Simulating high–Reynolds number flows through constricted passages by means of third–order upwind differencing. Numerical Methods in Fluid Dynamics II (Proc. Intl. Symp. on Computational Fluid Dynamics), pp 732–740

[39] Yabe T and Takei E (1988) A new higher–order Godunov method for general hyperbolic equations. J. Phys. Soc. Jpn., pp 2598–2601

[40] Ishigaki H (1982) Studies on the properties of turbulent jets (1st and 2nd reports). Trans. Jpn. Soc. Mech. Eng., Ser. B, 48:1692–1708

[41] Lee YJ, Hayashi T and Onuma Y (1990) The behavior of turbulence in hydrogen jet diffusion flames and isothermal hot air jets. Trans. Jpn. Soc. Mech. Eng., Ser. B, 56:359-365

[42] Komiya T and Onuma Y (1989) Studies on a triple jet diffusion flames (2nd Report, Numerical simulation). 27'th Symp.(Jpn.) on Combust., pp4-6

[43] Patel VC, Rodi W and Scheuerer G (1985) Turbulence models for near-wall and low Reynolds number flows: A review, AIAA J. 23:1308-1319

[44] Lee YJ and Onuma Y (1991) Modeling of turbulent jet diffusion flames (2nd Report, Application of the modified k-ε turbulence model to turbulent jet diffusion flames). Trans. Jpn. Soc. Mech. Eng., Ser. B. 57:339-345

[45] Takagi T, Okamoto T, Taji M, Nakasuji Y (1984) Retardation of mixing and counter-gradient diffusion in a swirling flame. 20'th Symp.(Intl.) on Combust., The Combustion Institute, pp 251-258

[46] Hirai S, Takagi T and Matsumoto M(1988) Prediction of the laminarization phenomena in an axially rotating pipe flow. Trans.ASME, J. Fluids Eng.110:424-430

[47] Hirai S, Takagi T and Higashiya T (1989) Numerical prediction of flow characteristics and retardation of mixing in a turbulent swirling flow. Intl. J. Heat Mass Transfer, 32:121-130

[48] Jones WP (1980) Models for turbulent flows with variable density, Lecture series, (ed. Kollmann W), Hemisphere, p379

[49] Janicka J (1986) A Reynolds-stress model for the prediction of diffusion flames. 21'st Symp.(Intl.) on Combust., The Combustion Institute, p 345

[50] Launder BE, Reece GJ and Rodi W (1975) Progress in the development of a Reynolds-stress turbulence closure. J. Fluid Mech. 68:537

[51] Daly BJ and Harlow FH (1970) Transport equations in turbulence. Phys. Fluids 13:2634

[52] Launder BE (1976) Turbulence,(ed. Bradshaw P), Springer-Verlag,p232

[53] Aoki H, Furuhata T, Tanno S, Miura T and Daikoku M (1991) Simulation of a spray combustion behavior for two kinds of slurry fuels (Effect of two stage air introduction and spray characteristics on combustion), ICLASS-91, Gaithersburg, MD, U.S.A., pp 499-506

[54] Aoki H, Furuhata T, Tanno S, Miura T and Ohtani S (1991) The effect of swirling flow on unburned ratio and NO concentration in a spray combustion system, Experimental Heat Transfer, Fluid Mechanics, and Thermodynamics 1991 (Keffer JF, Shah RK and Ganic EN, Eds.), Elservier Science Pub. Co. Inc., pp 575-582

[55] Magnussen BF and Hjertager BH (1976) On mathematical modeling of turbulent combustion with special emphasis on soot formation and combustion, 16'th Symp.(Intl.) on Combust., pp 719-729

[56] Scott CH, Smoot LD and Smith PJ (1984) Prediction of nitrogen oxide formation in turbulent coal flames, 20'th Symp.(Intl.) on Combust., p1391

[57] Abbas AS and Lockwood FC (1985) Prediction of soot concentrations in turbulent diffusion flames, J. Inst. Energy 58:112-115

[58] Launder BE and Spalding DB (1974) The numerical computation of turbulent flows, Comp. Methods Appl. Mech. Eng. 3:269-289

[59] Rajaratnam N (1976) Turbulent Jets, Elsevier Sci. Pub. Co.

Chapter 3
Spray Formation and Combustion

3.1 Diagnostics

3.1.1 Reviews

(a) Phase–doppler anemometry and spray measurements

Phase–Doppler anemometry is a measuring technique which developed out of laser Doppler anemometry to enable simultaneous measurements of velocity and size of particles whose diameters are larger than or comparable to the fringe spacing in the measuring volume. Based on general considerations on light scattering (see Durst [1]) and pertinent verification experiments, Durst & Zaré [2] demonstrated in 1975 that the general belief was incorrect that good laser Doppler signals could only result from particles that were much smaller than the fringe spacing. Using geometrical optics, fringe spacing in the scattered light field was correlated with the particle diameter. By means of a pair of detectors, phase–shifted signals were measured and used to verify the theoretical estimates of scattered fringe spacing.

In the years 1982 to 1984, Bachalo et al. [3], Saffman et al. [4] and Bauckhage & Flögel [5] reported extensions of the theoretical work by Durst & Zaré and computed the phase–diameter relationships using electromagnetic wave theory. The above authors also employed the technique for studying particulate two–phase flows. Since the inception of the phase Doppler technique, there has been a special interest in using it for droplet size and velocity measurements in various types of sprays. Commercial systems based on phase Doppler concept are nowadays readily available for studying particulate two–phase flows with low to moderate particle concentrations.

Recently, the phase Doppler technique has been extended to the measurement of droplet refractive index, in addition to the drop size and velocity. This extended technique is very useful for sizing of burning droplets in an internal combustion engine, where refractive indices of the drops are not known a priori and depend on their temperatures and compositions. Hence, the conventional phase Doppler system cannot be employed reliably in this application. On the other hand, the extended system allows not only to determine the drop size without an a priori knowledge of its composition but also provides a measure of the refractive index, which may be used to estimate the droplet temperature. Further work is underway to employ this method for in–situ measurements in internal combustion engines.   (Franz Durst and Amir Naqwi)

(b) The polarization properties of the scattered light to study the condensed phases in combustion systems

Sprays generated during the atomization of liquids are normally (but not always) formed by spherical droplets of size ranging from 1 to 1000 μm and therefore the Lorenz–Mie theory should exactly describe their light scattering and extinction properties. Based on this assumption a series of experiments has been carried out in the last few years by measuring the scattering and extinction coefficients inside hydrocarbon sprays, generated with different atomization procedures, in isothermal and burning conditions.

In a preliminary stage, the angular distribution of all the Stokes coefficients of scattered light was determined in the visible and it was found that the linearly polarized scattering coefficients Q_{VV} and Q_{HH} in the side region between $\theta = 80°$ and $\theta = 120°$ were the most promising features for the diagnostics of sprays formed by transparent liquids as water and light hydrocarbon mixtures [6]. Later on, the average size, number concentration, and volume fraction of droplets were determined inside different sprays by measuring the scattering coefficients and the polarization ratio $\gamma = Q_{HH}/Q_{VV}$ at $\theta = 90°$ [7]. The polarization ratio measurements were also applied to isothermal sprays formed by absorbing droplets and to sprays in burning conditions. In the first case it appeared that the forward scattering measurements, between $\theta = 30°$ and $70°$, allowed the determination of the droplet size, and in the second case it was possible to discriminate between conditions where submicronic soot particles were present in the scattering volume to those where the fuel droplets prevailed [7]. Furthermore, in the latter case, the change of the refractive index of the fuel droplets, due to liquid phase pyrolysis, was monitored with this technique [8]. More recently [9], it has been pointed out that the polarization peak in the backscattering region is peculiar of spherical objects around 1 μm. This feature allowed to detect the lower part of the droplet size distribution function inside a spray of heavy oil.

Although the method has shown an excellent space resolution and a sensitivity to the chemical nature of the scatterers, it suffered the limitation of giving a momentum of the size distribution function, since it was based on integrated light measurements of scattering by an ensemble of droplets. A first step toward the relaxation of this limit has been to measure the spectral behavior of the scattering coefficients due to droplets in a wavelength range in which their optical properties showed an appreciable dispersion, and then to obtain the size distribution function by inverting numerically the spectral scattering curve [10]. However, when a spray is dilute, a more direct way of obtaining the size distribution function is to measure the size of individual droplets over statistically significant number of measurements. Therefore, a new interferometric set–up for measuring, at selected scattering angles, both the polarization ratio and thus the diameter and/or the optical properties and the velocity of single droplets inside a spray, has been developed [11]. The measure of the imaginary part of the refractive index allowed to observe the absorptivity increase of heated–up droplets of hydrocarbon mixtures, which is related to the pyrolysis in the liquid phase of large aromatics compounds contained in the fuel [12]. Furthermore, the temperature of transparent droplets could be also determined by correlating the measured real part of the refractive index to the liquid density, and hence to its temperature [13].

To obtain the shape and the overall size distribution function inside practical combustion systems, time resolved measurements of the angular scattering coefficients and of the polarization ratio $\gamma = Q_{HH}/Q_{VV}$ were recently performed [9]. In this way, it was determined the size and the optical properties of the larger droplets, which passed individually through the scattering volume, and identified the average sizes and optical properties of the scatterers with typical size smaller than one micron, which passed as a cloud in the scattering volume.

To study unsteady diesel sprays either in evaporating regime or after ignition, the single point polarization ratio technique was extended to planar conditions in an optically accessible, high pressure, high temperature bomb [14]. The 2–dimensional approach permitted to find the instantaneous space distribution of the condensed phase discriminating between soot particles and fuel still in liquid phase [14,15].

(Antonio D'Alessio and Patrizio Massoli)

(c) 2–D soot imaging

The mechanism of soot formation in diesel flames has not been well understood because of the highly complex nature of the diesel flame in a high temperature and high pressure environment. Diagnostics which have high temporal and spatial resolution are desirable to improve our understanding of the mechanism of soot formation. Numerous studies have been devoted to the study of soot formation in diesel flames by means of various optical techniques, such as direct photography, shadowgraphy and Schlieren photography. However, these techniques lack spatial resolution along the line–of–sight direction.

In 1987, 1–D soot visualization along a laser beam was successfully applied to a diffusion flame[16]. Recently 2–D visualization techniques using a laser sheet have been applied to diesel flames in order to investigate the soot distribution with high spatial resolution[16–20]. Won et al.[18] obtained, using the 2–D elastic scattering technique, results that regions which show strong scattering from soot particles are composed of a large number of narrow strips or small clusters with a size of several millimeters, and that the intensity of scattered light by soot particles is high in the periphery near the flame head, whereas the scattering intensity in the inner region is weaker. They emphasized that the large–scale vortex structure in the flame head plays an important role in determining the soot distribution in diesel flames. Shioji et al.[19] also applied the 2–D elastic scattering technique to the flame in a direct injection diesel engine and found that it takes some time for soot particles to appear in the central region of a flame. They proposed two interpretations for this feature. One is that the temperature near the flame center is too low to be effective in soot formation, and the other is that soot clouds are formed only after a certain length of induction time, even though the temperature is sufficiently high. Dec et al.[20] employed the laser induced incandescence (LII) technique to visualize soot in a diesel flame, and found a general trend of a high soot concentration towards the leading edge of the combusting plume with a lower soot concentration extending upstream towards the injector. Although there are small differences in the observation of the 2–D soot concentration among these studies the trend is common that a high soot concentration is observed in the leading edge of the combusting plume.

(Takeyuki Kamimoto)

3.1.2 Phase–doppler anemometry and spray measurements

(a) The phase–doppler technique

A phase Doppler system is an extended version of laser Doppler anemometer (LDA) that employs at least two detectors in order to measure the phase shift between two signals. The transmitting and receiving optics of a typical phase Doppler anemometer (PDA) are shown in Fig. 3.1. The light beam from a laser source is split up into two beams, which are focused into a measuring volume in order to produce interference fringes. The particles crossing the measuring volume scatter light, which is collected by the detectors located in the Receiving Optics–1 and the Receiving Optics–2. The following relations yield the particle velocity and diameter using the ith Receiving Optics:

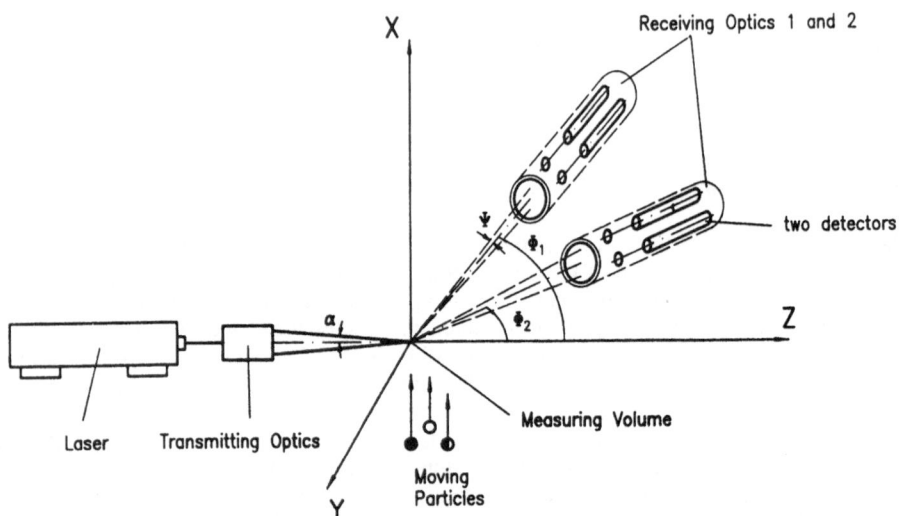

Fig. 3.1   Schematic layout of a phase Doppler system

$$u = \frac{\lambda}{2\sin\alpha} f_D \tag{3.1}$$

$$d_p = \frac{\Phi_i}{F_i(m)} \tag{3.2}$$

where λ, α and m are the wavelength of light, half–angle between the beams and the refractive index of the particle (relative to the surrounding medium) respectively. The signal frequency and the phase shift between two signals from the ith receiving optics are denoted by f_D and Φ_i respectively.

The transfer function $F_i(m)$ that relates the particle diameter to the measured phase Φ_i, may be expressed as follows:

$$F_i = -\frac{4\pi \, m \, \sin\alpha \, \sin\psi_i}{\lambda\sqrt{2\,(1+\cos\alpha\cos\psi_i\cos\phi_i)\,[1+m^2-m\sqrt{2\,(1+\cos\alpha\cos\psi_i\cos\phi_i)}]}} \tag{3.3}$$

where ϕ_i and ψ_i are the off–axis and elevation angles of the ith Receiving Optics respectively. As shown in Fig. 3.1, the off–axis angle is measured in the yz plane, whereas ψ_i represents the angle with which a detector is raised above or lowered below the yz plane. Equation (3.3) is valid for transparent particles, which act like spherical lenses for the incident light. Although fuel droplets are not completely transparent, the "spherical lens model" describes their behavior satisfactorilty.

The results of PDA measurements are unambiguous if the phase shift produced by the largest particle is smaller than 360°. In order to increase the size range, more than two detectors are needed, e.g. both the receiving units shown in Fig. 3.1 may be used simultaneously in order to obtain two phase signals necessary for overcoming the 360° phase ambiguity.

(b) Spray measurements

To illustrate applications of PDA in sprays, a summary of some recent investigations on
a Danfoss 60° oil burner nozzle are given here. Various tests were performed under
controlled environment using water as the flow medium, which was supplied to the
nozzle from a 120 liter tank whose pressure and temperature were maintained within
tolerance limits of ±1% and ±1°C respectively. The nozzle was mounted at the center
of a 300 mm diameter plexiglass tube, in which air pressure was regulated by a suction
type fan and flow straighteners were mounted at the air intake, so as to introduce the
secondary air axially.
    The PDA system consisted of a single receiving unit at an off–axis angle of 70°.
The beam angle and the elevation angle were adjusted, so as to measure water droplets
as large as 190 μm without phase ambiguity. The measured signals were digitized by a
100 MHz transient recorder and subsequently processed in a computer, where Fourier
transform was used to obtain the phase difference between the signals. At each
measuring location, 20,000 to 50,000 data points were taken in order to achieve a high
statistical reliability.
    Some of the results obtained in these investigations are presented in Fig. 3.2(a) &
(b), which show droplet frequency as a function of droplet diameter on a log–log scale.
In reference [21], more information is provided on these measurements. Although the
shapes of size distributions in the two diagrams in Fig. 3.2 are significantly different
from one another, the log–hyperbolic function exhibits an excellent fit to both the
measured distributions. Hence, the four parameters of the log–hyperbolic distributions
may be used not only to specify various moments of distributions at a given point but
also to describe the droplet migration trends within the spray cone.

(c) Extended PDA

In the previous example of PDA measurements, only one of the receiving units shown
in Fig. 3.1 was used. If both the receiving optics are empolyed, two phase signals are
available simultaneously. The ratio of these phase shifts, for small values of α and ψ_i
may be expressed as follows:

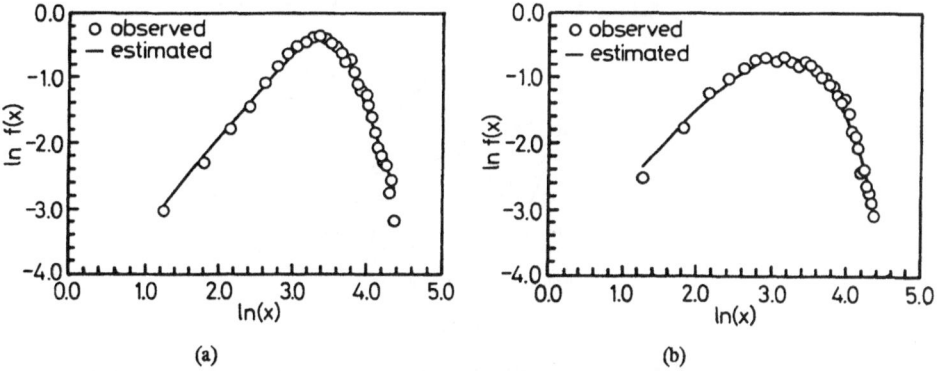

(a)   (b)

Fig. 3.2 The measured data and the fitted log–hyperbolic function at a distance of (a) 15 mm, (b) 25 mm from
    the nozzle.

$$\frac{\Phi_1}{\Phi_2} = \frac{\sin\psi_1}{\sin\psi_2}\sqrt{\frac{(1+\cos\phi_2)\,[1+m^2-m\sqrt{2}\,(1+\cos\phi_2)]}{(1+\cos\phi_1)\,[1+m^2-m\sqrt{2}\,(1+\cos\phi_1)]}} \qquad (3.4)$$

The above equation relates the measured phases to the particle refractive index, which is the main feature of extended phase Doppler anemometer (EPDA) [22]. It is clear from the above equation that the strongest dependence of the phase ratio on the refractive index is obtained if the off-axis angles ϕ_1 and ϕ_2 are largely different from one another. The available range of off-axis angles is indicated on Fig. 3.3(a) as the region containing horizontal lines on a plane of off-axis angle (scattering angle) versus particle refractive index. Only in this region scattered rays exist that are refracted by the particle and transmitted through it like the light rays through a lens. The vertical lines in the above diagram indicate the presence of internally reflected light. In Fig. 3.3(b), iso-intensity curves of the light refracted by a particle are shown with dotted lines. Solid lines are used to represent the iso-intensity curves of the light reflected from the surface of the particle, which is undesirable for the present application. In the case of fuel drops, the refractive index ranges from 1.2 to 1.6, so that the refracted light is at least 40 times

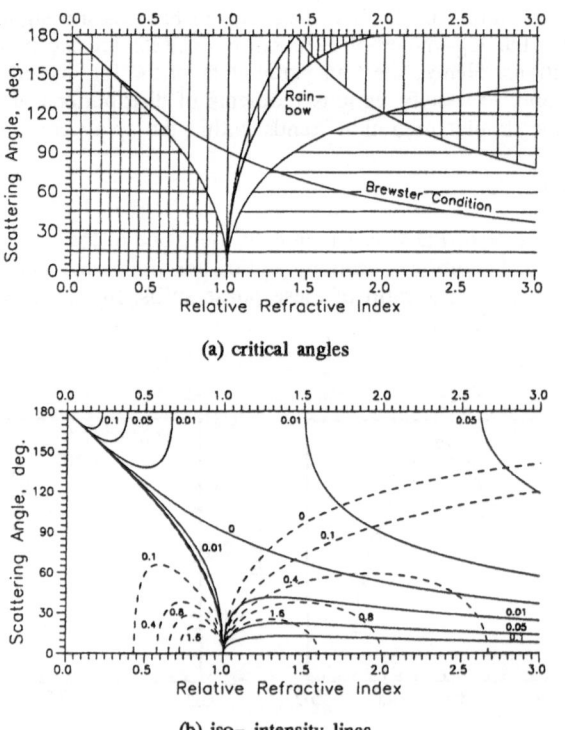

(a) critical angles

(b) iso- intensity lines

Fig. 3.3 Scattering characteristics as functions particle refractive index and off-axis angle

stronger than the reflected light for off–axis angles in the range 20°–70°. At smaller off–axis anlges, the performance of PDA is degraded due to presence of the light diffracted by the particles.

Based on the above considerations, verification experiments were conducted with an EPDA system having receiving units at $\phi_1 = 60°$ and $\phi_2 = 30°$. A mixture of glycerene and water was used in order to cover the refractive indices relevant to the fuel drops. The results, as shown in Fig. 3.4, exhibit a good agreement between experiments and theory. In Fig. 3.4(c), calculations based on Mie scattering theory are also included. Fluctuations in the rigorous calculations are comparable with the error bars in the experimental data.

(d) Concluding remarks

In this brief article, we have described the principle of phase Doppler anemometry and its extension to the measurement of droplet refractive index. The log–hyperbolic distributions are shown to be useful in reducing the large amount of phase Doppler data, taken in a spray. More recently, the principle of EPDA has been further refined and applied to fine droplets [23]. Furthermore, a number of projects on miniaturization of LDA and PDA instrumentation have been undertaken. Optical fibers [24] and semiconductor lasers and detectors are employed for this purpose [25]. These developments have significantly reduced the size of the equipment that is required in the vicinity of an IC engine.   (Franz Durst and Amir Naqwi)

3.1.3 The polarization properties of the scattered light to study the condensed phases in combustion systems

(a) Theoretical background and selected results

The solution for the scattering of a plane electromagnetic wave of wavelength λ by an isotropic and homogeneous sphere of diameter D and refractive index m=n–ik is exactly predicted, in the far field region, by the Lorenz–Mie theory [26]. It provides the expressions for the horizontally and vertically polarized scattering cross section in terms of infinite series:

$$C_{HH}(\alpha, m, \theta) = \left(\frac{\lambda}{2\pi}\right)^2 \sum_1^\infty \frac{2n+1}{n(n+1)} |\{a_n(m, \alpha)\, \pi_n(\theta) + b_n(m, \alpha)\, \tau_n(\theta)\}|^2$$

$$\text{(3.5)}$$

$$C_{VV}(\alpha, m, \theta) = \left(\frac{\lambda}{2\pi}\right)^2 \sum_1^\infty \frac{2n+1}{n(n+1)} |\{a_n(m, \alpha)\, \tau_n(\theta) + b_n(m, \alpha)\, \pi_n(\theta)\}|^2$$

where θ is the scattering angle, a_n and b_n are the scattering coefficients, and $\alpha=\pi D/\lambda$ is the size parameter. van de Hulst [26] has shown that the Lorenz–Mie theory is asymptotically equivalent to the ray optics theory for $\alpha \to \infty$ For finite values of a this assumption is only approximatively true because also the electromagnetic effects due to the surface waves have to be considered. In this case the angular polarized cross sections are given by the expressions:

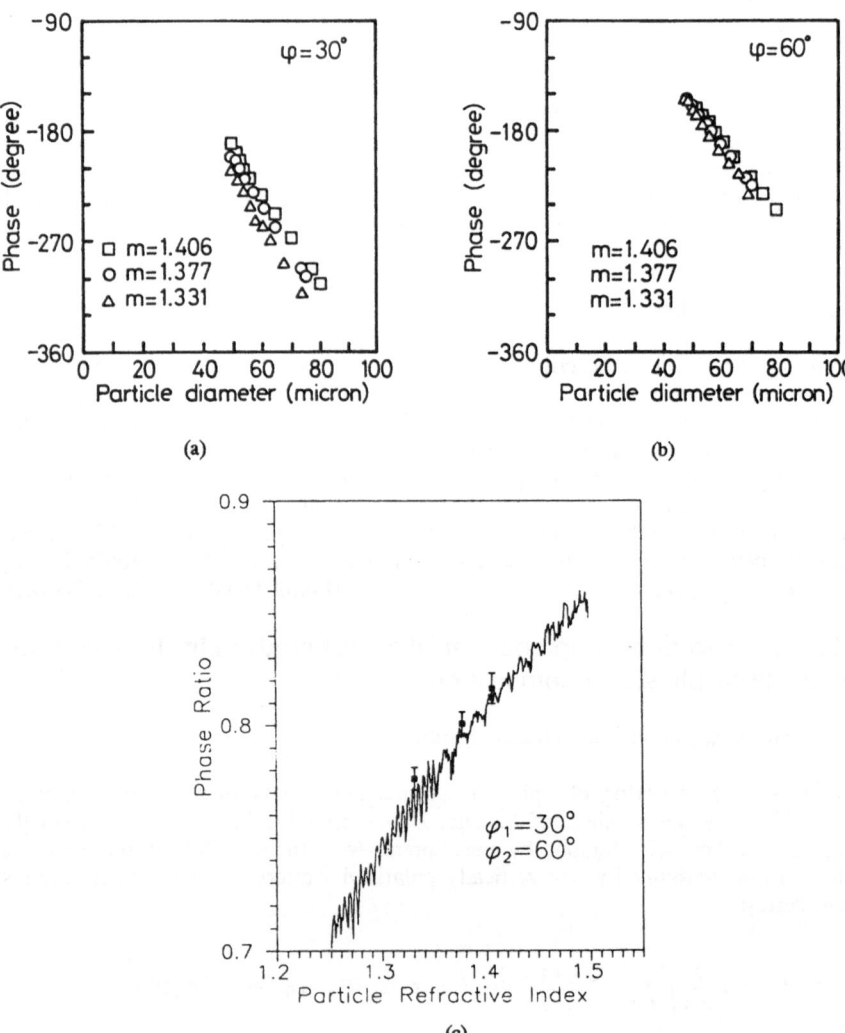

Fig. 3.4   Effect of refractive index on the phase Doppler data, (a) at φ=30°, (b) at φ=60°, (c) phase ratio

$$C_{HH}^{LM}(\alpha, m, \theta) = C_{HH}^{diff}(\alpha, \theta) + C_{HH}^{refr}(\alpha, m, \theta) + C_{HH}^{refl}(\alpha, m, \theta) + C_{HH}^{SW}(\alpha, m, \theta)$$
$$C_{VV}^{LM}(\alpha, m, \theta) = C_{VV}^{diff}(\alpha, \theta) + C_{VV}^{refr}(\alpha, m, \theta) + C_{VV}^{refl}(\alpha, m, \theta) + C_{VV}^{SW}(\alpha, m, \theta) \tag{3.6}$$

where the superscripts stand respectively for Lorenz–Mie, diffraction, refraction, reflection and surface waves.

In the following the principal features of equations (3.6), with particular emphasis on the polarization properties of the scattered light, will be outlined. On these basis, suggestions and experiments to study by optical diagnostic the condensed phases in combustion systems will be given. First it will be examined the behavior of the cross sections of single droplets. Then it will be reported the application of the polarization method to combustion environment, where simultaneously are present droplets and soot, and finally to instationary conditions where the 2–dimensional approach is necessary.

Diffraction is concentrated in a narrow forward lobe and it will not be considered here. Reflection is a purely geometrical optics effect and both the polarized scattering cross sections are proportional to the droplet surface area. The polarization ratio $\gamma = C_{HH}/C_{VV}$ depends upon the complex refractive index according to the Fresnel equations. Refraction is again a geometrical optics effect; the polarized cross sections are almost equal and they depend upon the size and the absorption cross section according to the expression $C_{ii} = AD^2 exp(-K_{abs}D sin\psi)$, where A is a proportionality constant, $K_{abs} = 4\pi k/\lambda$ is the absorption coefficient of the liquid, D is the particle diameter, λ is the light wavelength and ψ is the angle between the tangent at the impact point of the incident ray and the chord described by the refracted beam.

In the forward up to the limit angle, beyond which refraction disappears, the total scattering can be approximated with the sum of the reflection and refraction contributions ($C^{LM} = C^{refl} + C^{efr}$). For optically thin droplets ($kD \to 0$) refraction prevails over reflection, the polarized scattering cross sections follow a D^2–law and are almost equal. Optically thick droplets ($kD \to \infty$) have a negligible refraction contribution, their cross sections follow again a D^2–law but their polarization ratio, according to the Fresnel, law presents a deep minimum at the Brewster angle, with C_{HH} several orders of magnitude lower than C_{VV}. For intermediate optical path lengths the polarization ratio exponentially decays with kD between these two limits, whereas the intensity always follows the D^2–law. Therefore, in the forward region two simultaneous measurements of the droplet scattering cross sections (or intensities) in different polarization planes give the size and the imaginary part k of the refractive index. In the visible wavelength range and for liquid hydrocarbons, k varies from 10^{-6} for paraffins and light aromatic compounds to 10^{-1} for very heavy aromatic compounds. Therefore, its determination, by the measure of the polarized cross sections, permits to follow in non intrusive way the formation of heavier compounds during the heating of liquid hydrocarbons. This approach was recently employed to study the thermal behavior of single droplets composed of pure and commercial hydrocarbons mixtures heated up in a drop tube furnace [12]. No increase of k was observed for pure paraffins or light aromatic compounds. Liquid phase pyrolysis of polar compounds was instead observed when commercial light oils were used. In that case, a mass fraction between 10^{-4} and 10^{-5} of the initial droplets survives at temperature higher than 400°C as small absorbing droplets (Fig. 3.5).

In the side scattering region ($80° < \theta < 120°$) the surface waves contributions are detectable, since refraction is absent beyond the limit angle and reflection gives a lower contribution than in the forward. The scattered radiation due to the surface waves is preferentially polarized in the horizontal plane and is nearly proportional to the droplet diameter. By this contribution, the total Lorenz–Mie horizontally polarized cross section at $\theta = 90°$ is more than one order of magnitude larger than its geometrical counterpart and it is proportional to the first power of the droplet diameter. The total Lorenz–Mie vertically polarized cross section is instead very near to that predicted by geometrical optics and practically depends on the square of the droplet diameter. The different dependence of the cross sections upon the size allows to determine the diameter of the droplets from a measurement of the polarization ratio. Basing on this idea, and using a single drop interferometric apparatus, the diameter distribution and the velocity–diameter correlation of arrays of droplets were determined [11].

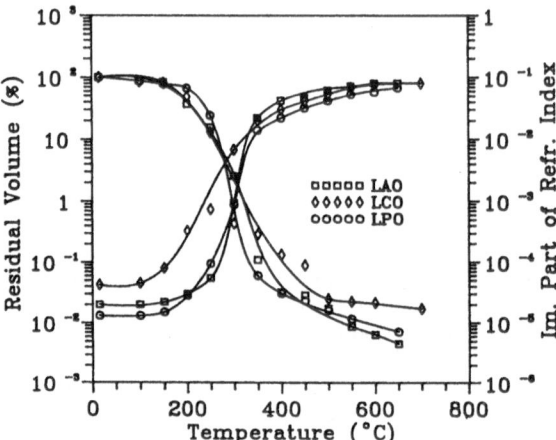

Fig. 3.5 Volume droplets percentage $(V_T/V_0)x100$ (full symbols) and imaginary part of the refractive index (empty symbols) for three different light oils vs the reactor temperature, from [12] with permission.

When the droplets are heated up, the real part n of the refractive index changes for the reduction of the liquid density and the diameter decreases due to the evaporation. A possible way to evaluate the droplet diameter is to measure the cross section C_{HH} at the scattering angle $\theta= 33°$ where it is almost independent of the refractive index and increases following a D^2 law for diameters larger than 5 μm [13]. In this way it is possible to obtain the drop size also if the refractive index of the liquid is unknown, as for heated droplets or multicomponent drops in vaporizing regime. Knowing the droplet size, the refractive index can be obtained by determining the primary rainbow position [27] or the ratio $C_{HH}(33°)/C_{HH}(60°)$ [13]. Being the real part of the refractive index connected to the liquid density, that is a function of the temperature, the determination of n permits the measure of the droplets temperature via scattering methods. Experiments were carried out on calibrated droplets of tetradecane injected in the drop tube furnace heated up to 200°C. The drop temperature as function of the furnace temperature are reported in Fig. 3.6 [13].

For dense particle clouds, single particle detection is not feasible, and ensemble scattering methods have to be used. The scattering coefficients from all the particles present in the scattering volume are simultaneously measured. For spherical particles in a single–scattering regime, the scattering coefficients are related to the cross sections by the equations:

$$Q_{HH} = N\int_0^\infty C_{HH}F(D)\,dD \propto N\int_0^\infty DF(D)\,dD$$

(3.7)

$$Q_{VV} = N\int_0^\infty C_{VV}F(D)\,dD \propto N\int_0^\infty D^2F(D)\,dD$$

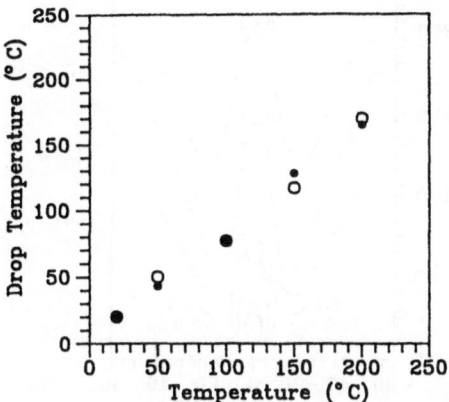

Fig. 3.6 Temperature of tetradecane droplets (D0=72μm) evaluated by the $C_{HH}(33°)/C_{HH}(60°)$ ratio (o) and the rainbow position (•)

where N is the number density of the particles and $F(D)$ is the size distribution function. Therefore, an absolute measure of the scattering coefficients Q_{HH} and Q_{VV} in the sideward directly gives the total periphery and the total surface area of the spray per unit volume, respectively, and the average diameter $D_{21}= D^2F(D)dD/DF(D)dD$, that is inversely proportional to the polarization ratio Q_{HH}/Q_{VV}.

It is worthwhile to remark that this method has its best sensitivity for the smallest droplets and the lower momenta of the size distribution function, where other techniques, like photography or diffraction, have their detection limits.

When the scatterer size is much smaller than the wavelength of the incident radiation (λ<<1), the elastic scattered field is equivalent to that emitted by an oscillating electric dipole induced by the incident field (Rayleigh effect) [26]. In the Rayleigh approximation, the scattering angular cross sections C_{VV} and $C_{HV}=C_{VH}$ are independent of the scattering angle θ, while the cross section C_{HH} varies with θ according to the expression:

$$C_{HH} = C_{VV}\cos^2\theta + C_{HV}\sin^2\theta \qquad (3.8)$$

The depolarized cross sections C_{HV} and C_{VH} are equal to zero for scatterers that do not present any optical or geometrical anisotropy. Consequently, the horizontal component C_{HH} at θ=90° is absent and the polarization ratio C_{HH}/C_{VV} is zero. This allows to separate the contributions due to submicronic (mainly soot in our cases) and supermicronic particles (mainly droplets in our cases) when both the classes of scatterers are simultaneously present in the scattering volume. In fact, even if due to any anysotropy of the practical scatterers γ is never zero, the droplets shown always an higher polarization ratio. With this method it was discriminated, inside oil flames, between zones where the concentration of soot particles was predominant to those where the fuel droplets prevailed (Fig. 3.7) [8].

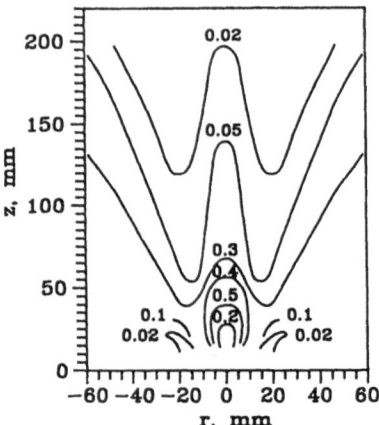

Fig. 3.7 Contours of the polarization ratio, γ, measured at θ=90° for flame of different light oils. The regions bounded by the curves with γ=0.05 indicate conditions where soot particles are dominant, from [8] with permission.

The scattering due to particles of dimension comparable with the wavelength of the incident radiation can't be brought back neither to that of an oscillating dipole (Rayleigh scattering) nor to the scattering due to supermicronic droplets. The angular pattern is dominated by surface waves and a relevant polarization peak is present in the backscattering region. This peculiarity represents a strong signature for the presence of micronic droplets in the scattering volume. The application of this scattering feature to sprays of heavy oil atomized by air assisted nozzle has shown that a large concentration of micron sized droplets are present in that conditions. In particular it was shown that the overall size distribution of sprays is bimodal and that the atomization efficiency had large influence on the ratio of the number concentration between the smaller particles, N_1, and those larger than 20μm, N_2, (Fig. 3.8) [9]. For dilute clouds of particles, time resolved measurements of the angular scattering coefficients and of the polarization ratio allow the determination of size and optical properties of large droplets, which pass individually through the scattering volume, and identify the average sizes and optical properties of the scatterers with typical size smaller than one micron, which pass as a cloud in the scattering volume [9]. This approach applied to sprays combustion has shown that, in the early region of oil flames, the smaller droplets decrease in size and become more absorbing than the original fuel, while the soot concentration increase strongly and the larger dropletsonly partially evaporate. Later on, in the soot oxidation zone ofthe flame, only soot particles and partially pyrolyzed larger particles are detected.

In unsteady phenomena, the determination of the spatial distribution of condensed phase requires to have simultaneously adequate time and spatial resolution measurements. This can be met only by multipoint measurements within very short characteristic times or better from a 2D quantitative imaging. Therefore, to study the combustion of unsteady diesel sprays in an optically accessible, high pressure, high temperature bomb, the single point polarization ratio technique has been extended to planar conditions [14,15]. The spatial morphology and the physical properties of the spray during the atomization and vaporization was obtained by determining the spatial map of the scattering coefficients in the horizontal and vertical polarization plane.

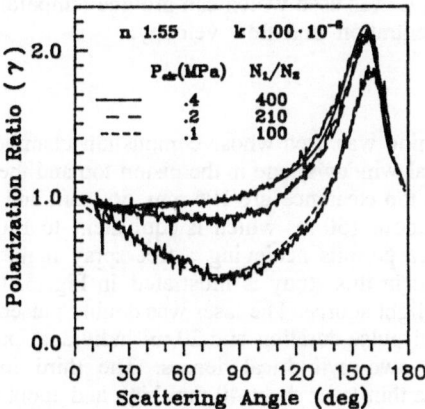

Fig. 3.8 Measured (noisy curves) and computed angular patterns of the polarization ratio for heavy oil sprays
at different pressure of atomization air

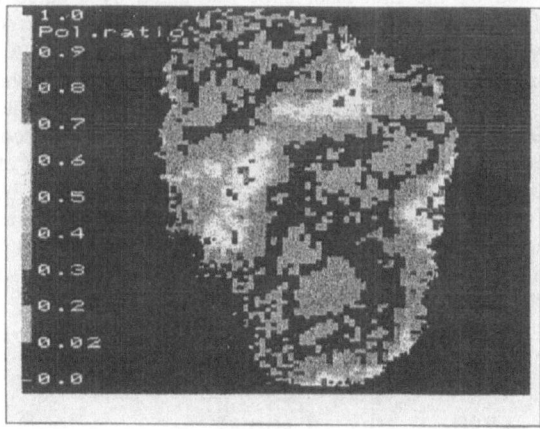

Fig. 3.9 Pattern of the polarization ratio, γ, detected, in pre–ignition condition, from a spray cross section at
30 mm from the nozzle and a time delay from injection beginning of 3.1 ms. The horizontal width
of the pattern is 6 mm, from [14] with permission.

Besides, the map of the polarization ratio permitted to find the instantaneous space
distribution of the condensed phase, discriminating between soot particles and fuel still
in liquid phase (Fig. 3.9) [14]. In that case the soot formation extended in the whole
field, very early after the autoignition time and it pervaded regions around the centerline
too.   (Antonio D'Alessio and Patrizio Massoli)

3.1.4 2–D soot imaging

The objective of this study is to investigate the mechanism of soot formation and
extinction processes in unsteady spray flames in a quiescent atmosphere. A 2–D elastic

scattering imaging technique was used which can provide temporal variations of spacial distributions of soot concentration and flow velocity.

(a) Experimental setup

A rapid compression machine was used whose combustion chamber is a pan–cake type with two transparent optical windows; one in the piston top and the other in the cylinder head. The bore, stroke and top clearance are 196 mm, 560 mm and 40 mm respectively. The compression time is about 150 ms which is equivalent to 180 rpm. The large size of the combustion chamber permits achieving a free spray flame.

The optical setup used in this study is illustrated in Fig. 3.10. Nd:YAG laser(SP: DCR–11) was used as the light source. The laser was double pulsed with a pulse interval of 100 μs, 3 nsec individual pulse duration and 20 mJ individual pulse energy. The laser beam was collimated by two cylindrical lenses. The third long focal cylindrical lens(f=1000 mm) formed a thin laser sheet 40 mm high and about 0.15 mm thick in the observation section. The laser sheet light was reflected by a prism located inside the combustion chamber and propagates through the mid–plane of the flame. The scattered light from soot particles was divided into two directions by a half mirror. The scattered light image generated by the first laser shot was taken by a gated intensified camera which was located opposite to the combustion chamber. The scattered image caused by the second shot was taken by another gated intensified camera. A band–pass optical filter with a central wavelength of 532 nm and FWHM of 4 nm was attached to each objective lens in order to reduce the background luminosity from the incandescent soot particles. The effect of background luminosity was further suppressed by the short gated duration of the cameras (0.5 micro–seconds). The individual gating of both cameras was synchronized to the individual laser pulse using a delay circuit. A single hole–type nozzle having an orifice diameter of 0.15 mm was used in all tests. The fuel was injected through an electro–hydraulically controlled unit injector at pressures of 55, 85, and 134 MPa into air at a pressure of 2.7 MPa and a temperature of 900 K.

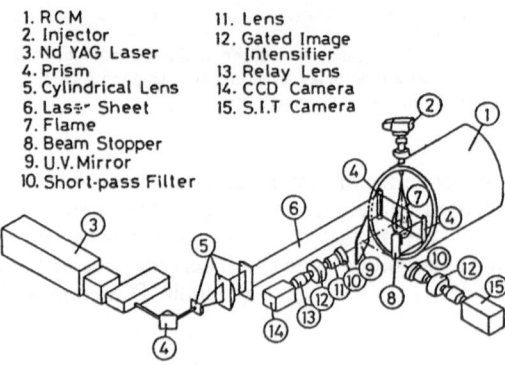

1. RCM
2. Injector
3. Nd YAG Laser
4. Prism
5. Cylindrical Lens
6. Laser Sheet
7. Flame
8. Beam Stopper
9. U.V. Mirror
10. Short-pass Filter
11. Lens
12. Gated Image Intensifier
13. Relay Lens
14. CCD Camera
15. S.I.T Camera

Fig. 3.10 Optical arrangement for 2–D soot imaging, from [93] with permission.

(b) Evaluation of the 2–D soot imaging technique based on the elastic scattering

When D_o denotes the particle diameter, and N number density, the elastic scattering intensity can be expressed as being proportional to $D_o^6 \times N$ if the Rayleigh theory is assumed for the scattering of soot particles. On the other hand a theoretical calculation of the LII intensity[28] shows that the intensity is approximately proportional to $D_o^{3.5} \times N$ [29]. This theoretical consideration suggests that the LII intensity is closer than the scattering intensity to the volume fraction, $D_o^3 \times N$, which needs to be measured. A preliminary imaging of the elastic scattering intensity and the LII intensity was performed simultaneously to compare these two intensity distributions and to evaluate the elastic scattering technique.

The method of LII was similar to that by Dec et al.[20], but differed in the use of the fundamental light of YAG laser at 1064 nm and a short wave pass filter with a cut-off wavelength of 400 nm. Laser output was 0.26 J per pulse, which resulted in a power level in the laser sheet about 4.3×10^8 W/cm². The gated time of the camera was 50 nsec. It was confirmed that the background luminosity was not imaged without a laser sheet.

Figure 3.11 shows typical results of simultaneous imaging. The region of strong scattering intensity coincides generally with the region of the strong LII signal. Since no liquid droplets are present within the observation field, Xc = 50 mm – 90 mm, their elastic scattering signal has little contribution to the scattering images. The general consistence of distributions of the elastic scattering signal and LII signal implies that the regions of high elastic scattering can represent qualitatively the regions of high soot volume fraction.

(c) Results and discussions

Selected 2–D soot images taken with a time interval of 0.1 ms at different times after start of injection for various injection pressures , are shown in Fig. 3.12 . The spatial

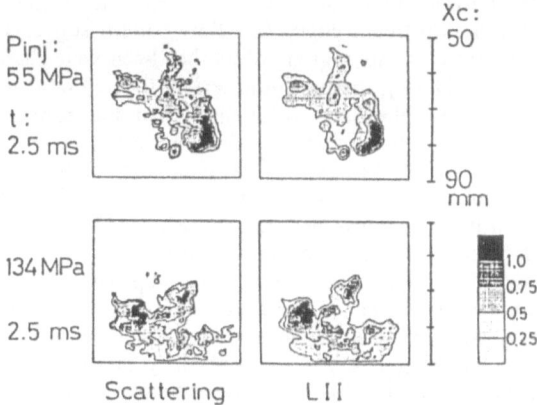

Fig. 3.11 Simultaneous 2–D soot imaging by both elastic scattering and LII technique (t: time after start of injection), from [93] with permission.

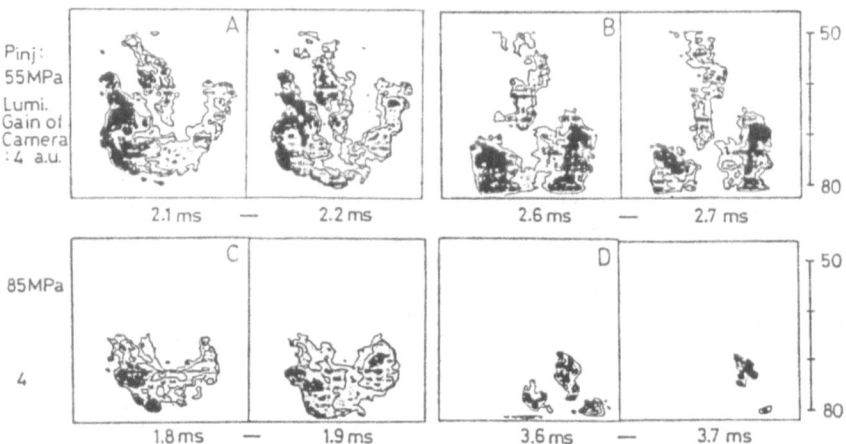

Fig. 3.12   2–D soot images at various times after start of injection and at various injection pressures, from [93] with permission.

distribution of soot is highly uneven. The soot clouds with high scattering intensity can be sharply distinguished inside the flame, and its spatial scale is on the order of several millimeters. The scattering intensity is higher in the narrow region along the periphery of the blunt flame head, which looks U–shaped. The soot clouds are usually concentrated in a region near the axis of the flame with a width of about 10 mm upstream of the U–shape sooting region. This trend holds on the whole regardless of the variation of injection pressure.

Figure 3.13 shows the velocity vector distribution of the soot clouds which was calculated using the spatial correlation between the image pairs in Fig. 3.12. The contours in the figures represent the outline of sooting regions. It is worth noting that the vectors are directed outward in the downstream edge of the U–shape sooting region, but toward the center in the upstream edge. This flow pattern suggests the existence of a large–scale vortex structure in this region, which has been well characterized for gas jets or unsteady sprays[30]. During the injection period, the velocity of soot clouds near the flame axis upstream above the flame tip is faster than the tip penetration velocity, but decreases rapidly after the end of injection. Also during the injection period, the soot clouds behind the flame tip seem to overtake the clouds in the flame tip, forcing them to move toward the periphery of the flame. This flow pattern indicates that the soot clouds are slowed down by the high drag force at the tip and are continually replaced by new later injected mixtures with a high momentum.

The most active soot formation occurs in the flame tip where both gas temperature and equivalence ratio are favorable for forming soot [31]. In the flame tip, the air entrainment seems very poor because the surrounding air is pushed away by the flame head. The active air–fuel mixing is expected to occur in the upstream edge of the U–shape sooting region where the high turbulent shear flow entrains the surrounding air. In this region, the equivalence ratio is also low. Therefore, a large fraction of soot in this region is probably transported from the downstream flame head and oxidized here, showing in high soot temperatures. There may be another possible mechanism of soot formation in the lean pre–mixture ombustion reported by Furutani et al[32].

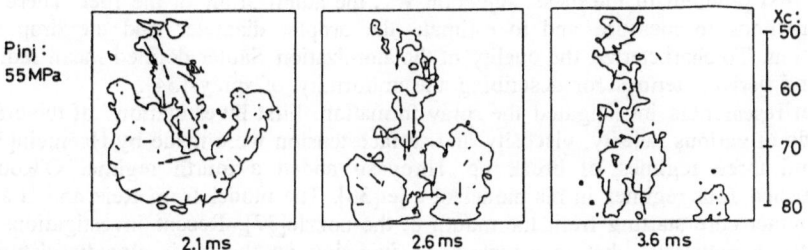

Fig. 3.13 2-D velocity vectors of soot clouds at various times and injection pressures, from [93] with permission.

(d) Conclusions

The following conclusions were derived from the experiments.

1. Soot formation occurs dominantly in the periphery of the flame tip, where both equivalence ratio and gas temperature favor soot formation. Soot oxidation hardly occurs in this region because of poor air entrainment.

2. Soot is conveyed by a flow in the large scale vortex in the flame head to an upstream edge of the soot formation region, and is actively oxidized by air which is entrained by turbulent eddies.   (Takeyuki Kamimoto and Young-Ho Won)

3.2 Spray and Ignition

3.2.1 Reviews

(a) Atomization and spray formation

Spray formation and atomization of liquid jets are subjects of many publications. In 1959 De Juhasz's "Spray Literature Abstracts" comprises more than 1300 litrature items. Today many hundred additional publications have appeared in this field. Spray formation and atomization have already been investigated from Rayleigh, who in 1859 published a fundamental mathematical analysis of the conditions for the breaking up of a liquid jet[33]. This work serves as the basis for most subsequent theoretical research work on sprays by many investigators.

In combustion engines the liquid fuel must first be injected into the combustion chamber and then be spread and atomized. In gasturbines the injection is stationary and in reciprocating engines it is instationary. Especially this processes cause many problems. R. Diesel had difficulties to inject and to atomize the fuel. His solution was to atomize the fuel with using compressed air, a so called pneumatic system. With the introduction of the hydraulic atomization many new difficulties arose. Therefore the diesel injection was subject of a great number of researches.

The investigations can be classified in four groups: atomization and determination of droplet diameters, spray formation, distribution of droplets in the combustion chamber and influence of the events in the hydraulic system on the spray formation and atomization.

The first problem of the diesel injection was the atomization of the fuel. There are many methods to measure and to estimate the droplet diameter and the drop size distribution. To characterize the quality of the atomization Sauter defined mean radii of drops and derived termes for describing the uniformity of sprays[34].

Other researchers investigated the spray formation. First investigations of the efflux of liquids of various density, viscosity and surface tension were made by Haenlein[35]. He found three regimes of break up. Hiroyasu added a fourth regime. O'Rourke distinguished four regimes in his model of a jet[36]. The model from Reiz and Bracco had an intact core starting from the mouth of the nozzle[37]. Recent investigations by Eifler[38] demonstrate, that in a real diesel injection jet the jet is already disturbed shortly after the nozzle.

Many researchers investigated the influence of the air motion in the combustion chamber of the spray formation and the distribution of the droplets.

Another problem is the influence of the hydraulic system of the spray formation and the atomization. The questions are: How is the effect of the instationary pressure? How are the influence of turbulence and cavitation in the nozzle, and how is the influence of the atomization?   Bergwerk investigated the inner flow in a nozzle and defined a cavitation number[39]. Ruiz and Chigier suggested, that there is an interaction between turbulence and cavitation[40].   (Fritz Eisfeld)

(b) Evaporation and impingement

In view of combustion in actual diesel engines, evaporation and impingement are both very significant processes, because the liquid fuel which is injected into the combustion chamber of a high–speed diesel engine burns as it is evaporating and impinging on the piston surface.

The evaporating spray includes a liquid phase region near the nozzle, the so–called the liquid intact core. The length of this intact core along the spray axis was investigated by the direct photography [41] and the laser shadowgraphy [42,43]. It was found in previous studies that the length is 15 – 25 mm, and is not affected much by the injection pressure, and the pressure and temperature of the surrounding atmosphere. The vapor phase of an evaporating spray was investigated intensively by the laser diagnostics which were currently developed. The statistical measurement of local fuel air ratio of an evaporating spray was performed by the first application of the spontaneous Raman spectroscopy [44]. The measured frequency distribution of the local fuel air ratio was well correlated to the ignition process of the spray investigated. Another sophisticated technique is the exciplex fluorescence technique which can visualize the liquid and vapor phase simultaneously [45]. This imaging technique was employed for the qualitative measurement of the two–dimensional distributions of the liquid phase and vapor phase concentration in an evaporating spray at a condition equivalent to that in actual engines [46].

Studies of an impinging unsteady spray are relatively limited. One of early example was the measurement of the external shape of a non–evaporating spray impinging on a flat wall [47]. The contours of fuel vapor concentration and temperature in an evaporating spray were obtained by the double exposure holographic interferometry [48]. A theoretical calculation of the evaluation of an impinging spray in the combustion chamber of a direct injection diesel engine was carried out using the KIVA code which introduced three new submodels to describe the spray–wall interaction processes [49]. The calculated result showed a good agreement to experimental data.

                                                                        (Hajime Fujimoto)

(c) Unsteady turbulent mixing

Most of fuel jets being used in engineering fields are turbulent ones whose physical characteristics vary statistically with time and space. The fundamental nature and turbulent mixing process of steady turbulent jets were studied numerously and they are summarized in textbooks [50,51]. Detailed structure and mixture formation process in steady turbulent jets have been studied by means of various modern laser diagnostics. Instantaneous planar measurement of the three–dimensional scalar gradient in a turbulent jet was made using laser Rayleigh scattering from two illumination sheets of different wavelength [52].The joint and independent pdf's of the mixture fraction and its gradient magnitude at a point on the center line of the jet provided useful information on turbulent mixing. Laser induced fluorescence techniques are also being used in the studies of jet mixing. An experimental study using a high resolution, LIF technique showed that the normalized scalar rms fluctuations decrease with increasing flow Reynolds number in the range of 3000 to 24000 [53].

Unsteady turbulent jets, on the other hand, have not been studied adequately, because it was recently that diagnostics which have fast response to the high frequency turbulence were developed. The concentration distribution in a methane jet injected into the combustion chamber of a DISC engine was measured by the Raman spectrometry [54]. The time history of the fuel concentration in a helium jet injected into a quiescent atmosphere was measured by a fast response concentration probe [55]. The authors obtained an empirical formula of the normalized concentration distribution for the fully developed jet region, and showed that the distribution expressed by the formula agreed well with results obatained in experiments on a diesel spray. This indicates that the concentration distribution in unsteady jets has a similarity in these jets despite the differences in the density ratio of injected medium to surrounding medium and injection conditions. The similarity was recently confirmed by a laser interferometry measurement of an unsteady helium jet [56]. Other works conducted on unsteady water jets also support this analogy [57,58]. The mechanism of mixing process in unsteady jets still needs to be pursued more in the future.   (Hideaki Tanabe)

(d) Application of the stochastic ignition theory

Using diesel engines or adiabatic compression vessels, many experimental and theoretical investigations of the spontaneous ignition in a fuel spray injected into compressed air have been carried out to observe the influences of 1) engine driving conditions, 2) the temperature and pressure of the compressed air, 3) the injected fuel quantity and character and 4) the particle size of injected fuel on the ignition delay. It is also important to observe the relation between the measurement method and the observed ignition delay, or to analyze the reaction mechanism during the ignition delay according to the classical ignition theory[59–63].

Even in a homogeneous mixture, the ignition never takes place homogeneously, but starts from several finite points[64], as the process is irreversible. Theoretically the induction period of ignition should be the period from the instant when the combustible substance is brought to an ignitable state till the formation of an ignition nucleus having a certain critical size from which the reaction can develop by itself to a propagating flame.

In general, however, because of the sensitivity of measurements apparatus the ignition is recognized when the ignition nucleus develops to a certain measurable size. The observed induction period of ignition, therefore, consists of two components, i.e. the ignition nucleus formation period and nucleus development period. The nucleus formation period fluctuates within a wide range and also depends on the quantity of the ignitable mixture[65,66].

The induction period of ignition in a fuel spray is usually assumed to be the period from the beginning of the fuel injection to an instant when the beginning of a pressure increase, temperature rise, an exothermal reaction or a light emission of ignition is recognized corresponding to the measurement apparatus. Such induction period, especially in diesel engines, is measured for investigating the ignition mechanism, but as the induction period fluctuates and its mean value depends on the quantity of air–fuel mixture around the fuel spray, the ignitability cannot be evaluated only by the mean induction period of ignition.   (Kunio Terao and Chihong Liao)

3.2.2 Atomization and spray formation

(a)Models of spray formation

The analysis by Raleigh is strongly valid for non viscouse fluids only. The disintegration is the result of the capillar forces. The thesis is that the wave length with the highest amplification in the jet leads to the disintegration. Rayleigh found that this wave length is 4.508 times the diameter of the nozzle hole.

Haenlein found three regimes for the disintegration of a jet. Figure 3.14 demonstrates the mechanism of disintegration and Fig. 3.15 shows the regimes as a function of the efflux velocity. The mechanisms of disintegration are as follows: 1. drop formation only by the surface tension of the liquid, 2. drop formation by surface tension reinforced by air interaction, 3. wave formation by air interaction. The regimes are the laminar Rayleigh regime, the first wind and the second wind regime followed by a sudden and complete disintegration of the jet. Hiroyasu has added an atomization regime with a steadily decreasing break up length.

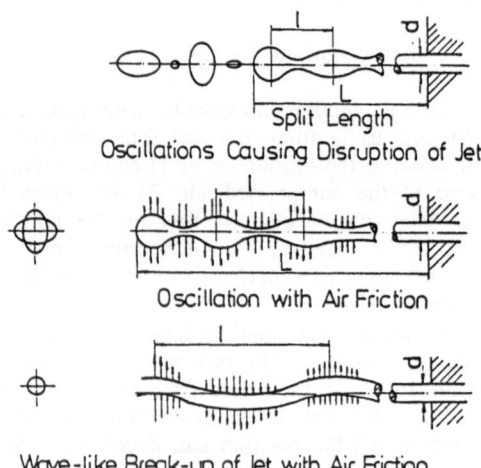

Split Length

Oscillations Causing Disruption of Jet

Oscillation with Air Friction

Wave-like Break-up of Jet with Air Friction

Fig. 3.14   Mechanisms of break up

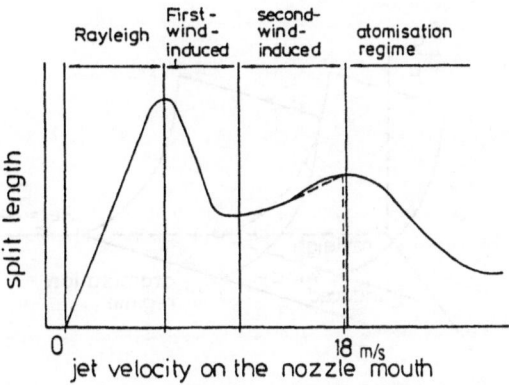

Fig. 3.15 Regimes of disintegration

Table 3.1 Characteristic stresses for an orifice nozzle spray

Type of stress	Force per unit area of interface
Ambient phase	
impact or inertia stress	$\rho_g v^2/2$
shear stress	$2\eta_g v_o/d_o$
Phase being dispersed	
impact of inertia stress	$\rho_\ell v^2/d_o$
Surface normal stress	$4\sigma/d_o$

There is also a relationship between the Reynolds number and the jet disintegration. Important for the disintegration of a jet is also the density of the gas in which it is injected. Ranz[67] analyzed the spray formation considering the characteristics of an orifice nozzle spray. Table 3.1 shows the characteristic stresses for an orifice nozzle spray.

The limits of the regimes of the spray formation are a function of the Weber number. It limits the Rayleigh regime to the following:

$$We_\ell > 8 \qquad We_g = \rho\, v_o^2\, d_o/\sigma \quad < 0.4 \tag{3.9}$$

the induced regime

$$by \quad 0.4 < We_g < 13$$

and the beginning of the atomization regime

$$by \quad We_\ell > 8 \qquad We_g > 13 \quad \text{by other authors is } We_g > 40 \tag{3.10}$$

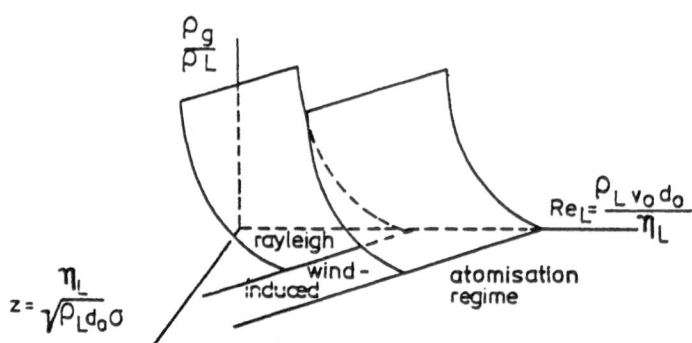

Fig. 3.16 Break up regiomes as function of Re_1^2

Figure 3.16 shows the break up regimes as function of $(\rho_g/\rho_\ell)+Z^2 = Re_\ell^2 = $ const.

$$\text{with} \quad Re = v_o \, d_o \, \rho_\ell/\eta_\ell \qquad\qquad\qquad\qquad (3.11)$$

and the Ohnesorge number

$$Z = We^{1/2}/Re_{do} = \eta_\ell \, /(\rho_\ell \, d_o \, \sigma)^{1/2} \qquad\qquad\qquad (3.12)$$

η_ℓ : viscosity of liquid d_o : diameter of orifice
ρ_ℓ : density of liquid σ : surface tension
v_o : jet velocity We : Weber number $= \rho_\ell \, v_o^2 \, d_o/\sigma$

There is a great number of possible physical break up mechanisms. In the following list events are given which could start the jet break up:

-growth of surface waves by aerodynamic interactions caused by infinitesimal
  disturbances in the nozzle,
-micro turbulences of the flow in the nozzle hole,
-macro turbulence caused by the shape of the nozzle and instationarities,
-change of the velocity profil after the mouth of the nozzle,
-disturbances of the fluid column caused by pressure pulsations,
-boundary layer relaxation in the free jet,
-cavitation and falling out of gases solved in the fuel

All these effects are effective and increase the instability. But if all of these mechanisms lead to a break up of the jet is a question still to be investigated.

(b) The fluid core

In many theories and intact core of the liquid jet is assumed. In Reitz and Bracco[39] this intact core length reaches up to the atomization regime shown in Fig. 3.17. Ruiz and Chigier[40] suggested that the turbulence generated cavitation bubbles inside the nozzle.

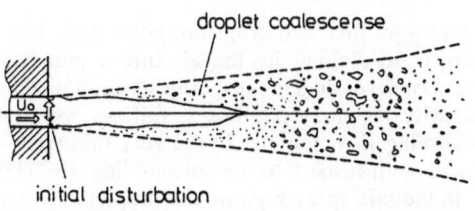

Fig. 3.17   Jet model by Reitz and Bracco

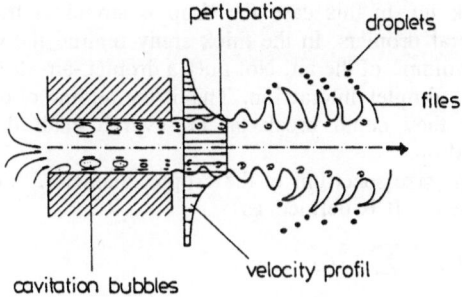

Fig. 3.18   Jet model by Ruiz and Chigier

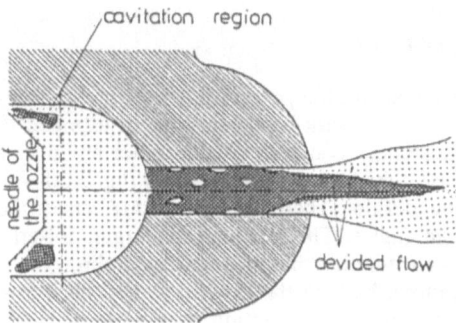

Fig. 3.19   Jet model by Eifler

In the mouth are fluid and gas bubbles and no intact core. These cavitation bubbles implode and generate eruptions shown in Fig. 3.18. Eifler has proved that in a real cavitation bubbles grow and no intact core is found in the mouth. His model of a jet is demonstrated in Fig. 3.19.

(c) The atomization

By the break up of a liquid jet files and drops are generated. The files are disintegrated into drops and the drops are divided by the air forces into smaller droplets. On the otherhand droplets could coalesce to a greater drop. A good method to estimate the stage of atomization is to classify the spray regimes as follows: very thin sprays, thin sprays, thick sprays and the churning flow regime. In the very thin spray regime the volume of the droplets is very small with respect to the surrounding air. There are no interactions between the droplets. In the thin spray regime the droplets have a significant mass with respect to the air, but the volume of the droplets is small. There are hardly any interactions between the droplets, but there are interactions droplet–gas–droplet. The small droplets move from the border of the jet in the core region. This effect is caused by the air motion. Then the free jet sucks surrounding air. In this regime the relative velocity of droplets is very high and the droplets break up. In the diesel engine there is mostly stripping break up. In this case the drop is stretched by viscouse forces and disintegrates into several droplets. In the thick spray regime the volume of liquid is of the same order as the volume of the air. Not only a droplet–air–droplet interaction takes place but also a droplet–droplet interaction. This is the regime of droplet collision. If the droplets collide, then they could break up in several small droplets or they could coalesce to a greater drop.

Important is also the size spectrum of the droplet diameter. A characteristic value is the Sauter mean diameter. It is defined as

$$D_{32} = \sum r_i^3 n_i \ / \ \sum r_i^2 n_i \tag{3.13}$$

r_i : radius of droplet of one order of magnitude
n_i : number of droplets of one order of magnitude

The SMD depends on the orifice diameter, the fluid pressure, the gas pressure and the viscosity of the fluid.

(d) The real diesel injection jet

In a diesel engine the cone angle, the penetration length and also the atomization are of special interest. Many research work has been done to estimate the cone angle. But the results from several researchers contradicts each other as far as the dependence of the fuel pressure is concerned, which is demonstrated in Fig. 3.20. Another question is how the cone angle is to be defind. There are great differences between measured and analytical estimated cone angles. The real angle is larger. Eifler[38] has found that this is an effect of imploding cavitation bubbles and exploding gas bubbles. With a high-speed rotating mirror camera he took the flow in a glass nozzle which had a hole of 0.3 mm. He could prove that there is cavitation (see Fig. 3.21). In injection into oil he visualized shock waves generated by imploding bubbles (see Fig. 3.22). The implosions and explosions spread the spray. Using high–speed cinematography with frequencies up to 1,000,000 f/s he could demonstrate that there are eruptions on the jet and sometimes fine files shut out of the jet with a velocity of twice the tip penetration velocity (see Fig. 3.23). Another problem is the atomization when the needle of the nozzle begins to close. Then there are long files and very big drops as shown in Fig. 3.24. Pressure pulsations in the hydraulic system are important for spray formation. The result is the

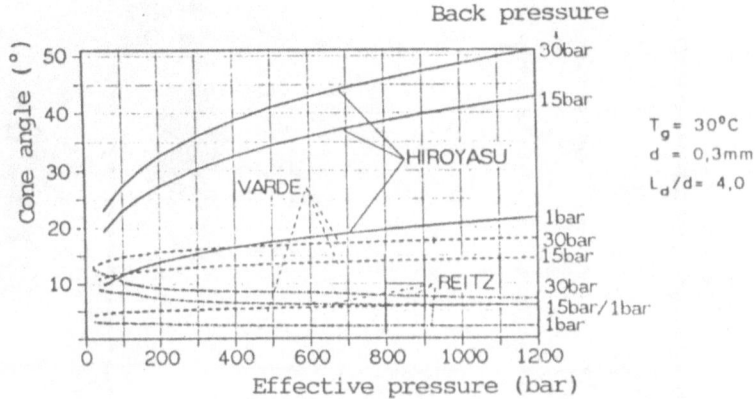

Fig. 3.20   The cone angle as function of p_1

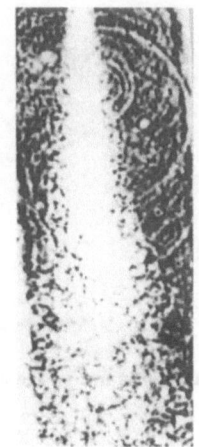

Fig. 3.21 Cavitation in a nozzle   Fig. 3.22 Shock waves   Fig. 3.23 Eruptions

generation of branches in the spray as shown in Fig. 3.25, but also a better break up. This is also a result of analytical research work by Janson (see [38]). Recent investigations on high-speed films from diesel injection jets demonstrate that there are vortices, sources and sinks on spray. This is a result of picture analysis by Wagener[68] too. Furthermore on the tip of the jet new fuel particles flow out permanently which are pushed away and decelerated. Furthermore it is demonstrated that there is an effect of the walls on the spray formations. Janson analyzed that a pressure wave goes on ahead of the jet. When the wave has reached the wall the jet spreads more (see Fig. 3.26). This could be observed by low gas pressure.   (Fritz Eisfeld)

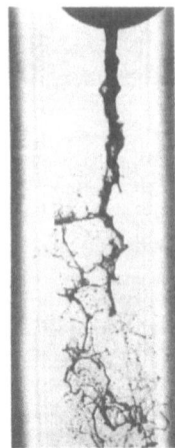

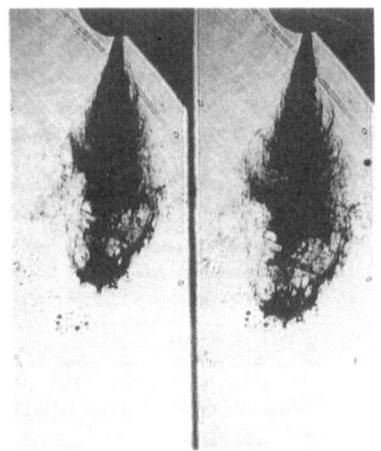

Fig. 3.24   Atomization by injection end   Fig. 3.25   Formation of branches

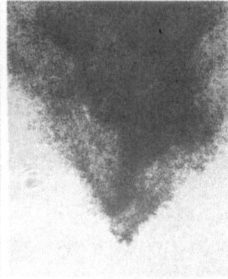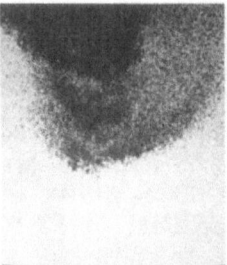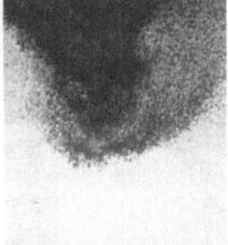

Fig. 3.26   Penetration of a free jet and a jet near the wall

3.2.3 Evaporation and impingement

(a) Fuel concentration distribution in an impinging fuel spray

Detailed internal structures of non-evaporating and evaporating sprays which impinge
upon a flat wall were investigated [69–71]. Gas oil was injected into a quiescent
atmosphere in a constant volume bomb through an orifice which has a diameter of 0.2
mm. Figure 3.27 is a typical result showing the measured spatial distribution of droplet
density C_f in a non–evaporating unsteady spray. C_f was calculated by applying the
computer tomography to the analysis of instantaneous shadow pictures. The figure shows
that C_f in the peripheral region of the impinging spray increases with the increase in the
surrounding gas density and the decreases in the impinging distance and the fuel
injection pressure. The figure also shows that C_f is higher near the impinging wall and
at the same time lean mixtures distribute dominantly in the wall spray. Also it is noticed

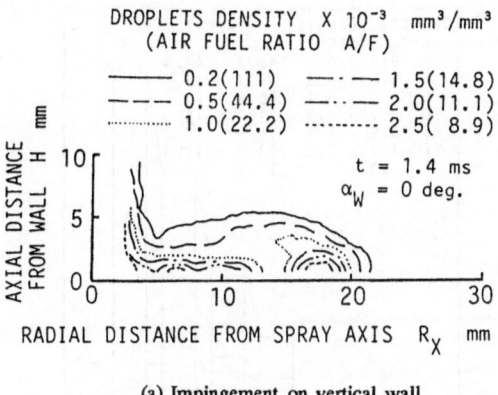

(a) Impingement on vertical wall

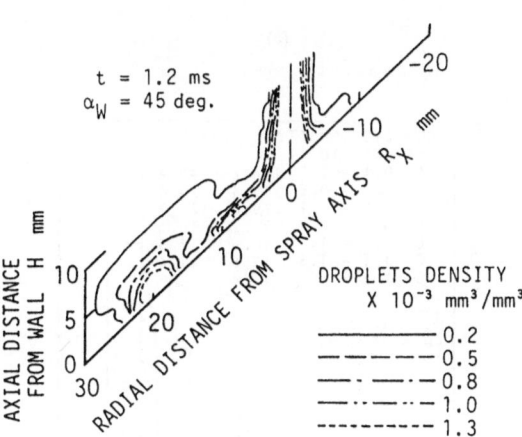

(b) Impingement on inclined wall

Fig. 3.27 Spatial distribution of droplets density in non–evaporating unsteady spray impinging on flat wall (Impinging distance: 24 mm, ρ_a: 18.5 kg/m^3, nozzle diameter: 0.2 mm, mean injection pressure: 14.0MPa, injection period: 1.2 ms), (a) from [69] with permission, (b) from [70] with permission.

that some of C_f contours exhibit a cross sectional profile like a mushroom. These experimental observations permit to draw a picture that droplets drifting in a region near the spray tip are pushed forward and upward by the succeeding droplets, and this results in the generation of a large vortex, the so–called roll–up vortex or the wall jet vortex, in the penetrating spray tip.

The spray tip penetration and the spray thickness on the wall were both functions of the surrounding gas density, fuel injection pressure, and time after the impingement. However the penetration and the thickness were hardly affected by the impinging distance when it was changed from 20 to 34 mm.

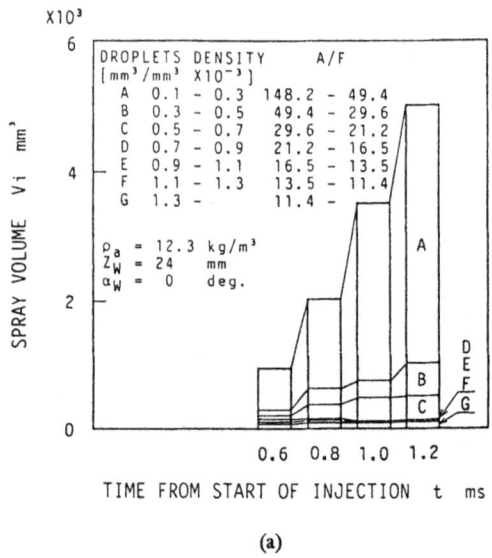

(a)

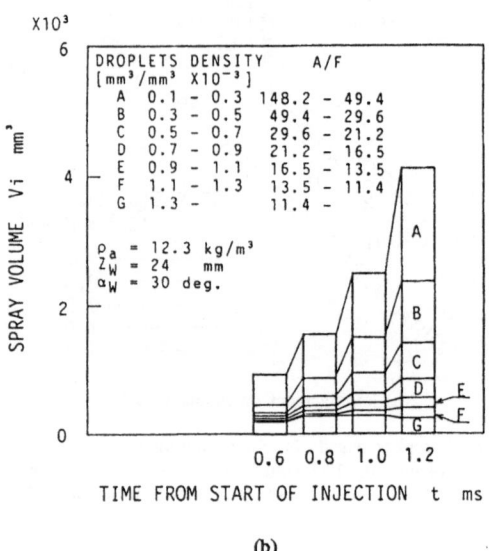

(b)

Fig. 3.28 Spray volume by droplets density in non–evaporating unsteady spray impinging on flat wall
(Experimental conditions are as same as those in case of Fig. 3.27), from [70] with permission.

Figure 3.28 displays spray volume V_f as a function of time after the start of injection.
Letters A to G in bars stand for different levels of C_f respectively. That is, A
corresponds to the leanest C_f and G the richest. It is interesting to note that the increase
in spray volume with time owes mostly to the increase in lean mixtures represented by

A to C, and that the volume of rich mixtures, D to G, scarcely varies with time. This figure revealed the fact that the reason for the larger volume of impinging spray against that of the free spray is the increase in lean mixtures for the impinging spray. Similar trends to that observed in Fig. 3.28 were found even at different experimental conditions where gas density, impinging distance, and an angle of impingement were varied.

The exciplex 2–D fluorescence technique [45] was used to image the liquid and vapor phase concentration in an evaporating impinging spray. A fuel which is composed of 90 % n–decane, 9 % Naphthalene, and 1 % TMPD, by mass fraction, was injected through a nozzle orifice with a diameter of 0.2 mm toward a wall placed in a nitrogen atmosphere at a pressure of 2.5 MPa and a temperature of 700 K. The surface temperature of the wall was kept 550 K. A thin sheet of laser light from a Nd: YAG laser (wavelength: 355 nm) irradiated an evaporating spray through the spray axis. The exciplex fluorescence from the liquid phase and the monomer fluorescence from the vapor phase were both imaged by an image intensifier from the direction perpendicular to the laser sheet. These two fluorescences were imaged separately by optical filters whose centerwavelengths were 532 nm and 390 nm respectively.

Figure 3.29 shows instantaneous distributions of the liquid phase and the vapor phase concentration. The magnitude of the vapor phase concentration is not quantitative but qualitative at present. The vapor phase evolves with time along the wall surface at a velocity faster than that of liquid phase. The liquid phase, or liquid film on the wall surface stops to develop after 0.5 to 0.75 ms after the start of injection. The thickness of the liquid phase on the wall remains constant, 1 mm to 2 mm irrespectively of time and impinging distance. When looking at the contours of the fluorescence intensity of the vapor phase, we notice that immediately after the impingement the highest intensity are observed in two regions; one near the wall the other in the peripheral near the leading edge of the vapor phase. The high fluorescence intensity near the wall disappears first, followed by a gradual extinction in the peripheral region. It is interesting to note that after the end of injection a high fluorescence intensity appears again in a thin layer along the wall. This is probably due to the late evaporation of the liquid film which was formed on the wall during the period of fuel injection.   (Hajime Fujimoto)

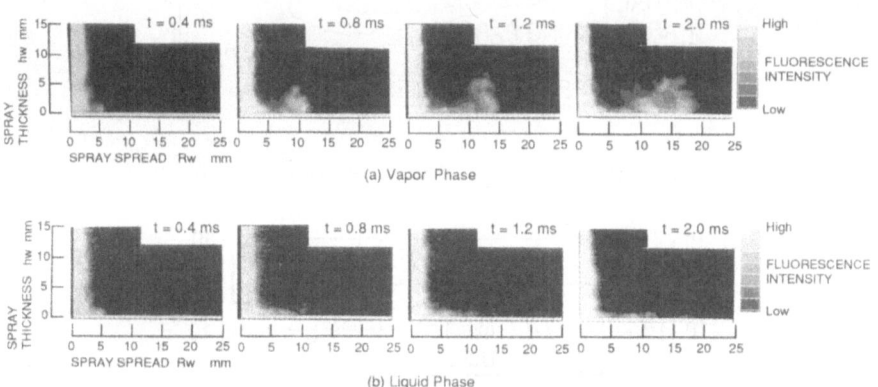

Fig. 3.29 Spatial distributions of fluorescence intensity of vapor and liquid phases in evaporating unsteady spray impinging on flat wall (Impinging distance: 24 mm, ρ_a : 12.3 kg/m^3, T_g: 700 K, P_g: 2.5 MPa, T_w: 550 K, T_{fuel}: 373 K, nozzle diameter: 0.2 mm, mean injection pressure: 16.4 MPa, injection period: 1.8 ms), from [71] with permission.

3.2.4 Turbulent mixing process in unsteady gas jets

(a) Experimental apparatus and procedure

Helium was injected into a quiescent atmosphere through a solenoid type injection valve for a period of 20 ms at a constant injection pressure of 0.35 MPa. The helium concentration was measured by a fast response hot wire concentration probe. Static and dynamic pressures were measured using a semiconductor pressure transducers incorporated in a pressure probe. Injection was repeated several hundred times to obtain statistical turbulent characteristics such as ensemble mean, variance and intermittency factor. Conditional mean and variance of concentration were obtained to investigate the mixture formation process. Effects of wall impingement on mixing was also investigated.

(b) Results and discussion

Figure 3.30 is a schematic illustration of a fully developed unsteady free jet, showing that the jet is composed of a steady part and an unsteady part. Fig. 3.31 shows a typical example of the history of ensemble mean values of velocity, pressure and concentration measured on the axis of the helium jet. It is observed that the static pressure rises at first, then decreases and reaches a constant, which is lower than the surrounding gas pressure. The pressure keeps constant during the injection period, and then increases to

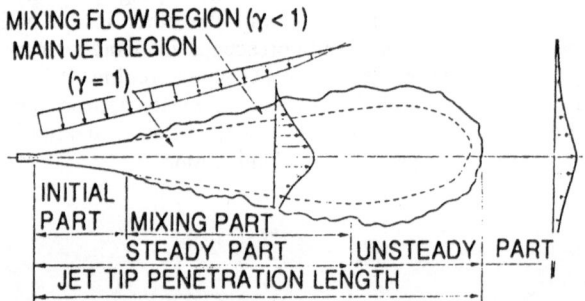

Fig. 3.30 Schematic illustration of fully developed unsteady free jet

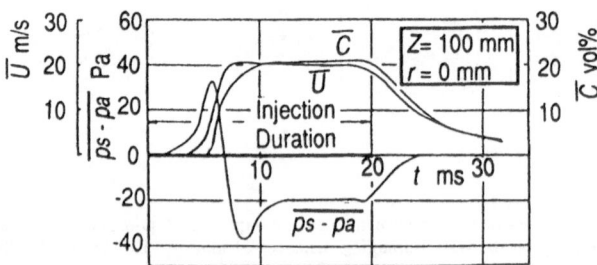

Fig. 3.31 Time history of velocity, $\bar{U}$, concentration, $\bar{C}$, and static pressure, $\bar{P}_s - \bar{P}_a$ in unsteady free jet

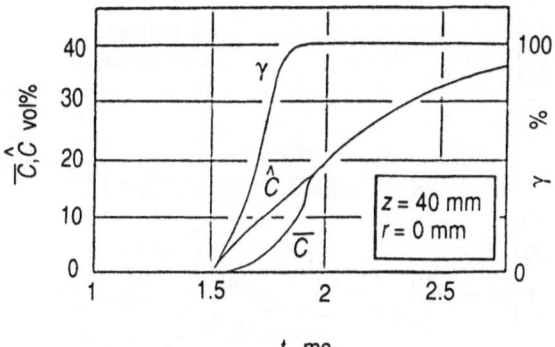

Fig. 3.32 Time history of conditional mean concentration, $\hat{C}$, ordinary mean concentration, $\bar{C}$, and intermittency factor at unsteady free jet's tip

the surrounding pressure after the end of injection. The axial velocity starts to increase after the pressure rise and keeps constant during injection period. It decreases gradually after the end of injection. The time history of helium concentration is similar to that of axial velocity. The first period is the unsteady one, in which the velocity, pressure and concentration vary with time. The second period is steady one, in which the jet characteristics are identical to those of steady jets. The last period after the end of injection is also unsteady.

The first period corresponds to the jet tip growth process. The surrounding air around the jet tip is pushed forward by the jet tip prior to its arrival, because the static pressure in the jet tip is higher than that in the surrounding air as indicated in Fig. 3.31. The jet tip catches up the surrounding air, then entrains air into the jet, because the tip moves faster than the surrounding air. The mixture formation process in the jet tip is thus specific to the unsteady jet. In the steady part, the negative pressure gradient governs the air entrainment. Fig. 3.32 shows a typical example of the time histories of ordinary ensemble mean concentration, conditional mean concentration and intermittency factor at a position on the axis of the jet. When the jet tip arrives at the measuring point the concentration rises sharply, because the mixture eddies exist on the interface between the jet and surrounding. This is a stochastic phenomenon, and the intermittency factor, γ, gives a probability when the measuring point is included in the jet region. The figure shows that the conditional mean concentration is higher than the ordinary ensemble mean in the early period after the jet tip arrives at the measuring point. This means that in the jet tip, the nibbling entrainment takes place and small scale mixing hardly occurs.

In the practical application of unsteady fuel jets, the jet impinges on the combustion chamber wall. Figure 3.33 indicates a typical example of the concentration contour and wall pressure distribution of an impinging unsteady jet. Broken lines in the figure denote those of the corresponding steady jet. It is shown that the jet tip roles up in the wall impinging jet. The reason for this can be interpreted by the flow pattern indicated by arrows in the figure. The entrained surrounding air flows into a region between the jet tip and the wall as arrows indicate, and the mixture flows outward in the jet tip. This flow makes the jet tip separate from the wall and rotate in counter clockwise. Fig. 3.34 shows a typical example of radial distribution of ordinary mean and conditional mean concentration, and intermittency factor. Ordinary mean concentration decreases gradually as radial distance increases. Conditional mean, on the other hand, is higher than

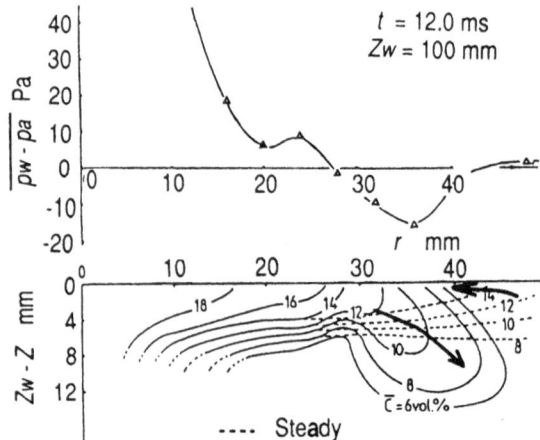

Fig. 3.33   Instantaneous concentration contour and wall pressure distribution of unsteady wall impinging jet, from [94] with permission.

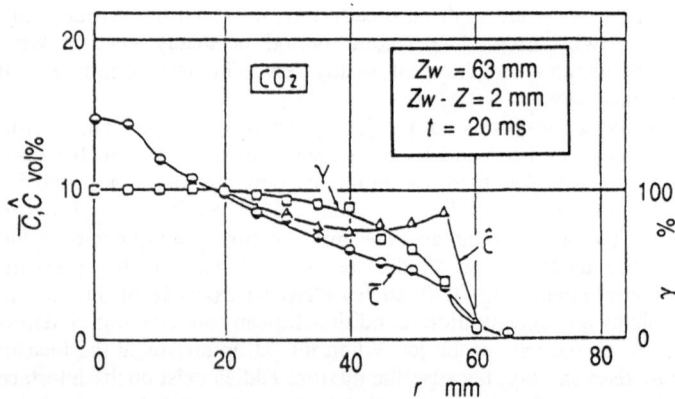

Fig. 3.34   Instantaneous radial distribution of conditional mean concentration, $\hat{C}$, ordinary mean concentration, C, and intermittency factor in unsteady wall jet

unconditional one and decreases sharply at the edge of the jet tip. This means that the nibbling entrainment takes places in the jet tip and small scale mixing occurs a little, as described in the case of unsteady free jets.

(c) Conclusion

Mixture formation process in unsteady gas jets was studied using a fast response probe to measure concentration, velocity and pressure in the jet.The experiment showed that the unsteady jet is composed of a steadty part and an unsteady jet tip part, and that a nibbling air entrainment occurs in the unsteady jet tip region.   (Hideaki.Tanabe)

3.2.5 Application of the Stochastic Ignition Theory

In order to clarify the ignition mechanism in the fuel spray developing unsteadily and heterogeneously, the ignition in an n–octane spray injected into high–temperature air behind reflected shock waves in a shock tube is statistically investigated, applying the stochastic ignition theory recently developed for the homogeneous mixture [65,66]. The theory, experimental method and results are explained here.

(a) Stochastic ignition theory

Even though in a homogeneous mixture, the induction period of spontaneous ignition fluctuates within a wide range not at random but under a certain probability. Repeating the same experiment many times, a histogram of the ignition induction period is obtained, from which the probability density $q(t)$ can be obtained. The ignition probability $w_0(t)$ per unit time in a whole combustible mixture can be calculated from $q(t)$ according to the following equations:

$$P(t) = \int_t^\infty q(t)dt \qquad\qquad\qquad (3.14)$$

$$w_0 P(t)dt = -dP \qquad\qquad\qquad (3.15)$$

$$w_0 = - \ d \ lnP(t)/dt \qquad\qquad\qquad (3.16)$$

where $P(t)$ is the probability of ignition whose induction period is longer than t.

According to the stochastic ignition theory the well–known explosion limits in hydrogen–oxygen mixture having so–called explosion peninsula which have been explained by the chain–branching kinetics as well as the thermal explosion theory are recognized to be the state having a certain constant ignition probability[65,66].

In a heterogeneous mixture, the partial ignition probability in each part is generally different. Dividing the space of the mixture into many small sections by an arbitrary method, the ignition in each section can be observed separately at the same time. Repeating experiments under the same condition many times and measuring the induction period of ignition in each section, the histogram of the induction period in each section as well as that in the whole space can be obtained. If the number of ignitions which take place in the whole space during a short period Δt at an arbitrary instant t is expressed to be ΔN, while that in a small section j to be ΔN_j, then the following relationship is obtained from eq.(3.15):

$$w_j - w_0 \bar{N} \ \Delta N_j/\Delta N, \qquad\qquad\qquad (3.17)$$

where $w_j(t)$ is the partial ignition probability in the section j. According to this equation, the partial ignition probability $w_j(t)$ can be deduced from the histogram of the induction period in each section, the histogram in the whole space and the ignition probability $w_0(t)$ in the whole space.

(b) Experiments [72]

A shock tube of stainless steel with 50 mm innerdiameter, as shown in Fig. 3.35 is prepared for the experiments. Shock driver gas, helium, is filled in the high pressure tube at 2.3 MPa, while the low pressure tube is filled with dry air at 33 kPa and room

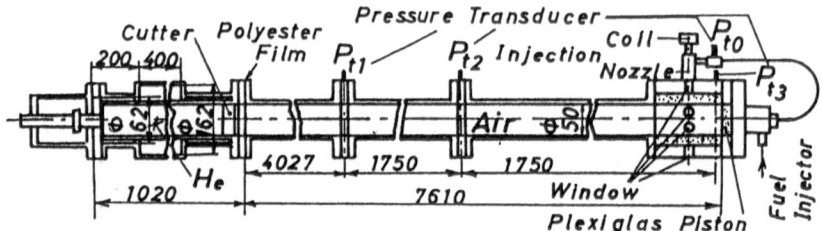

Fig. 3.35   Shock tube having a fuel injection system at the tube end, from [66] with permission.

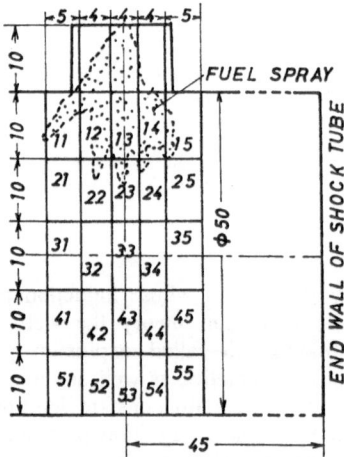

Fig. 3.36   Optical division of the fuel injection space at the tube end, from [66] with permission.

temperature of 293.5 ± 0.5 K. By breaking the polyester film set between both the tubes by a cutter, a shock wave of Mach number 2.86 propagates through the air in the low pressure tube and reflects at the tube end. Behind the reflected shock waves the air is compressed to 2.3 MPa and heated to 1270 K.

A piston of plexiglas having the same diameter as the innerdiameter of the shock tube and a thickness of 15 mm is set on the end plate of the shock tube. A stainless steel plunger of 6 mm diameter is connected to the back side of the plexiglas piston. Every time when the incident shock is reflected at the end of the low pressure tube, the plexiglas piston is pushed a few mm against the plunger, which in 0.4 ms compresses the liquid n– octane fuel in the plunger room and the injection pipe to a pressure higher than 17 MPa. The fuel is then injected into the air behind the reflected shock waves through an injection nozzle (DN30SDN180,ZEXELL) set at the end of the shock tube. The injection pressure of the nozzle is set at 12 MPa and the injection quantity of the fuel at each time is 3.0 ± 0.04 mg.

Setting a transparent window on each of the both sides of the fuel injection space at the shock tube end, we can take Schlieren or shadow photographs of the fuel spray. Further, the fuel injection space can be optically separated into 25 sections using several photomultipliers, as shown in Fig. 3.36. The ignition in each section can thus be

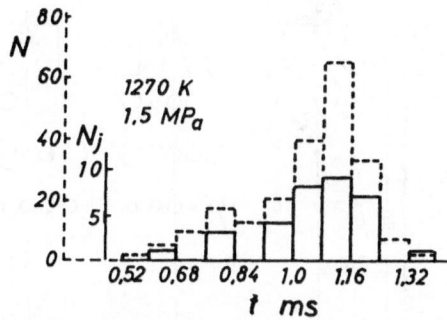

Fig. 3.37   Histograms of ignition induction period. Solid line: N_j in section 15 and broken line: N in the whole space, from [66] with permission.

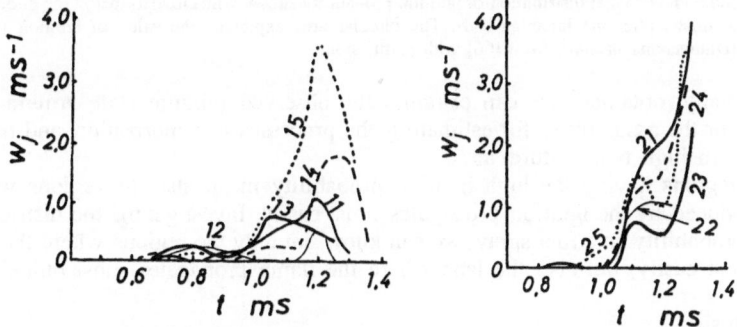

Fig. 3.38   Ignition probability w_j in each section. Number on each curve means the section number shown in Fig. 3.36, from [66] with permission.

observed separately by the photomultipliers and the induction period of ignition in each section is also measured as the period from the start of fuel injection to the instant when the first light emission from ignition is observed by one of the photomultipliers.

(c) Distribution of the ignition probability

By the method above explained the histogram of the induction period of ignition in each section as well as that in the whole space can be obtained, as shown in Fig. 3.37 According to eqs.(3.16) and (3.17) the ignition probability $w_j(t)$ in each section and $w_0(t)$ in the whole space can be deduced from these histograms. The results are illustrated in Fig. 3.38. From $w_j(t)$ in all sections, the spatial distribution of the ignition probability in the whole fuel injection space at different moment can be obtained, as shown in Fig. 3.39.

The results in Figure 3.39 suggest that ignition first takes place in the upper regions near the injection nozzle during a short period in the early stage after the onset of injection, but afterwards the most inflammable regions move downwards, increasing the ignition probability. The process is similar, with some delay, to that of fuel spray spreading into the air. The distribution of the ignition probability, therefore, should mainly depend on the vaporization, diffusion and mixing of the fuel with air of a high temperature. Using the temporal and spatial distribution of ignition probability

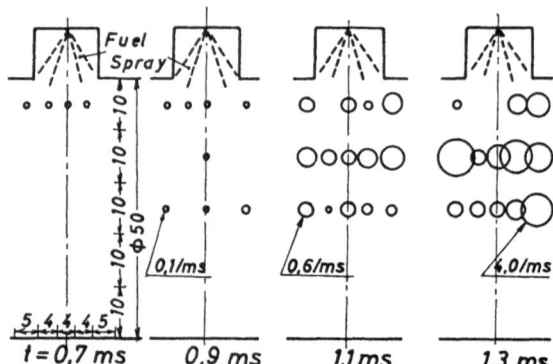

Fig. 3.39   Side-view spatial distribution of ignition probability shown with circular area in the n–octane spray
at time t after the injection start. The circular area expresses the value of ignition probability.
(Dimensions in mm), from [66] with permission.

experimentally obtained, we can examine the observed mixture state around the fuel
spray[44] or the assumption for estimating the processes of vaporization and mixing of
the fuel with high temperature air.

The regions having the high ignition probability mean also the regions where the
flame produced by the ignition propagates most easily. Investigating the distribution of
ignition probability in a fuel spray, we can know not only the regions where the ignition
occurs most easily, but also the lane where the flame propagates most quickly.

(d) Conclusion

Applying the stochastic ignition theory, the spontaneous ignition in an n–octane spray
injected into air behind reflected shock waves in a shock tube is experimentally
investigated. Observing the temporal and spatial distribution of ignition probability
obtained by the experiment, the most probable ignition position as well as the lane of
flame propagation are estimated.   (Kunio Terao and Chihong Liao)

3.3 Spray Combustion

3.3.1 Review

(a) Combustion with high injection pressure

Diesel combustion is dominated by the mixture formation in the engine combustion
chamber, which includes the characteristics of injection system, the nature of air swirl
and turbulence generated by an inlet port, and the combustion chamber geometry [73].
As an effective technique which allows simultaneous reduction of soot and NOx
emissions, high pressure injection combined with other techniques (reducing the nozzle
hole diameter, retarding the fuel injection timing, changing the injection rate profile, et
al.) has attracted a great attention. Extensive studies have been conducted on the analysis
of spray characteristics [74–77], the combustion mechanism and the soot formation [75,

77] and the engine combustion characteristics [74,77,78]. It is found that the decrease in the mean diameter of the spray and the increase in the spray angle are realized with the employment of the high pressure injection, which may increase the air entrainment rate into the spray. This is assumed to result in a shortening of the combustion duration and a reduction of soot emission. The effect of high pressure injection seems to be more appreciable with a small orifice diameter [75]. And a combination of high pressure injection with pilot injection is effective in reducing all NOx emission, smoke and combustion noise [77]. All these show the promising prospect of the high pressure injection to the diesel combustion improvement.   (Hiroyuki Hiroyasu)

(b) Numerical Simulation of Transient Spray Combustion

Spray combustion is encountered in various combustors such as furnaces, gas turbine combustors and diesel engines. The spray combustion occurs in a steady or unsteady state, at a low- or high- pressure atmosphere. Spray combustion includes complex processes such as drop formation, collision, coalescence, secondary breakup, evaporation and interaction with turbulence, turbulent mixing and chemical reactions. Experimental and numerical studies of spray combustion have been conducted so far in various aspects to elucidate spray flow, mixing and combustion as reviewed in ref.[62,79,80]. Here is described a short note concerned with problems to be solved in the numerical simulation of the spray combustion.

(1) Initial spray condition (drop velocity and its direction, and drop size near the nozzle exit) is needed to begin the numerical simulation of the spray flow and combustion. The experimental findings on initial conditions in full cone sprays, like Diesel sprays, are very few because the central core of the spray near the nozzle exit is too dense to approach by an optical manner. The mechanism of drop formation is not well established to predict the initial spray conditions. Drop collision, coalescence and break up might occur in the dense spray. No experimental evidence of those phenomena has not been revealed. Researches on this matter are collected in ref.[81].

(2) Interaction of turbulence and drop motion is to be considered. The turbulence induces random walk (diffusion) of drops and the existence of drops would promote or reduce the turbulence formation in the flow.
Especially, the latter phenomena (promotion or reduction of turbulence due to existence of drops) are not well elucidated.

(3) Evaporation of drops at supercritical pressure is an up-to-date problem which is encountered in practical diesel sprays.

(4) The processes and mechanism of spray combustion are still under controversy. One idea is that the drops evaporate at the early stage of the spray flame and most of the fuel vapor burns in a gaseous turbulent diffusion flame. This is true when the drops do not disperse widely and evaporation rate is high as in diesel spray [42],[82] and axial-flow air blast spray flame [83]. In this flame, turbulent gaseous diffusion combustion dominates the spray combustion. Onother idea is that the spray combustion should be the ensemble of drop combustion or group combsution of drops. This might be true when the drops are well dispersed prior to combustion to conduct premixed-like combustion

(5) The knowledge of the gaseous turbulent diffusion combustion is applicable to the spray combustion such as turbulence model and combustion model. Combustion model incorporated by chemical kinetics with elementary reactions is needed to predict formation and destruction of pollutant such as CO, NOx and soot. (Toshimi Takagi)

(C) Flame structure of steady spray flame

Steady spray flames take either of the following forms depending on the ratio of the characteristic time for air entrainment and mixing to that for droplet vaporization; (1) Gas–phase turbulent diffusion flames, (2) Clusters of two–phase diffusion flamelets of single drops or small–scaled droplet groups, and (3) Intermediate flames at various stages between (1) and (2).

Pure flames belonging either to category (1) or (2) are barely observed, and only intermediate flames at various stages between them are observed. As the typical examples of such flames, the direct photographs and the optically observed structures of the flames of No.2 and No.6 fuel oils injected in the same condition have been reported by Brena de la Rosa et al.[84] The structure of intermediate flames is very much complicated since they are usually nonuniform complexes of gas–phase diffusion flames and two–phase diffusion flamelets.

Onuma and Ogasawara [83] studied the structure of intermediate spray flames close to the gas–phase turbulent diffusion flames. They stabilized air–blast nonswirling kerosene spray flames using a small pilot flame and observed their structure by intrusive diagnostic techniques, finding that such spray flames had a structure similar to the gaseous diffusion flames produced by turning the kerosene jet into a propane one. It has been elucidated by their study that the flame of low–speed spray of a volatile fuel like kerosene has a structure similar to a gaseous turbulent diffusion flame except in the core region immediately following the fuel injector tip.

Liquid fuels are often injected into a preheated air stream or a hot recirculating gas stream to stabilize a flame. In this case, a fraction of droplets evaporate in contact with the hot gas entrained, and the flame of fuel vapor is formed outside of the spray boundary. Mizutani et al.[85] observed the structure of such kerosene spray flames around their upstream fronts by flow visualization and intrusive diagnostic techniques. It has been elucidated that the flame front is located at the position where the fuel vapor concentration reaches a fixed value so that the flame front position is governed by the heat flux convected by the gas entrained into the spray, and that the flame front position is dominated by the flame propagation mechanism rather than by the ignition delay mechanism.

Another method frequently used to stabilize spray flames is the swirl in combustion air stream and/or in liquid jet. Edwards and Rudoff [86] produced the flame of a 60 deg. hollow–cone spray by injecting kerosene from a Simplex air–blast atomizer placed at the center of the combustion air port, the swirl number of the air stream being 1.0. The luminous flame zone, cold two–phase flow, annular cold air flow and external recirculation zone were visualized by the combination of schlieren, direct and laser–extinction photographies, whereas, by LDA technique, the mean flow field was divided into the fuel jet, main–air jet, preserved (equilibrated) burnt gas flow, external recirculation zone and internal recirculation zone. Furthermore, the mean velocity of droplets belonging to every size class was determined by phase/Doppler interferometry, finding that, in the internal recirculation zone, the reverse flow of gas and forward flight of droplets took place alternately. The concept of "droplet group combustion" has been resulted from the study on the structure of intermediate spray flames at various stages between gas–phase turbulent diffusion flame and two–phase diffusion flamelet cluster. This concept has thereafter been developed into the stage that the nonuniform complexes of gaseous diffusion flames and two–phase diffusion flamelets are observed by combined optical techniques. Further discussions on this subject are to be made in section 3.3.3.

(Yukio Mizutani)

3.3.2 Combustion with high injection pressure

(a) Spray characteristics at high pressure injection

Spray characteristics in a high pressure vessel was investigated for obtaining the basic information on the combustion at the high pressure injection. Figure 3.40 shows the effect of the injection pressure on the Sauter mean diameter of the whole spray injected from the single hole nozzle with a hole diameter of 0.2 mm. The injection pressure is almost constant during the injection period. Diesel fuel is used as a test fuel. Ambient gas in the high–pressure vessel is nitrogen gas at room temperature and pressure of 1.2 MPa whose gas density is the same as the high–temperature and high–pressure air at 773 K and 3.0 MPa.

The tip penetration of the spray at any injection pressure is about 80 mm. The Sauter mean diameter is decreased with the injection pressure up to about 250 MPa, whereas the Sauter mean diameter is scarcely decreased above the injection pressure of about 250 MPa. This results shows that there is a limit in decreasing the Sauter mean diameter by only increasing the injection pressure.

Figure 3.41 shows the Schlieren photographs of the spray at an injection pressure of 275 MPa. The penetrating velocity of the spray tip estimated from the tip penetration

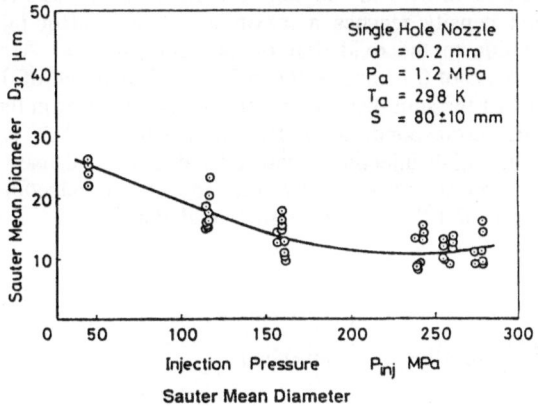

Fig. 3.40 Sauter mean diameter

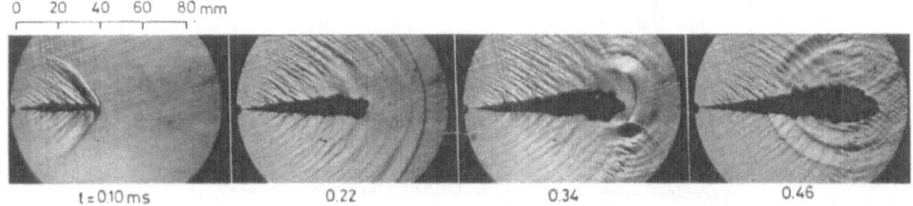

Fig. 3.41 Schlieren photographs of Mach wave around a spray (Pinj=275 MPa, d=0.2 mm, Pa=1.2 MPa, Ta=298 K)

versus time curve is about 440 m/s at 0.05 ms after the start of injection, which exceeds the sound velocity of the ambient gas of 355 m/s. Compression waves with a cone shape, that is the Mach waves, are observed around the spray at 0.10 ms after the start of injection. At 0.22 ms, the front compression wave is detached from the spray tip, but the Mach wave is still existed around the spray of the nozzle side. An influence of the Mach wave on the spray characteristics is not clear yet.

(b) Combustion mechanism in a high pressure vessel

Characteristics of the spray flame at the high pressure injection were investigated in quiescent high pressure and temperature air achieved in a rapid compression machine [75]. The rates of heat release are shown in Fig. 3.42. Two types of unit injectors of ACE and TIT were used. The fuel injection duration was kept at 2.6 ms. It is found that at the highest injection pressure of 250 MPa, the rate of heat release is the highest during the early stage of diffusion combustion and drops most rapidly, which shows a fast burning. High–speed Schlieren photographs of the spray flames taken at the same condition as that of Fig. 3.42 are shown in Fig. 3.43. The dark region near the nozzle orifice is assumed to result from the extinction of incident laser light caused by the scattering of the unvaporized liquid droplets. This dark region disappears immediately after the end of the injection. On the other hand, there also exists a dark region near the flame tip which is likely to be caused by the absorption of the soot particles. As seen in this picture, the soot is formed in a region near the flame tip 0.5 ms after the intiation of ignition. The soot density reaches a maximum at the end of injection, and, then, decreases gradually due to the oxidation of the soot particles. A comparison of the pictures between the case of 0.17 mm – 160 MPa (ACE) and that of 0.17 mm – 97 MPa (TIT) makes it clear that the soot disappears comparatively faster in the former case than in the latter one, which corresponds to the fast burning for the ACE type injector shown in Fig. 3.42. At the ultra–high injection pressure condition, the density of the dark region is the lowest and disappears very rapidly after the end of injection. This shows the reduced soot formation and the increased soot oxidation rate.

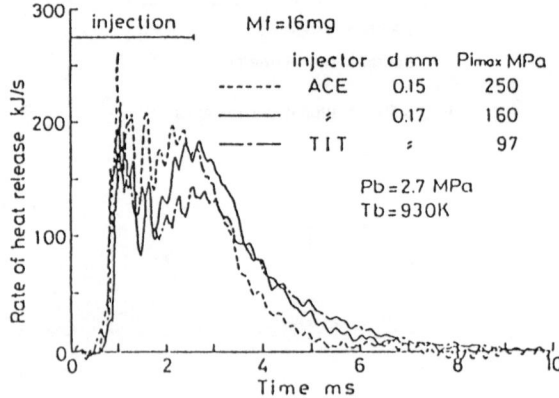

Fig. 3.42 Effect of maximum injection pressure on rate of heat release, from [75] with permission.

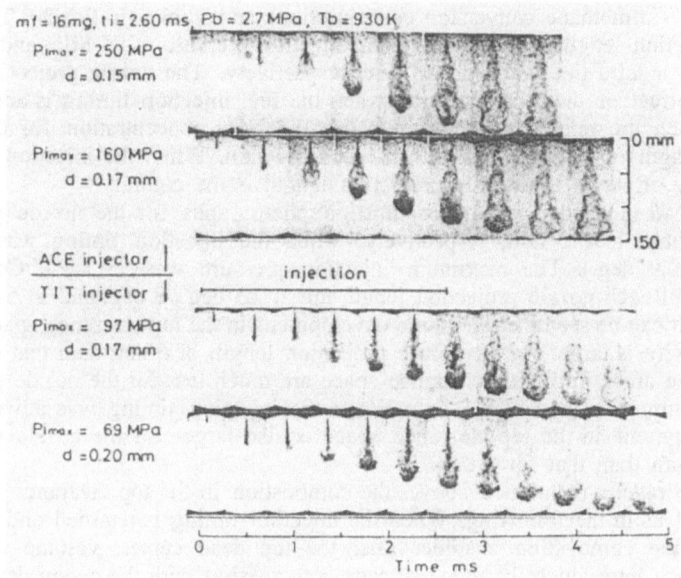

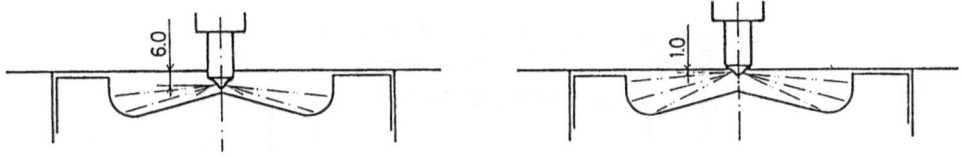

(1) Nozzle projection length 6 mm   (2) Nozzle projection length 1 mm

(c) Combustion mechanism in a D.I. diesel engine

An experiment on the combustion process and the emission characteristics of an engine at the high pressure injection was made in a single cylinder, four stroke cycle, D. I. diesel engine with a bore of 135 mm and stroke of 130 mm. Intake swirl was almost quiescent. The piston had a cavity of a shallow dish type with a diameter of 90 mm. Amount of fuel injected was 70 mg/cycle, and engine revolving speed was 800 rpm. A nozzle used was a 6–hole nozzle whose hole diameter was 0.13 mm. The maximum injection pressure was set at 95 MPa and 130 MPa.

Two different spatial distributions of the fuel sprays in the combustion chamber were obtained by adjusting the projection length of the nozzle tip from the cylinder head, that is, 6 mm and 1 mm, as shown in Fig. 3.44.

Figure 3.45 shows the effect of fuel injection timing on the exhaust emissions such as NOx, THC (in methane conversion concentration) and smoke (in Bosch). The results of both projection lengths for the maximum injection pressure of 95 MPa and 130 MPa are shown by a solid line and a dotted line, respectively. The nozzle projection length has differnt effect on the NOx emission when the fuel injection timing is advanced or retarded. When the injection timing is retard, the NOx concentration for the nozzle projection length of 6 mm is lower than that for 1 mm. When the injection timing is advanced, the effect of the nozzle projection length is the contrary.

Figures 3.46 (1) and (2) show combustion photographs for the nozzle projection length of 6 mm and 1 mm, respectively, when the injection timing was retarded (θinj=2.4 – 2.9 deg.). The maximum injection pressure was 95 MPa. Combustion photographs of each nozzle projection length are at 2.3 deg., 8 deg. and 11.5 deg. after the ignition. It can be seen that the flame development in the top clearance space outside the piston cavity is larger for the nozzle projection length of 6 mm than that for 1 mm. The non flame areas in the top clearance space are much less for the nozzle projection length of 6 mm than those for 1 mm. When the injection timing was advanced, the flame development in the top clearance space is also larger for the nozzle projection length of 6 mm than that for 1 mm.

From the results mentioned above, the combustion in the top clearance space can be considered as in the following. When the injection timing is retarded and the flame develops in the combustion chamber after the top dead center, volume of the top clearance space into which the flame spreads is increasing with the piston descent. The wider spread of the flame in the top clearance space (a case for nozzle projection length of 6 mm) induces more intensive mixing of burning gases and enough air. Such a combustion in the top clearance space reduces NOx emission due to the decrease in high

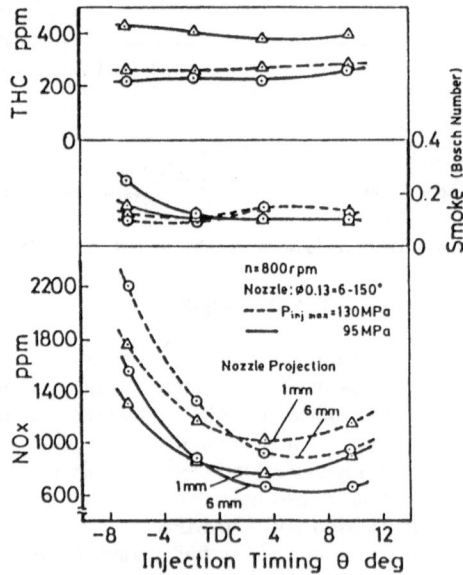

Fig. 3.45 Exhaust emission characteristics, from [78] with permission.

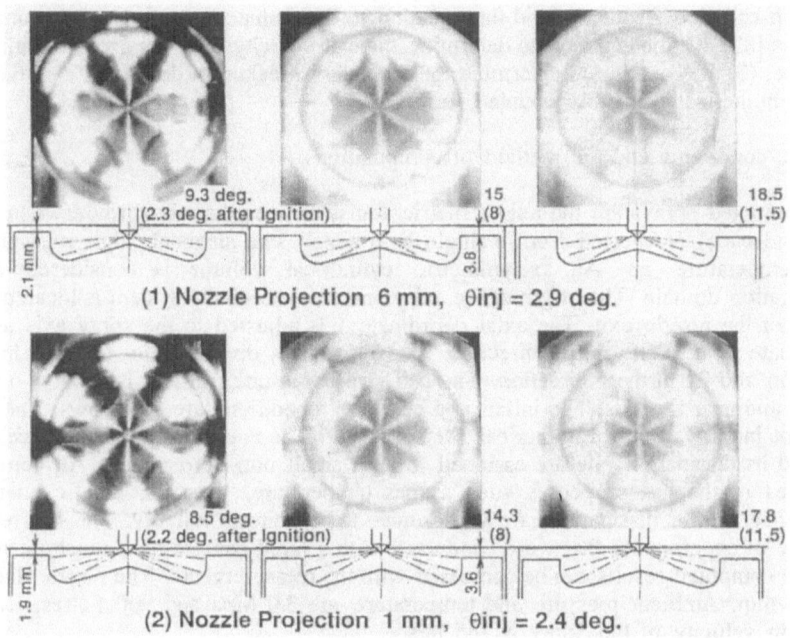

(1) Nozzle Projection 6 mm, θinj = 2.9 deg.

(2) Nozzle Projection 1 mm, θinj = 2.4 deg.

Fig. 3.46 Combustion photographs for retarded injection timing, from [78] with permission.

temperature spots in the combustion gas. On the contrary, when the injection timing is advanced, the flame spreads into the top clearance space whose volume is still small. The wider spread of the flame in the top clearance (a case for nozzle projection length of 6 mm) results in the increase in NOx emission due to the increase in high temperature spots in the combustion gas.   (Hiroyuki Hiroyasu)

3.3.3 Numerical Simulation of Transient Spray Combustion

Evaporation, ignition and combustion of transient sprays were computed and comparisons were made with measurements taken in a rapid compression machine [42] operated at high pressure and high temperature in order to elucidate the process and the internal structure of the combusting transient spray.

(a) Fundamental equations

The fundamental equations are the transient Eulerian equations for the gas in cylindrical coordinates and the Lagrangian equations for the drops [87]. The equations include (1) the conservation equations of mass, momentum, energy and species concentration for the gas, (2) the exchange rate equations of heat, mass and momentum between gas and drops, and (3) reaction models for gas species. Reaction models are the most indeterminable. In the present computations, one–step irreversible reaction scheme is selected for simplicity. k–ε turbulence model was used for evaluating turbulent transport of momentum and scalars. Other auxiliary equations are (1) the equation to determine the drop collision frequency and the equation to determine the coalescence probability of drops [87], (2) the equation to determine the exit velocity of drops given the injection pressure, (3) the equation to determine the secondary breakup of drops. All equations are solved numerically in fully coupled form.

(b) The conditions and the method of computations

The computed sprays are the axisymmetric transient ones that are formed when liquid fuel (tridecane) is injected from a single hole nozzle into stagnant, high–pressure and high–temperature air. An axisymmetric cylindrical volume is considered as the computation domain. The origin of the axisymmetric coordinate system is located at the center of the nozzle exit. The axial coordinate x is adjusted to the spray axis and the coordinate y is in the radial direction. The domain is divided into 44 cells in the x direction and 21 in the y direction. The cell size is not uniform and is smaller near the nozzle and near the axis. The initial and boundary conditions are as follows. The gas is stagnant initially. But, for numerical reasons, the kinetic energy of the turbulence of the gas and its dissipation rate are assumed to have small non–zero values. All dependent variables of the gas properties such as gas temperature, pressure, kinetic energy of turbulence, and its dissipation rate are assumed to be uniform initially. They correspond to those of experiments that were conducted with a rapid compression machine [42] so that the computed results can be compared with the measurements. The nozzle diameter is 0.16 mm. Ambient pressure and temperature are 3.0 Mpa and 900 K, respectively. Injection velocity of the Spray is 185 m/s.

(c) Results and discussions

Figure 3.47(a) shows the computed isotherms at different times during injection. Fig. 3.47(b) shows the experimental Schlieren photographs [42]. The Schlieren photographs indicate that ignition occurs at about $t = 1$ ms. The calculated isotherms also indicate that the temperature rise initiates at about $t = 1$ ms . The location of the ignition

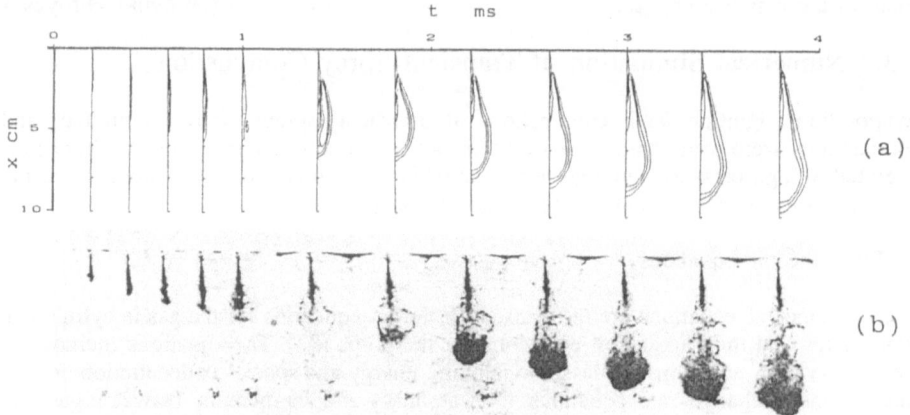

Fig. 3.47   Comparison of the computed equi–temperature contours (a) and Schlieren photographs (b), from [82] with permission.

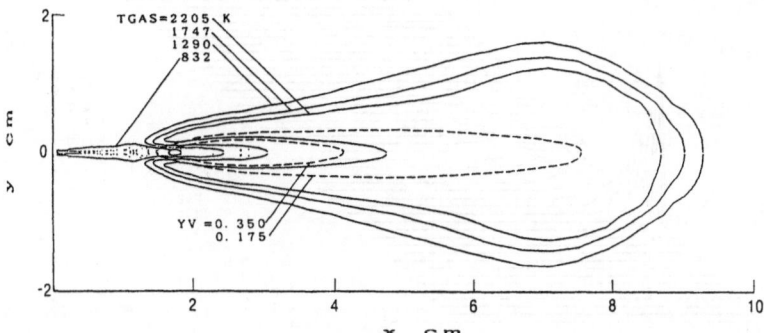

Fig. 3.48   Schematic picture of the spray flame (t = 3.5 ms), from [82] with permission

(rapid temperature rise) is just upstream of the tip of the spray. After ignition, the radial extent of the spray increases due to the expansion of the gas. The computed configuration of the spray and the penetration distance correspond fairly well to the measured ones.

Figure 3.48 shows the schematic picture of the spray flame structure based on the calculation at t = 3.5 ms when considerable time has passed after ignition. Small dots indicate the positions of fuel drops. The drops disappear nearly up to x = 3 cm due to the evaporation. Existence of the drops is limited to local region near the nozzle exit. The equi–concentration line of the evaporated fuel vapor is indicated by the dotted lines. High concentration of the vapor is observed at x = 2 – 4 cm near the central axis but the fuel vapor region extends up to x = 8 cm. Gaseous turbulent diffusion combustion seems to dominate in the large portion of the flame. Low temperature region corresponds to the drop–existing region and high–vapor–concentration region near the nozzle exit. High temperature region is observed at the circumferential part at x = 2 – 5 cm and at the central part in the downstream.

Figure 3.49 shows the radial profiles of the local heat release rate per unit volume q due to chemical reactions at many axial distances to get the map of the heat release rate at t = 3.5 ms which corresponds to the condition of Fig. 3.48. The figure indicates that the heat release is very large near the flame front at about x = 1.5 cm. The flame front is denoted by the front of the equi–temperature contours in Fig. 3.48. The high heat release rate near the flame front is caused by the premixed type combustion. Fuel vapor evaporated in the upstream of the flame front is premixed with the surrounding air prior to the combustion. In the downstream, heat release rate is positive at the circumferential region and near the tip of the flame where the heat release rate is not intense but the region of the heat release extends widely to the off–axis region. The heat release at the circumferential region is caused by the turbulent diffusion combustion.

In Fig. 3.50, radial profiles of gas temperature (TGAS), species gas concentration (YI), soot concentration (YSO, mass fraction), local equivalence ratio (FAI) are shown at t = 3.5 ms. Plots do not denote the calculation grids but indicate which line denotes which quantity. Fig. 3.50 indicates that profiles of the fuel vapor, O_2, H_2O, CO_2 and temperature are similar to those of the typical turbulent gas diffusion flame. That is to say, the fuel vapor or oxygen concentration is high at the central part or circumferential

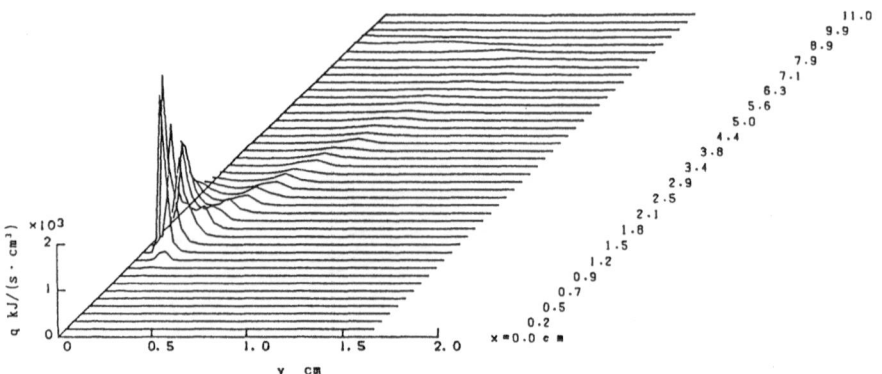

Fig. 3.49   Local heat release rate by reactions (t = 3.5 ms), from [82] with permission.

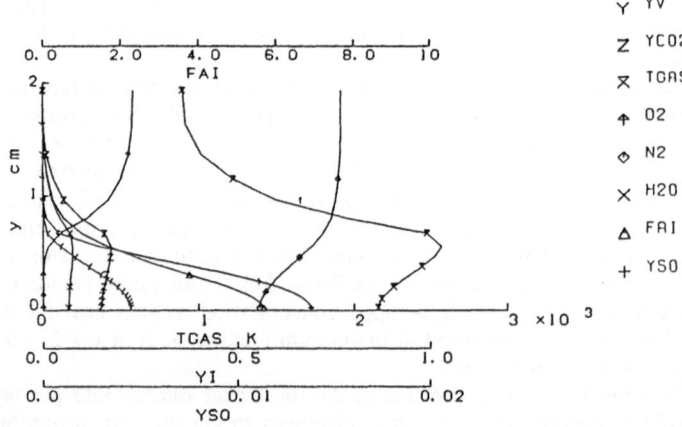

Fig. 3.50   Radial profiles of various properties (t = 3.5 ms, x = 4.4cm), from [82] with permission.

part, respectively. CO_2 and H_2O concentration and temperature are high at the boundary where fuel vapor and oxygen meet and equivalence ratio is about unity.

(d) Summary

(1) The computations predict the ignition delay, the transient configuration of the spray flame and the drop penetration distance reasonably well.
(2) The drops evaporate in the early stage of the spray combustion. The drop penetration distance is predicted well by taking account of the secondary breakup of drops. The secondary breakup is liable to occur at high pressure and high temperature atmosphere.
(3) Ignition occurs at a hot spot in the off axis region followed by the rapid spread of

the combustion and fast release of heat.

(4) A flame front is formed downstream from the nozzle exit. Near the flame front, the local heat release rate is very intense where premixed combustion occurs. Gaseous turbulent diffusion combustion dominates combustion of the fuel in the core of spray.

(Toshimi Takagi and Tatsuyuki Okamoto)

3.3.4 Flame Structure of Steady Spray Flames

(a) Concept of Droplet Group Combustion

If a premixed spray flame (the burner port diameter: 52.7 mm, the two–phase flow velocity: 4.3 m/s and the fuel–to–air mass ratio: 0.05) stabilized by a central pilot flame is observed by human eyes, a rather smooth luminous flame front is seen accompanied by a thin blue flame zone (typically the direct photograph [88] taken with 1/15 s exposure are shown in Fig. 3.51(a)). If, however, the same flame is recorded by high–speed photography at a speed of 5000 frames/s, the continuous flame front disappears and a complex–shaped luminous cluster of two–phase diffusion flamelets is observed as shown in Fig. 3.51(b). It is supposed from these pictures that a spray flame has a very complicated structure and that the spray is divided into several droplet groups corresponding to their number density patterns and burns by the so called group combustion mechanism. In addition, it is also supposed from the time–averaged flame front structure that blue flames exist between the two–phase luminous diffusion flamelets. Generally, sprays are apt to produce nonuniform density patterns [89]. If fuel vapor is produced in the region where droplets are sparse, a blue flame propagates there at a high speed [90], and the dense droplet regions left behind it burn in diffusion combustion mode with a luminous flame.

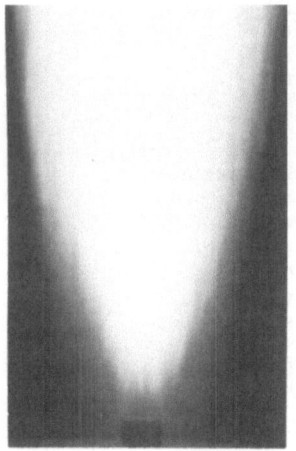
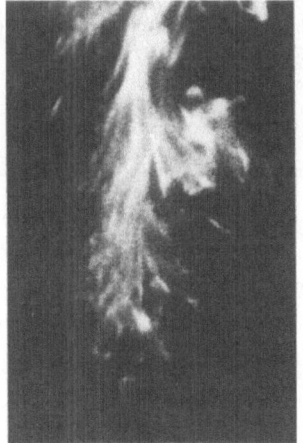

∟⌐ 10 mm

(a) Long exposure (1/15 s)   (b) Short exposure (100μs)

Fig. 3.51   Variation of flame image with exposure time, from [88] with permission.

As for droplet group combustion, Chiu et al.[91] theoretically analyzed the combustion of a spherical cluster of diameter D, which consisted of monodisperse droplets of diameter d, showing that four kinds of combustion mode should appear as illustrated in Fig. 3.52. They defined the group combustion number, G, as a criterion to judge which mode of combustion appeared; i.e.,

$$G = 1.5 L_e \ (1 + 0.276 S_c^{1/3} R_e^{1/2}) \ n_T^{2/3} \ (d/l) \tag{3.18}$$

where L_e was Lewis number, S_c Schmidt number, R_e the droplet Reynolds number, n_T the total number of droplets contained in the cluster and l denoted the mean interdroplet distance. G corresponds to the ratio of the gross evaporation rate of droplets to the gas exchange rate across the cluster boundary.

Figure 3.52(a) shows the single droplet combustion mode, where every droplet burns surrounded by each envelope flame due to the rapid diffusion of oxygen to the cluster center $(G<10^{-2})$. Fig. 3.52(b) shows the internal group combustion mode, where, in the central portion, droplets burn with a common group flame due to an insufficient oxygen supply, whereas, in the peripheral region, single droplet combustion mode is maintained $(10^{-2}<G<1)$. Fig. 3.52(c) shows the external group combustion mode, where the oxygen supply is far insufficient against the gross evaporation rate of droplets so that the region of single droplet combustion mode completely disappears. In this case, a group flame is formed outside the cluster $(1<G<10^2)$. Fig. 3.52(d) shows the external sheath combustion mode, where droplet evaporation is confined to the peripheral region since the temperature does not rise in the central region due to the extremely high density of droplets. Naturally a group flame is formed outside the cluster $(10^2<G)$[88].

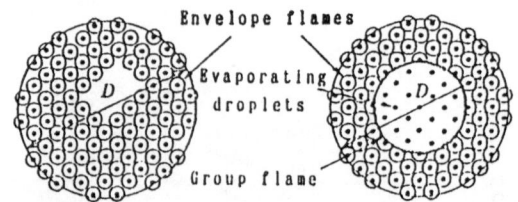

(a) Single droplet combustion mode   (b) Internal group combustion mode

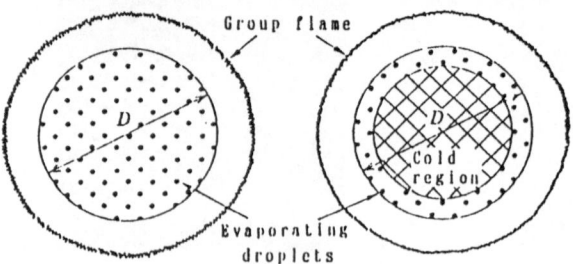

(c) External group combustion mode   (d)External sheath combustion mode

Fig. 3.52   Four modes of droplet group combustion

In the case that the spray has a nonuniform density pattern, the group combustion concept should not be applied to the spray as a whole, but should be applied to each droplet cluster which is produced by the density nonuniformity and the preferential flame propagation.

(b) Structure of Intermediate Spray Flames

The structure of intermediate spray flames is very much complicated and quite different from the model of Chiu et al.[91] where monodisperse droplets are arranged at regular intervals within an isolated spherical cluster. Then, aiming at the elucidation of the detailed structure of intermediate spray flames, the light emission signals in the OH– and CH–radical emission bands, the Mie scattering signal from droplets and droplet velocity signal were monitored simultaneously at the time–averaged flame front [92]. In order to simulate various intermediate spray flames ranging from a gas–phase turbulent diffusion flame to a two–phase diffusion flamelet cluster, propane was mixed with the combustion air at various fractions in place of the fuel vapor.

The flames adopted were a spray flame similar to the one shown in Fig. 3.51(a) and a gas–liquid coburning flame where propane was mixed with the combustion air (the velocity of two–phase mixture: 5.5 m/s and the overall fuel–to–air mass ratio: 0.065). Using the optical system shown in Fig. 3.53, the light emission signals from a point, Mie scattering and droplet velocity signals were monitored simultaneously. The monitoring points were located at the time–averaged flame front and at 10 mm downstream from the front, both at a radial position, r = 16 mm. In the figure, PM_D and PM_P are the photomultipliers to detect the Doppler and Mie–scattering signals of LDA, respectively, being offset by 5 deg. from the incident beam direction in the horizontal plane to restrict the control volume. PM_U and PM_V are the photomultipliers to detect the light emission signals in OH– and CH–bands, respectively, the control volume of which was restricted to 2 mm diam. and 4 mm depth by the light guide assembly in the figure.

Figure 3.54 shows the recorded signals monitored at the flame front. (a) is a gas–liquid coburning case where the fraction, ϕ_l/ϕ, of liquid fuel (kerosene) in the whole fuel is 0.23 and the mean diameter, $\bar{d}$, of droplets is 17 μm, whereas (b) is a spray combustion case where $\bar{d}$ = 100 μm. The recorded signals are, from top to bottom, emission intensities, and I_{OH}, at OH– and CH–bands, respectively, the intensity, I_P,

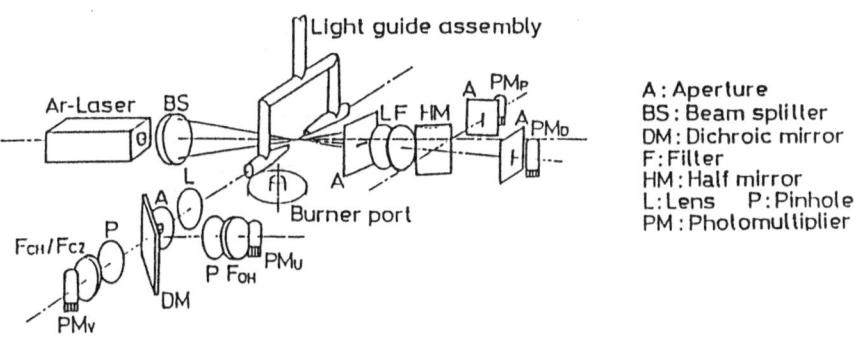

Fig. 3.53 Optical monitoring system, from [92] with permission.

of Mie scattered light and droplet velocity, V_p. In case (a) of gas–liquid coburning, I_{OH} and I_{CH} well synchronize with each other. Generally, CH–radical emission is overlapped by the solid emission from soot. For a blue flame of a fixed equivalence ratio, the intensity ratio of OH– to CH–band emission is kept almost constant, whereas, for a luminous flame accompanied by the solid emission from soot, the correlation between the emission signals at both bands is deteriorated. Therefore, the synchronism of the emission signals at both bands implies that thin blue flames are predominant to luminous flames accompanied by the solid emission of soot. In addition, I_{OH} and I_{CH} are well synchronous with I_p. This fact implies the preferential propagation of blue flames through the droplet–existing region. (Note that the monitoring spot is located at the flame front.)

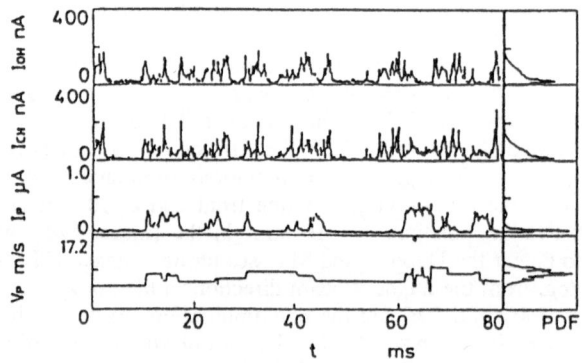

(a) $\phi_\nu/\phi=0.23$ and $\bar{d}=17\mu m$ (Gas–liquid coburning)

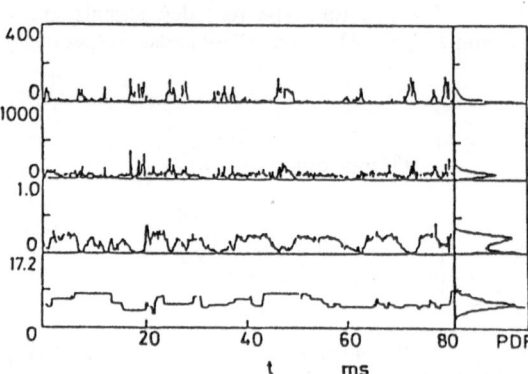

(b) $\phi_\nu/\phi=1.00$ and $\bar{d}=100\mu m$ (Spray combustion)

Fig. 3.54 Comparison between the signals of light emissions, Mie scattering and droplet velocity, monitored at the flame front, from [92] with permission.

In case (b) of spray combustion, on the other hand, the synchronism between any signals is deteriorated, which implies that the relative appearance frequency of luminous flames is increased. Furthermore, I_{OH} and I_P are almost on an opposite phase, which implies that combustion takes place in the external group or sheath combustion mode where a group flame is formed outside the droplet cluster (see Fig. 3.52(c) and (d)). These arguments become clearer if the correlations between I_{OH}, I_{CH} and I_P are calculated. The coherences, $C_{OH,CH}$ and $C_{OH,P}$, along with the phases, $P_{OH,CH}$ and $P_{OH,P}$, which are defined as the correlation coefficients and phase differences, respectively, calculated as functions of Fourier frequency, f, are superior to the simple cross correlation coefficients, since the mean droplet velocity divided by f results in a length scale so that the correlation coefficient and spatial gap between any pair of signals are recognizable as functions of length scale.

The coherences, $C_{OH,CH}$ and $C_{OH,P}$, and the phases, $P_{OH,CH}$ and $P_{OH,P}$, between I_{OH}, I_{CH} and I_P are plotted against Fourier frequency, f, in Figs. 3.55 and 3.56. As for Fig. 3.55 (a) on gas–liquid coburning, although the mean value of $P_{OH,CH}$ is always zero, its fluctuation gradually increases as f is increased, becoming a correlationless state as f exceeds 6 kHz. $C_{OH,CH}$, on the other hand, decreases linearly from 0.9 to 0.2 between f = 0~6 kHz. This fact implies that the smaller is the scale of a flamelet, the stronger is the character of diffusion flame where the local equivalence ratio varies widely or that of luminous flame where the solid emission of soot is dominant. In Figure 3.56 (a), $P_{OH,P} \rightarrow 0$ and $C_{OH,P} \rightarrow 0.3$ as $f \rightarrow 0$; $P_{OH,P}$ fluctuates around π and $C_{OH,P}$ decreases from 0.4 to 0.1 between f = 0.2 ~2 kHz; and the correlation is lost for $f > 2$ kHz. Since the mean value of droplet velocity, V_P, is read as 8 m/s from Fig. 3.54 (a), a droplet cluster and a flame overlap each other if they are observed at a time scale exceeding 1/0.2 = 5 ms or at a length scale exceeding 8/0.2 = 40 mm. If, however, they are observed at a smaller scale, they appear alternately. At scales smaller by an order of magnitude, the correlation cannot be determined due to the discrepancy of control volumes between the light emission and droplet Mie scattering.

Although the situation is almost the same for the case of spray combustion shown in Figures 3.55(b) and 3.56(b), $P_{OH,P} \rightarrow \pi$ in this case as $f \rightarrow 0$ so that a droplet cluster

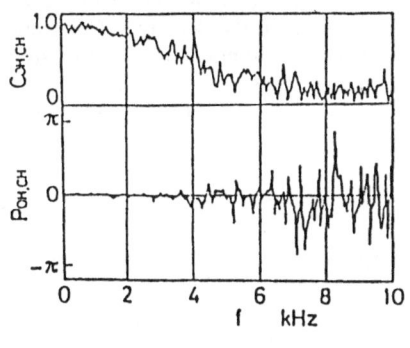

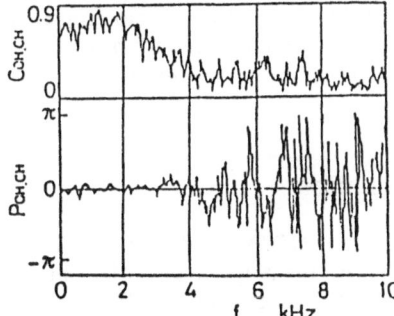

(a) ϕ_v/ϕ=0.23 and $\bar{d}$=17μm
(Gas–liquid coburning)

(b) ϕ_v/ϕ=1.00 and $\bar{d}$=100μm
(Spray combustion)

Fig. 3.55 Coherences and phases between OH– and CH–band emission signals monitored at the flame front

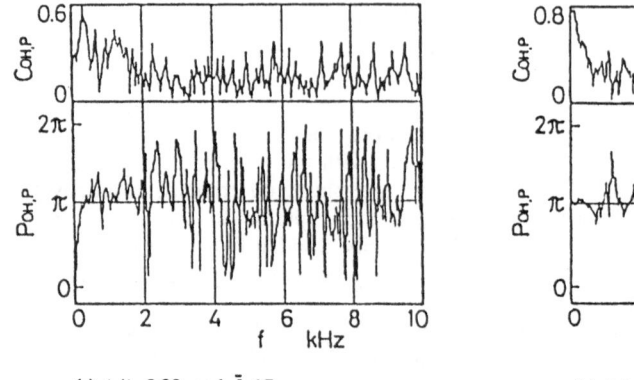 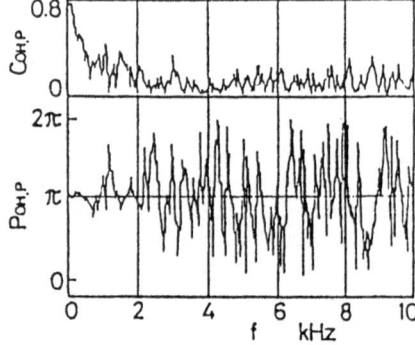

(a) ϕ_v/ϕ=0.23 and $\bar{d}$=17μm   (b) ϕ_v/ϕ=1.00 and $\bar{d}$=100μm
(Gas–liquid coburning)   (Spray combustion)

Fig. 3.56   Coherences and phases between OH–band emission and Mie scattering signals monitored at the flame front

and a flame does not overlap each other at any scale. In addition, $C_{OH,CH}$ has a value around 0.7, far smaller than unity for $f < 2$ kHz, which suggests that diffusion combustion that the local equivalence ratio varies widely is dominant in spray combustion. $C_{OH,P}$ rapidly decreases from $f = 0$ to 1 kHz, whereas the fluctuation in $P_{OH,P}$ rapidly increases from 1 kHz. This fact implies that droplet clusters burn in external group or sheath combustion mode at large scales, whereas, at a time scale less than 1/(1 kHz)=1 ms or at a length scale less than (5.4 m/s)/(1 kHz) = 5.4 mm, they burn with a luminous or diffusion flamelets emitting a solid emission from soot. Here, 5.4 m/s is the mean value of the droplet velocity, V_P, being read from Fig. 3.54(b). Although the frequency at which the correlation is lost is slightly lower than that of the gas–liquid coburning case, the corresponding length scale does not differ much due to the lower droplet velocity[92].

All the above–mentioned facts seem to suggest that liquid fuel sprays make group combustion being divided into droplet clusters corresponding to the density pattern, and that they assume a complicated flame structure where a number of small–scaled luminous or diffusion flamelets are embedded in larger–scaled group flames.

                                                                (Yukio Mizutani)

References

[1] Durst F, (1973) Scattering phenomena and their applications in optical anemometry. Zeitschrift für Angewandte Mathematik und Physik (ZAMP), 24: 619– 643
[2] Durst F and Zaré M (1975) Laser doppler measurements in two–phase flows. LDA–Symposium, Copenhagen, Congress Proceedings, pp 403–429
[3] Bachalo WD and Houser MJ (1984) Phase/doppler spray analyzer for simultaneous measurements of drop size and velocity distributions. Optical Eng. 23 pp 583–590
[4] Saffman M, Buchhave P, and Tanger H (1984) Simultaneous measurements of size, concentration, and velocity of spherical particles by a laser doppler method. Proc. 2nd

Int. Symp. on Appl. Laser Anemometry to Fluid Mechanics, Lisbon

[5] Bauckhage K and Flögel H (1984) Simultaneous measurements of droplets size and velocity in nozzle sprays. Proc. 2nd Int. Symp. on Appl. Laser Anemometry to Fluid Mechanics, Lisbon

[6] Beretta F, Cavaliere A, and D'Alessio A (1983) Experimental and theoretical analysis of the angular distribution and polarization state of the light scattered by isothermal sprays and oil flames, Combust. Flame. 49:183

[7] Beretta F, Cavaliere A, and D'Alessio A (1984) Ensemble light scattering diagnostics or the study of fuel sprays in isothermal and burning conditions. 20'th Symp.(Intl.) on Combust., The Combustion Institute, Pittsburgh, pp 1249–1258

[8] Beretta F, D'Alessio A, and Noviello C (1986) Spray vaporization and soot formation in flames generated by light oils with different chemical composition. 21'st Symp.(Intl.) on Combust., The Combustion Institute, Pittsburgh, pp 1133–1140

[9] Arnone L, Beretta F, D'Alessio A, Ossler F, and Tregrossi A (1992) Ensemble and time resolved light scattering measurements in isothermal and burning heavy oil sprays, to be presented at the 24'th Symp.(Intl.) on Combust., Sydney

[10] Beretta F, Cavaliere A, D'Alessio A, Massoli P, and Ragucci R (1986) A Spectral scattering method for determining size distribution functions and optical characteristics of droplets ensembles in fuel sprays. 21'st Sympo.(Intl.) on Combust. The Combustion Institute, Pittsburgh, pp 675–683

[11] Massoli P, Beretta F, and D'Alessio A, (1989) Single droplets size, velocity, and optical characteristics by the polarization properties of scattered light. App.Opt., 28, 6: 1200–1205

[12] Massoli P, Beretta F, and D'Alessio A (1990) Pyrolysis in the liquid phase inside single droplets of light oil studied with laser light scattering methods. Combust. Sci. and Tech., 72: 271–282

[13] Massoli P, Beretta F, D'Alessio A, and Lazzaro M, Temperature and size of single transparent droplets by light scattering in the forward and rainbow regions. submitted for publication on Applied Optics.

[14] Cavaliere A, Ragucci R, D'Alessio A, and Noviello C (1991) Digital imaging of condensed phases fields in ignited unsteady sprays. Comb. Sci. and Tech., 77: 73–93

[15] Ragucci R, Cavaliere A, Ciajolo A, D'Anna A, and D'Alessio A (1992) Structures of diesel sprays in isobaric combustion conditions. to be presented at the 24'th Symp.(Intl.) on Combust., Sydney

[16] Miake–Lye RC and Toner SJ (1987) Laser soot– scattering imaging of a large Buoyant diffusion flame. Combustion and Flame, 67, 9–26: 9

[17] Cavaliere A, Ragucci R, D'alessio A, and Cardamone P (1990) Morphological analysis of diesel sprays structure in ignition regimes. Proceeding of Workshop and Exposition on Fluid Mechanics, Combustion and Emission in Reciprocating Engines, Naples, Italy, 89

[18] Won YH, Kamimoto T, Kobayashi H, and Kosaka H (1991) 2–D Soot visualization in unsteady spray flame by means of laser sheet scattering technique. SAE technical paper series 910223

[19] Shioji M, Yamane K, Sakakibara N and, Ikegami M (1990) Characterization of soot clouds and turbulent mixing in diesel flame by image analysis. Proc. of Intl. Symp. on Diagnostics and Modeling Combustion in Reciprocating Engines(COMODIA 90), Kyoto, Japan, pp 613–6180

[20] Dec JE (1992) Soot distribution in a D. I diesel engines using 2–D imaging of laser–induced incandescence, elastic scattering, and flame Luminosity, SAE technical paper series 920115

[21] Bhatia JC, Domnick J, Durst F, and Tropea C, (1988) Doppler anemometry and the log–hyperbolic distribution applied to liquid sprays. Part. Part. Syst. Charact. 5: 153–164

[22] Naqwi A, Durst F and Liu X (1991) Extended phase–doppler system for characterization of multiphase flows. Part. Syst. Char., 8: 16–22

[23] Naqwi A, Ziema M, and Durst F (1992) Fine particle sizing using an extended phase doppler anemometer. Proc. 5th Euro. Sym. Part. Char., pp 267–279

[24] Bopp S, Tropea C, and Durst F (1990) In–cylinder velocity measurements with a mobile fiber optic LDA system. SAE technical paper series 900055

[25] Domnick J, Durst F, Müller R and Naqwi A (1991) Improved optical systems for velocimetry and particle sizing using semiconductor lasers and detectors. in the book Applications of Laser Techniques to Fluid Mechanics (Ed.: Adrian, Durão, Durst, Maeda, Whitelaw), Springer–Verlag, pp 317–330

[26] van de Hulst HC (1991) Light scattering by small particles. Dover publications, New York

[27] Roth N, Anders K, and Frohn A (1991) Refractive index measurements for the correction of particle sizing methods. App.Opt., 30, 33: 4960–4965

[28] Melton LA (1984) Soot diagnostics based on laser heating. Applied Optics 23–13: 2201

[29] Won Y–H, A study of formation and extinction of soot in unsteady spray flames, Dr. Thesis, Tokyo Institute of Technology

[30] Cossali GE, Brunello G, and Coghe A (1991) LDV characterization of air entrainment in transient diesel sprays, SAE technical paper series 910178

[31] Flenklach M, Ramachandra MK, and Matula RA (1984) Soot formation in shock-tube oxidation of hydrocarbons. 20'th Sympo.(Intl.) on Combust. The Combustion Institute, pp 871–878

[32] Furutani M, Ohta Y, Terada K, and Takahashi H (1991) Soot formation in low-temperature compression ignition, 9th Internal Combustion Engine Symp., Tokyo, Japan, pp 433–438 (in Japanese)

[33] Rayleigh (1878/79) On the instability of jets, Proc. London Math. Soc., 10: 4–13

[34] Sauter J (1926) Die Größenbestimmung der im Gemischnebel von Verbrennungskraft–maschinen vorhandenen Brennstoffteilchen, VDI–Z, 70: 1040–1042

[35] Haenlein A (1931) Über den Zerfall eines Flüssigkeitsstrahls, Forschung, 2:139–149

[36] O'Rourke JK (1981) Collective drop effects on vaporizing liquid sprays, Ph. Thesis, Princeton University

[37] Reitz RD (1978) Atomization and other breakup regimes of a liquid jet, Ph. Thesis, Princeton University

[38] Eifler W (1990) Untersuchungen zur Struktur des instationären Dieseleinspritz-strahles im Düsennahbereich mit der Methode der Hochrequenz–Kinematographie, Thesis, Universität Kaiserslautern

[39] Bergwerk W (1959) Flow pattern in diesel nozzle spray holes, Proc. Instn. Mech. Engrs., 173–25: 655–660

[40] Ruiz F and Chigier N (1985) The mechanics of high speed atomization, Proc. of ICLASS, 1, S 6B 13/1, London

[41] Kuniyoshi H, Tanabe H, Sato, GT, and Fujimoto H (1980) Investigation on the characteristics of diesel fuel spray, Trans. SAE 89–800968

[42] Kamimoto T, Yokota H, and Kobayashi H (1987) Effect of high pressure injection on soot formation process in a rapid compression machine to simulate diesel flames, SAE technical paper series 871610

[43] Browne KR, Partridge IM, and Greeves G (1986) Fuel property effects on fuel/air mixing in an experimental diesel engine, SAE technical paper series 860223

[44] Scheid E, Pischinger F, Knoche KF, Daams H-J, Hassel EP, and Reuter U (1986) Spray combustion chamber with optical access, ignition zone visualization and first raman measurements of local air-fuel rates, SAE technical paper series 861121

[45] Melton LA (1983) Spectrally separated fluorescence emission for diesel fuel droplets and vapor, Applied Optics, 22-14: 2224-2226

[46] Bardsley MEA, Felton PG, and Bracco FV (1988) 2-D visualization of liquid and vapor fuel in an I.C. engine, SAE technical paper series 880521

[47] Fujimoto H., TanabeH. Kuniyoshi H, and Sato GT (1983) Investigation on combustion in medium-speed diesel engine using model chamber, Proc. 15th CIMAC, Paris, D13,3:1471-1490

[48] Lakshminarayan PA and Dent JC (1983) Interferometic studies of vaporizing and combusting sprays, SAE technical paper series 830224

[49] Naber JD and Reitz RD (1988) Modeling engine spray/wall impingement, SAE technical paper series 880107

[50] Abramovitch GN (1963) The theory of turbulent jets, MIT Press, Massachusetts

[51] Rajaratnum N (1976) Turbulent jets, Elsvier, Amsterdam (1976)

[52] Yip B and Long MB (1986) Instantaneous planar measurement of the complete three-dimensional scalar gradient in a turbulent jet, Optics letters, 11: 64-66

[53] Miller PL and Dimotakis PE (1991) Reynolds number dependence of scalar fluctuations in a high Schmidt number turbulent jet, Phys. Fluids A3(5) 1156-1163 1151.

[54] Johnston SC, Robinson CW, Rorke WS, Smith JR, and Witze PO (1979) Application of laser diagnostics to an injected engine, SAE Trans., 790079

[55] Tanabe H, Suzuki N, Sorihashi T, Fujimoto H, and Sato GT (1982) Experimental study on transient gas jet, Proc. XIX FISITA, No. 82026

[56] Hamamoto Y, Tomita E, Tsunashima Y, Nsunge FC, and Ikeda K (1990) Measurement of gas jet concentration by laser interferometry, J. Marine Engineering Society of Japan, 25-8: 498-504

[57] Shirakashi,M. and Wakiya, M. (1986) A study of turbulent structure in an impulsively started jet by means of an image analysis, Trans. JSME. 52-475,B 1032

[58] Cho IY, Fujimoto H, Kuniyoshi H, Ha J-Y,Tanabe H, and Sato GT (1990) Similarity law of entrainment into diesel spray and steady spray, SAE technical paper series 900447

[59] Jost W (1939) Explosions-und Verbrennungsvorgaenge in Gasen, Verlag von Julius Springer, Berlin

[60] Lewis B and von Elbe G (1961) Combustion, flames and explosions of gases, Academic Press Inc., New York and London

[61] Wolfer HH (1938) Der Zuendverzug im Dieselmotor, VDI- Forschungsheft 392: 15-24

[62] Miwa K, Ohmija T. and Nishitani T (1988) A study of the ignition delay of diesel fuel spray using a rapid compression machine, JSME Int. J. II,31-1:166-172

[63] Kamimoto T and Kobayashi H (1991) Combustion process in diesel engines, Prog. Energy Combust., Sci., 17: 163-189

[64] Martinengo A, Wagner HG, and Zunft D (1959) Untersuchungen ueber Selbst-zuendungsreaktionen von Kohlenwasserstoff-Luftmischungen durch adiabatische Verdichtung, Z. Phys. Chemie, N.F. 22: 292-304

[65] Terao K (1991) Combustion and detonation waves, IPC, Tokyo

[66] Liao C, Terao K, and Utaka Y (1992) Ignition probability in a fuel spray, Jpn. J. Appl. Phys., 31-7: 2299-2303

[67] Ranz WE (1958) Some Experiments on Orifice Sprays, The Canadian J. of

Chem. Engs., 175–181

[68] Wagener J (1992) Ein digitales Bildverarbeitungssystem zur Auswertung hochfreequenter Bildfolgen von Dieseleinspritzstrahlen, Thesis, Universität Kaiserslautern

[69] Katsura N, Saito M, Senda J, and Fujimoto H (1989) Characteristics of a diesel spray impinging on a flat wall, Trans. SAE 89–890264

[70] Fujimoto H, Senda J, Nagae N, and Hashimoto A (1990) Characteristics of a diesel spray impinging on a flat wall, Proc. COMODIA90, Kyoto, pp193–198

[71] Senda J, Fukami Y, Tanabe Y, and Fujimoto H (1992) Visualization of evaporative diesel spray impinging upon wall surface by exciplex fluorescence method, SAE technical paper series 920578

[72] Terao K, Liao C, and Utaka Y (1991) Proc.1st Intern. Conf. on Combust. Techn. for Clean Environment, Vilamoura, Engines II, pp.26–32

[73] Hiroyasu H (1985) Diesel engine combustion and its modeling, COMODIA 85, pp 3–75

[74] Kato T, Tsujimura K, Shintani M, Minami T, and YamaguchiI.(1989) Spray characteristics and combustion improvement of D.I. diesel engine with high pressure fuel injection, SAE technical paper series 890265

[75] Yokota H, Kamimoto T, Kosaka H, and Tsujimura K (1991) Fast burning and reduced soot formation via ultra–high pressure diesel fuel injection, SAE technical paper series 910225

[76] Nishida K, Ochiai M, Arai H, and Hiroyasu H (1991) Characteristics of diesel sprays at high–pressure injection, preprint of the Japan Society of Mechanical Engineers, 904–4: 68–70 (in Japanese)

[77] Shimada T, Shiji T, and Takeda Y (1989) The effect of fuel injection pressure on diesel engine performance, SAE technical paper series 891919

[78] Hiroyasu H, Nishida K, Yoshikawa S, Kown S, and Arai M (1991) Combustion process in a D.I. diesel engine with high pressure combustion chamber on NOx emission, Trans. of Japan Society of Automotive Engineers, 22–4: 53–58 (in Japanese).

[79] Faeth GM (1987) Mixing, transport and combustion in sprays, Prog. Energy Combust. Sci., 13: 293–345

[80] Chigier NA (1976) The atomization and burning of liquid fuel sprays, Prog. Energy Combust. Sci., 2: 97–114

[81] Lefebre AH (1989) Atomization and sprays, Hemisphere Publising Corporation, New York.

[82] Takagi T, Fan CY, Kamimoto T. and Okamoto T (1991) Numerical simulation of evaporation, ignition and combustion of transient sprays, combustion science and technology, 75: 1–12

[83] Onuma Y and Ogasawara M (1975) Studies on the structure of a spray combustion flame, 15th Symp. (Int.) on Combust., pp.453–465

[84] Brenã de la Rosa A, Sobiesiak A, and Brzustowski TA (1988) The influence of fuel properties on drop–size distribution and combustion in an oil spray, 21st Symposium(International) on Combustion, pp 557–566

[85] Mizutani Y, Yasuma G, and Katsuki M (1977) Stabilization of spray flames in a high–temperature stream, 16th Symposium (International) on Combustion, pp 631–638

[86] Edwards CF and Rudoff RC (1990) Structure of a swirl–stabilized spray flame by imaging, laser doppler velocimetry and phase doppler anemometry, 23rd Symposium (International) on Combustion, pp 1353–1359

[87] O'rourke PJ (1981) Collective drop effects on liquid spray, Ph. D. Thesis, Princeton University

[88] Nakabe K, Mizutani Y. and Hirao T (1988) Burning characteristics of premixed

sprays and gas–liquid coburning mixtures, Combust. Flame 74–1: 39–51

[89] Mizutani Y and Nishimoto T (1972) Turbulent flame velocities in premixed sprays, Part I. Experimental study, Combust. Sci. and Tech. 6–1/2: 1–11

[90] Mizutani Y and Nakajima A (1973) Combustion of fuel vapor–drop–air systems: Part I–Open burner flames, Combust. Flame 20–3 : 343–350

[91] Chiu HH, Kim HY, and Croke EJ (1982) Internal group combustion of liquid droplets, 19th Symposium (Int.) on Combustion, pp.971–980

[92] Nakabe K, Mizutani Y, and Hirao T (1991) An experimental study on detailed flame structure of liquid fuel sprays with and without gaseous fuel, Combust. Flame 84–1/2: 3–14

[93] Won YH, Kamimoto T, and Kosaka H (1992) A study on soot formation in unsteady spray flames via 2–D soot imaging, SAE technical paper series 920114

[94] Tanabe H and Sato GT (1990) Experimental study on unsteady wall impinging jet, SAE technical paper series 900605

Chapter 4
Kinetics

4.1 Chemical Kinetics and Modeling of Combustion

4.1.1 Recent development of reaction kinetics and modeling study

(a) Introduction

Numerical modeling based on chemical kinetics is a powerful technique for the analysis of many combustion phenomena including turbulent diffusion combustion, as is reviewed from time to time [1].

In order to apply chemical kinetics to real combustion, thermochemical information (in order to compute heat capacities, enthalpies, equilibrium constants, etc.), knowledge of the reaction mechanism and relevant reaction rates, and transport properties are required as basic data. Among them, the reaction mechanism and relevant rate constants are most important and thus will be discussed in the present chapter. In order to make up a reliable reaction mechanism, we must be very careful not to miss any important elementary reactions and branching channels, which is sometimes very difficult. The recent trial of developing a computer modeling in which the reaction mechanism is automatically provided seem to be very promising [2,3].

The modeling may be classified to the following three categories. The first category includes a kinetic modeling which only deals with the reaction progress, in other words, a zero–dimensional model with kinetics. This type of modeling is not applicable to many practical problems, but still it is useful for the sake of simulating many experimental situations such as those encountered in shock tubes, flow reactors and static reactors. The second category of modeling is addressed to a relatively simple flow field such as one– or two–dimensional laminar flow field, where the contribution of reaction, heat and transport can be taken into account relatively easily. The third category is a modeling in a more complicated flow field such as turbulent flow field.

At present, in these complicated flow fields, the direct simulation based on the Navier–Stokes equation may be possible only by using a very simplified kinetic model. In practical systems, the direct simulation seems to be still too early, considering the large range of relevant time constants and scales. The entire field of turbulent combustion and its relation to chemical kinetics modeling are of immense importance in applied combustion, but it is not yet clear how the connection is best accomplished. Making up a reduced reaction model is expected to be one of such trials [4].

(b) Experimental approach to the reaction mechanism and rate constants

Experimentally, the reaction mechanism is constructed on the basis of comprehensive knowledge of chemical reactivities, elementary reactions and their rate constants. The identification of elementary reactions and their rate constants are widely developed owing to recent advent of various physicochemical methods adequate to study elementary reactions, one of which is the combination of the shock tube technique and pulsed laser photolysis and will be described in section 4.2 by Matsui. As the results, tabulations of rate constants over a wide temperature range are now available, for example, in the literature [5], and they are still expanding very rapidly.

The kinetic modeling itself is also very useful so as to extract information concerning elementary reactions from the experiments such as conventional shock tubes. Historically, shock tube studies have been the most productive, primarily those in which the ignition delay time or induction time are reported. By searching for good models and rate constants which can reproduce these time measurements, the unknown elementary reactions and their rate constants can be determined. Kinetic models are also often used to describe the propagation rate and structure of laminar flames under laboratory conditions, though these require molecular transport properties at the same time. Additional classes of experimental problems often used to yield kinetic information include stirred reactors [6], plug flow reactors [7] and static reactors [8]. In each case, it is rather simple to incorporate the boundary and initial conditions into a kinetic model and then derive the evolution of the model system.

(c) Theoretical approach to reaction mechanism and rate constants

Ab initio calculations of thermochemical values, potential surfaces, and the theory and numerical calculations of reaction dynamics are now contributing very remarkably to the modeling study.

Owing to the recent development of methods of handling electron correlation such as fourth order Möller–Plesset perturbation theory, now we can estimate heats of formation of free radicals and potential energy barriers to chemical reactions with the accuracy up to the order of 1 to 2 kcal mol^{-1}, being aided by some empirical correlations in some cases. A large amount of potential surfaces relevant to combustion reactions are also calculated by means of *ab initio* methods by many theoretical chemists [9]. Study of reaction progress on the *ab initio* reaction surface is sometimes very helpful to understand the branching ratio leading to various products, and the specific temperature dependence of some reaction channels. One example is the $R+O_2$ reaction [10]. More empirical relationships, in principle, based on the reaction surface, are sometimes very helpful. The problem of finding a correlation among $O_2 + R$ and $NO + R$ reactions will be discussed in section 4.2 by Matsui.

In order to estimate rate constants, the conventional transition state theory is adequate for reactions with potential energy barriers. By combining the knowledge of energy barrier height and the transition state structure, the rate constant and its temperature dependence can be estimated, though in some cases, the tunneling and re–crossing in the transition region require adequate corrections. The $CO + OH$ reaction is one important example of the reactions which show a temperature dependence deviated from the Arrhenius equation and are successfully treated by the transition state theory [11]. The transition state theory is also employed in treating the rate constants of a series of reactions altogether, such as O [12] and H [13] atom reactions with alkanes.

Variational transition state theory can deal with the reactions with no intrinsic potential energy barriers and has been applied successfuly to several combustion reactions [14]. Classical trajectory methods have been also applied widely to reactions of combustion interest.

(d) Modeling and computer simulation

Chemical kinetic reaction systems are generally solved numerically, using high speed computers. The kinetic rate equation systems are often very stiff, with widely differing time constants, which places significant demands upon the solution algorithms to be used. Fortunately, there are now many flexible, convenient and portable software packages available which can solve these problems [1]. One of these trials will be shown in section 4.5 by Sano.

The essential element in determining the difficulty and cost of solution of a given kinetic problem is the total number of equations to be integrated. For each spatial point or computational node, one differential equation must be solved for each chemical species. The number of chemical reactions does not directly influence the cost of the computation. For relatively simple fuels such as H_2 or CO, the detailed kinetic mechanisms are quite small; a reasonable mechanism for H_2 includes about 10 chemical species, with about 15 for CO oxidation. When the number of species is small, multidimensional spatial grids can be employed. For practical hydrocarbon fuels the reaction mechanisms become very large as will be shown later, making it necessary to simplify the geometry of a problem to one dimension or even to consider only spatially homogeneous conditions. It is interesting to note that reaction mechanisms of current interest such as those for hexane, include more than 300 species per spatial zone.

4.1.2 Kinetic models of individual fuels

(a) Hydrocarbon fuels

In past years, many combustion modeling studies have considered the oxidation of methane and methanol [15,16]. These fuels have considerable practical importance, with methane being the primary component in natural gas and methanol as an important alcohol fuel. In addition, since these fuels are structurally very simple, kinetic models for their combustion are relatively small and require only modest computer resources. However, interest has grown in combustion of larger and more realistic hydrocarbon fuels, particularly in systems such as internal combustion of larger fuel molecules. In due course, the sizes of commonly used reaction mechanisms have increased, because reaction mechanisms are built in a hierarchical manner, with mechanisms for each fuel containing the mechanisms for smaller fuels; this process leads inevitably to increasingly larger and more complex mechanisms. In addition, the accompanying growth in the capacity and power of computers in the past few years has made it possible to include much larger reaction mechanisms in practical computations.

Reaction mechanisms for propane [17], n–butane [18], n–pentane [6], heptanes [19,20] and octanes [2,21] have been developed in past few years. These mechanisms have been used in studies of laboratory experiments and in analysis of practical combustion systems such as internal combustion engines, pulse combustors and others, as discussed below. The kinetic principles used in the development of reaction mechanisms for these fuels are applicable to other fuels as well, and it is currently possible to develop a reaction mechanism for practically any paraffinic hydrocarbon fuel, with any number of carbon atoms and any branched molecular structure. The rules for all of the important reaction steps have been established and tested, and application of these rules for any fuel of interest can be implemented with a high degree of confidence. For example, although reaction mechanisms for the oxidation or ignition of n–dodecane have not been reported in the combustion literature, if this fuel were important to a specific research program, it would be straightforward to establish a reaction mechanism for its combustion that would be reliable over a wide range of experimental operating conditions.

Nearly all of the fuels treated in current modeling studies are paraffinic hydrocarbons or are derived from paraffin fuels. Simple alcohol fuels, ketones, aldehydes, ethers, and epoxide species based on paraffinic structures can also be treated using current modeling techniques. In contrast to the situation for paraffinic and related fuels, however, reaction mechanisms for the combustion of aromatic hydrocarbon species are much less mature. There have been relatively few detailed combustion experiments carried out for species such as benzene, toluene, xylene, and others involving an aromatic ring structure. Accordingly, few kinetic reaction mechanisms have been proposed or analysed for aromatic species. Information on the major kinetic steps have not yet been established; for example, it is not yet known what are the major ring–breaking reactions or the major intermediate species in the reaction of aromatic hydrocarbons. This is a very serious limitation in the development of practical combustion models, since aromatic species are an important component in many practical oxidations and its detailed kinetic modeling represent very important subjects in current hydrocarbon kinetic research. Some of recent experimental studies will be described in section 4.3 by Fujii.

The understanding of the reactions of unsaturated aliphatic compounds such as acetylene, aromatic compounds and corresponding radicals altogether, that is, the understanding of reactions preceding the soot formation process which is the subject of chapter 5, is also very immature [22], and should be developed in future, considering its importance for the development of soot formation and destruction models.

(b) Practical applications of hydrocarbon models

The ultimate goal of combustion modeling is the description of practical systems, although it is always interesting to study chemical systems under carefully controlled conditions. Flame models can often be applied directly to practical problems, and there are models in which laminar flame models are used as a basis space to describe all of the conditions which can be encountered in practical systems.

Kinetic modeling has been successfully applied to ignition problems in some practical combustion systems [23–25] and initiation of detonations [26,27]. It has been used extensively in the analysis of the critical conditions leading to engine knock [28,29] and of hydrocarbon ignition at relatively low temperstures [30,31]. These studies will be described more in detail in section 4.4 by Westbrook and Pitz. There is every reason to believe that the same reaction mechanisms should apply to a wide variety of chemical engineering problems at temperatures below 800K, but extensions of current modeling capabilities to such conditions have not yet been developed.

(c) Other fuels and pollutant species

In addition to the hydrocarbon fuels already discussed, reaction mechanisms for other classes of fuels and pollutant species have been developed recently. The work of Miller and Bowman [32] has shown how the production and destruction of oxides of nitrogen (NOx) in combustion systems can be described quantitatively, including reaction pathways involving NH_i and CH_i species, HCN, and NCO. Their work also explains how NOx reduction can be achieved using post–reaction addition of species such as ammonia, urea, and cyanuric acid.

Combustion of other energetic species has been studied by numerical modeling techniques. Combustion of silane has been simulated [33,34] and reviewed recently by Koda [35]. Oxidation of ammonia [36], and ignition of alkyl nitrates and nitroalkanes [37] have all been the subjects of kinetic modeling studies. Another large class of kinetic modeling problems of increasing importance considers the oxidation of chlorinated hydrocarbon species [38], an important type of toxic chemicals for which incineration and other combustion treatments have been proposed. For all of these non–hydrocarbon species, kinetic modeling offers the opportunity to develop experimental combustion strategies designed to optimize overall system performance with a minimum of difficulty, cost, and time.   (Seiichiro Koda and Charles K. Westbrook)

4.2 The Rate Constants of Elementary Combustion Reactions and Empirical Rate Laws

4.2.1 Measurement of rate constant

Within the last few decades, appreciable advance of experimental techniques in the studies of elementary reaction processes on combustion has been achieved: highly sensitive detection systems for many key species with sufficient time resolution are now available. For example, new informations on the reaction mechanisms have been achieved with the improved techniques such as laser spectrometry (LIF, CARS, LMR, REMPI, SEP and etc.) or mass spectrometry (E–I mass, P–I mass, TOF).

As an example, some of the reaction systems recently studied in the Department of Reaction Chemistry, The University of Tokyo, are summarized in Table 4.1. In this case, reactions at low temperature range are examined mostly by using VUV to visible LIF and E–I mass spectrometry. Improving the reliability of kinetic data is an important and urgent task: thus crucial examinations have been often tried, for example, by combining two or more techniques to confirm the conclusions on the target reactions.

Informations on the oxidation processes at high temperature range (>1000 K) are being accumulated recently by laser flash photolysis combined with shock tube technique [39,40].

Rate constants of about a magnitude different from the previous indirect studies have been reported [41]. Non–Arrhenius temperature dependence was found for many atom–molecule reaction systems; the experimental results were in accordance with the predictions of TST calculation with the *ab initio* potential energy surfaces [41–44]. These high temperature data obtained by using flash photolysis/shock tube technique are also consistent with those of high temperature static cell/flash photolysis studies [45].

Reaction mechanisms for producing some minor species that play central roles for important problems in combustion (such as formation of highly active radicals like CH_2, CH, C_2, or C_2H) have not been well understood. They are believed to be responsible for the prompt NOx formation: measured profiles for CH radical in combustion is not in agreement with that of numerical simulations even though profiles of OH are consistent each other.

Also, reaction mechanisms for oxidation of aromatic compounds or formation of PAH in combustion, as well as those of soot formation have to be explored in details, although there have been different types of proposals for the mechanisms [22].

Understanding cyclization reactions starting from very simple hydrocarbon radicals is still a challenging task. For example, C_2H is believed to be a key radical species in the initial stage of soot formation in combustion of C_2H_2: there have been numerous studies on this reaction process. However, the measured rate constants are not in agreement each other as shown in Fig. 4.1.

Table 4.1   Summary of the elementary reactions and the experimental methods recently performed in the Department of Reaction Chemistry of The University of Tokyo

type of reaction	monitored species	method
(1) [RH + X → R + HX]		
R = CH_3CO, X = O	CH_3CO, CH_3CHO, CH_3	FT/PIM
alkyl radicals	O	ST/ARAS
(C_1–C_5), X = O		
CH_2F, CF_3, X = O	O	ST/ARAS
HS, X = O, H	O, H	ST/ARAS
CH_3CO, X = H	CH_2CO, CH_4, CH_3CHO	ST/EIM
SiH_nF_{3-n}, X = H	H	SC/VUV–LIF
(n=3,2,1)		
(2) [R + O_2 → products]		
R = CH	CH, OH	SC/LIF
CH_3CO	CH_3CO, CH_3CHO, CH_3	FT/PIM
C_1–C_3 hydroxy–	reactants	SC/PIM
alkyl radicals		
CH_3	O	ST/ARAS
HOCO	HOCO	SC/PIM
SiH_3	SiH_3, SiH_3O, Si_2H_4	SC/EIM
	Si_2H_6	LIF, VUV–LIF
(3) [R + NO → products]		
R = C_2H_4OH (α and β)	reactants	SC/PIM
CH_3O	CH_3O, CH_3ONO	SC/LIF
NH	NH, H	SC/LIF,
		VUV–LIF
CH	CH, NH, CN, OH, NCO	SC/LIF
(4) C_2H + C_2H_2	C_4H_2, H, D	VUV–LIF
C_2H + H_2, D_2		ARAS, EIM

[FT: microwave discharge–fast flow reactor, ST: excimer laser photolysis/shock tube, SC: static cell or slow flow reactor, PIM: photoionization mass spectrometry, ARAS: atomic resonance absorption spectroscopy, EIM: electron impact ionization mass spectrometry, LIF: laser induced fluorescence in the visible to ultraviolet region, VUV–LIF: laser induced fluorescence in the vacuum ultraviolet region]

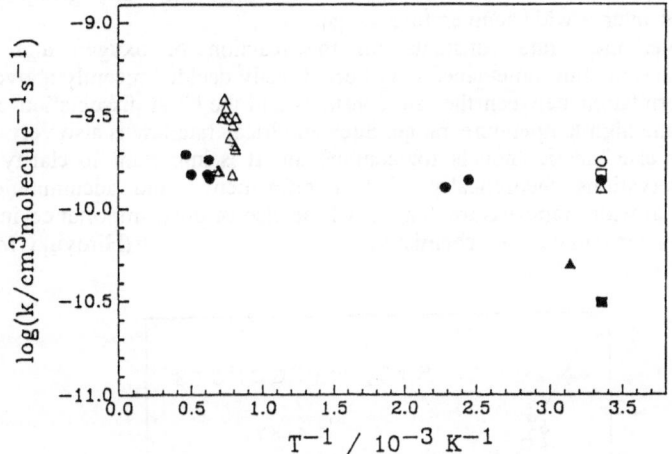

Fig. 4.1   Summary of the rate constant for the reaction $C_2H + C_2H_2$
$\rightarrow C_4H_2 + H$

o, • :Koshi *et al* [44],   Δ :Shin and Michael [46],
    ▲ :Lange and Wagner [47],   □ :Stephens *et al* [48],
    ■ :Laufer and Bass [49]

4.2.2 Empirical rate laws

Modeling of the oxidation processes even for relatively simple hydrocarbons (such as
n-heptane) is still not an easy task. The recent trials of developing a computer pragram
in which the reaction mechanisms are automatically provided [2] will be of great
importance for establishing modeling of the practical combustion systems with fuels of
general use: in order to construct useful softwares for automatic generation of reaction
mechanism of oxidation processes, accumulation and refinement of the kinetic data will
be inevitable: specially, general principles of the reactivity for the oxidation of
hydrocarbons have to be clarified.

Group additivity rules of the reaction rate constant have been carefully tested for
series of alkane and OH radical at low temperature ranges below 1000 K [50–53].

Also, correlation between the activation energies (or the rate constants) and the bond
dissociation energies for simple absraction reactions have been tested very long for a
number of systems important in the combustion reactions: reaction rate constants and
branching fractions for the radical–O_2 reactions have been systematically examined in
relation to the ionization potentials IP (and electron affinities) [54–56]. As an example,
some of the measured rate constants (high pressure limit value) for the reactions of
various radicals with O_2 are shown in Fig. 4.2. It is interesting that many hydrocarbon
radicals have good correlation with IP for the reaction with O_2.

In contrast, no apparent dependence for the reactions with NO has been reported. As
reliable kinetic data even at high temperature range are being accumulated, it is possible

to compare the rate constants for some substances within the same group to test empirical relations over a wide temperature range.

As an example, the rate constants for the reaction of oxygen atom with alkanes(C_1–C_5) and some fluoromethanes have been directly decided recently above 1000 K [57]: a linear correlation between the rate constants and the bond dissociation energy was confirmed at the high temperature range. Such empirical rate law is also very useful in constructing precise kinetic models for combustion. It is important to clarify these experimental observations theoretically: further refinement and accumulation of experimental data at wider temperature ranges will be also of great importance in order to improve models for combustion chemistry.   (Hiroyuki Matsui)

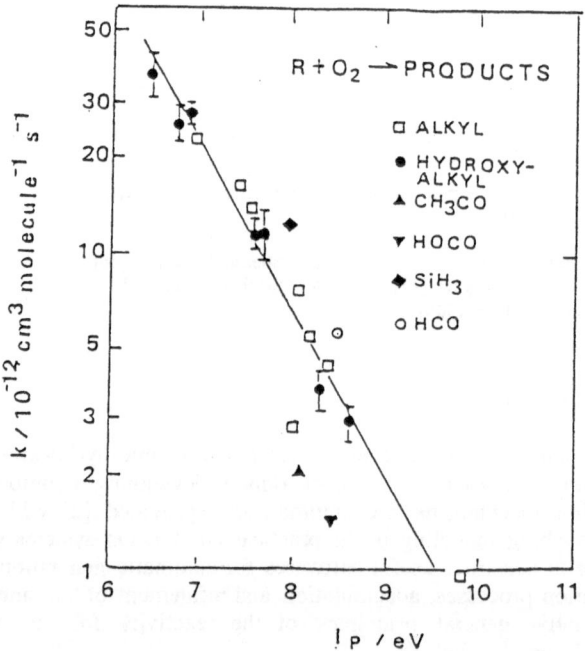

Fig. 4.2   The high pressure limit rate constant for the reaction of radical
with O_2 plotted against their ionization potentials (room temperature)

4.3 Combustion and Oxidation Mechanism of Aromatic Compounds

4.3.1 Kinetic study of benzene oxidation and some problems

Recently there is considerable interest in the kinetics and mechanism of oxidation of aromatic compounds because of the increasing use of aromatics as fuel components. This reaction involves the breakage of aromatic rings by attack of oxygen, which is very important in understanding the combustion kinetics of aromatic compounds. Up to data, however, in contrast with the studies on the oxidation of paraffin hydrocarbons, there have been only few kinetic studies on the oxidation of aromatic hydrocarbons.

The oxidation of gaseous benzene, which is the most typical compound among aromatic hydrocarbons, has been studied kinetically by several authors, but these studies were performed at relatively low temperatures, i.e. below 1000 K before 1970. Previously Fujii et al studied the pyrolysis of benzene from 1300 to 1800 K using a shock tube and proposed a kinetic scheme of soot formation as follows [58],

$$C_6H_6 \to C_6H_5 \to \text{biphenyl} \to \text{polyphenyl} \to \text{soot (at low temperature)}$$

$$\text{acetylene} \to \text{polyacetylene} \to \text{soot (at high temperature)}$$

Braun–Unkhoff et al [59] studied the thermal decomposition of toluene and C_6H_5NO behind shock waves by monitoring hydrogen atom by ARAS in the temperature range 1300–1800 K and proposed a scheme including linear C_6H_5 as a route of the decomposition of phenyl radical, $C_6H_5 \to \text{liner-}C_6H_5 \to C_4H_3 + C_2H_2$. Frenklach et al [60] studied the kinetic modeling of soot formation in shock tube pyrolysis of acetylene and proposed a complicated scheme including phenyl radical and liner–C_6H_5. However, the reactions including in the scheme seem to be very speculative. Also, the authors studied the oxidation of rich mixtures of benzene from 1300 to 1700 K using a shock tube. By using gas analysis, UV absorption by benzene and IR emission by CO and CO_2, they found that the initiation reaction was $C_6H_6 \to C_6H_5 + H$ and CO was a major early product. To elucidate the experimental results, they proposed kinetic scheme including an overall CO formation reaction with phenyl radical and molecular oxygen, $C_6H_5 + O_2 \to 2CO + C_4H_4 + H$ [61]. After that the kinetic model was applied to the ignition delays of the lean mixtures of benzene with oxygen from 1100 to 1600 K in reflected shock waves and good agreement between simulation and experiments was obtained [62].

Glassman et al [63,64] studied the oxidation of benzene, toluene, ethyl benzene, and o-xylene at about 1200 K using a high temperature adiabatic turbulent flow reactor. From the products analysis using gas chromatograph, they found that the overall rate was dominated by the oxidation of phenyl radical in above cases and proposed a detailed mechanism of CO formation, $C_6H_5 + O_2 \to \text{phenoxy} + O \to CO + \text{lower hydrocarbons}$. Their detailed mechanism seems to be roughly consistent with Fujii's overall scheme. However, the rate constant for each reaction step was not reported. Hippler et al [65] studied the oxidation of benzyl radical by UV absorption in shock waves over the temperature range 1200– 1500K, and observed no direct attack of benzyl by O_2. Hsu, Lin and Lin [66] studied the oxidation of benzene under fuel–lean condition from 1600 to 2300 K in a shock tube using a stabilized cw CO laser to monitor CO production. They found that the rate of CO formation in the early stage of the oxidation depends very sensitively on the rate–limiting unimolecular decomposition of benzene: $C_6H_6 \to C_6H_5 + H$. The rate constant of this reaction obtained was similar to that by Asaba et al [58]. Recently, Thyagarajan et al [67] measured the ignition delay of benzene in the temperature range 1200–2200 K behind shock waves and compared the data with those calculated using a mechanism consisting of 70 reactions among 30 species. Here, the rate constants of major reactions were compiled from Kiefer et al [68]. The agreement between experiments and simulations was good.

In the present stage, the mechanism for high temperature oxidation of benzene remains largely speculative. In addition, the mechanism of the oxidation of lower hydrocarbons, especially unsaturated C_2 and C_4 hydrocarbons, is very important for the oxidation of aromatics.

4.3.2 A Study of reaction mechanism of benzene oxidation at high temperature by means of shock tube method

In order to evaluate the kinetics more precisely, Fujii *et al* studied the high temperature oxidation of benzene using two kinds of mixtures C_6H_6–N_2O and C_6H_6–O_2–H_2 highly dilute in Ar compared with the previous studies. These systems seem to involve larger amounts of atoms and radicals than C_6H_6–O_2 system in the early stage of reaction behind shock waves. Experiments were carried out using two type shock tubes. Simultaneous measurements of two infrared emissions (3.39 μm for C–H stretching, 4.25 μm for CO_2, 4.52 μm for N_2O and 4.87 μm for CO) isolated with interference filters were performed behind reflected shock waves in the temperature range of 1500–2400 K at pressures of 3–4 atm. Also time profiles of O atom concentration in incident shock waves were measured using atomic resonance absorption spectroscopy (ARAS) at 130.5 nm in the temperature range 1500–2000 K at pressures of 1 atm. The measured concentration profiles were compared with those by simulations. Diagram of main elementary reactions used in the simulation are shown in Fig. 4.3. Here, reaction (i) is initiation reaction for C_6H_6–N_2O system and reaction (ii) is formation of phenyl radical by decomposition of benzene and hydrogen abstraction from benzene by H, O and OH. The overall oxidation process of lower hydrocarbons, reaction (iv), may have some problems, but they are not considered to be very severe in the present experiment. Accordingly, those reactions are seemed to be almost well established except for the ring–opening–oxidation (iii). The detailed reaction scheme and the rate constants of the elementary reactions have been given elsewhere [69]. Here, the rate constants of low hydrocarbons were mainly referred to Warnatz [70].

$$N_2O \rightarrow O + N_2 \qquad \qquad \text{(i)}$$
$$C_6H_6 + O\ (H,\ OH) \rightarrow C_6H_5 \qquad \text{(ii)}$$
$$C_6H_5 + O_2 \rightarrow CO + \text{lower hc} \qquad \text{(iii)}$$
$$\text{lower hc} \rightarrow CO,\ H_2O \qquad \qquad \text{(iv)}$$
$$H_2\text{–}O_2 \text{ reaction} \rightarrow H,\ O,\ OH,\ H_2O \qquad \text{(v)}$$
$$CO \rightarrow CO_2 \qquad \qquad \text{(vi)}$$

Fig. 4.3 Diagram of mechanism for benzene oxidation

(a) C_6H_6–N_2O system [69]

Typical infrared emission profiles of 3.39 and 4.87 μm are shown in Fig. 4.4. The 3.39 μm emission was found to be due to C_6H_6 and C_6H_5 at the initial stage of the reaction from the comparison with simulations. In order to evaluate the infrared emission profiles, the maximum decrease rate of 3.39 μm emission, and the maximum growth rate for 4.25 μm emission were estimated from the observed and simulated profiles. Examples of the comparison are shown in Fig. 4.5. The calculated values at lower temperature, especially for the CO_2 formation, were much smaller than experiments. It was found that the calculated rate of N_2O consumption was also smaller than the experiments at lower temperature. From those results, it seemed that another reaction paths for C_6H_6 (and/or C_6H_5) + O (and N_2O) $\rightarrow$ CO $\rightarrow$ CO_2 exist in the system, especially at low temperature under the experimental conditions.

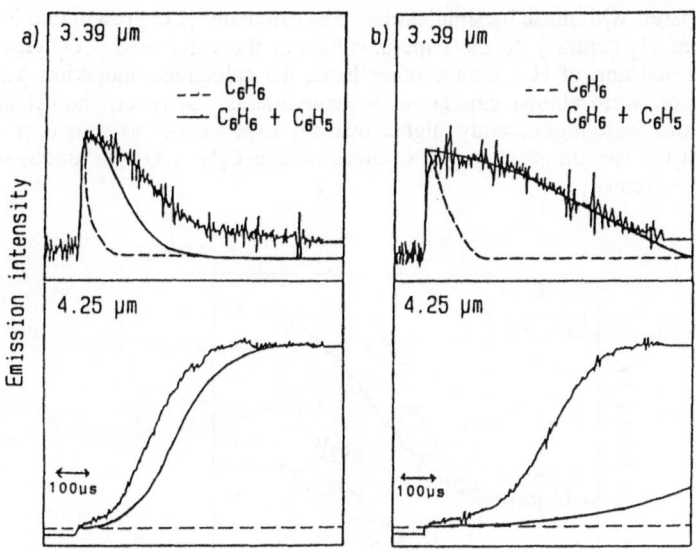

Fig. 4.4 Examples of IR emission profiles for C_6H_6–N_2O system
a) 1810 K b) 1645 K
Smooth lines are calculations

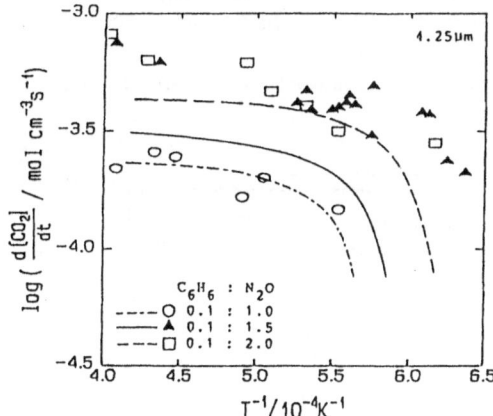

Fig. 4.5 Maximum growth rate of CO_2 from 4.25 μm emission
Lines are calculations

(b) C_6H_6–O_2–H_2 system [71]

Typical infrared emission profiles and O–atom absorption profile are shown in Figs. 4.6 and 4.7, together with those by simulations. The growth rate of CO became fast slightly by adding H_2, and the CO_2 growth rate became low considerably by adding H_2. When the amount of H_2 addition was as large as $H_2/O_2 = 1.0$ under the present condition, however, the CO growth rate became low. The measured emission profiles of CO and CO_2 were compared with those by simulations. The calculated CO growth rate increased with addition of H_2 contrary to the experiments, but the calculated CO_2 growth rate decreased with addition of H_2. On the other hand, the calculated induction periods of O atom formation were almost similar to the experiments, however, the calculated O atom growth rates were significantly higher than the experiments as shown in Fig. 4.7. This means that the consumption step of O atom, such as $C_6H_5 + O \rightarrow$ products, was lack in the reaction scheme.

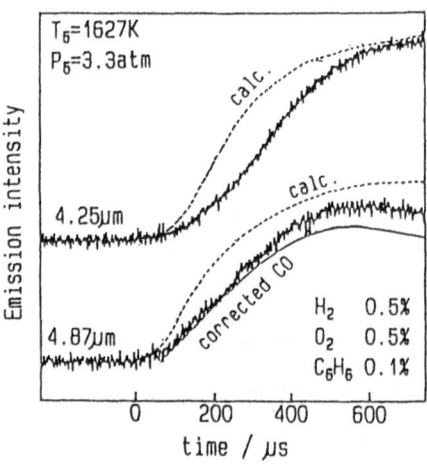

Fig. 4.6 Example of IR emission profiles for C_6H_6–H_2–O_2 system

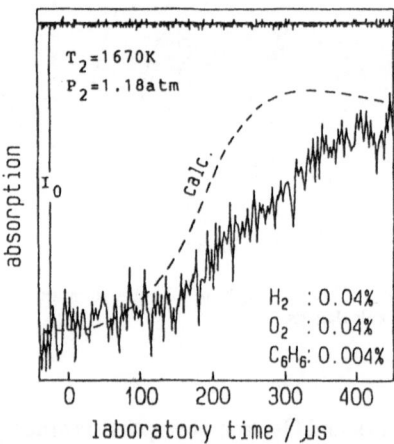

Fig. 4.7 Example of 130.5 nm absorption in the C_6H_6–O_2–H_2 system

(c) Discussion of the reaction scheme

The simulated maximum decrease rate of 3.39 μm emission was about five times higher than the experiments in C_6H_6-O_2 system as shown in Fig. 4.8. This result was compared with that in C_6H_6-N_2O system in Fig. 4.8. The figure shows that the experimental decrease rates of both systems are almost same, but the simulated rates are very different. The similar tendency about the growth rates of CO_2 in both system were also observed. As the conclusion, it is considered that the breakage reaction of benzene ring via the reaction of phenyl radical with O atom or OH radical, such as C_6H_5 + O (or OH) → CO + low hc, is not negligible in the reaction mechanism of high temperature oxidation of benzene.   (Nobuyuki Fujii)

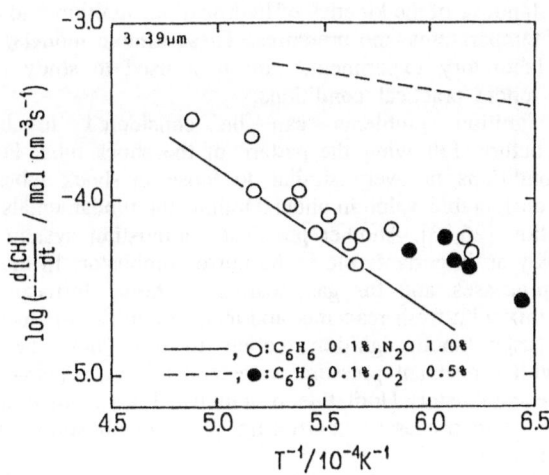

Fig. 4.8   Comparison of 3.39 μm emission decrease rate for C_6H_6-N_2O and C_6H_6-O_2 systems

4.4 Applications of Simulation Work to Some Combustion Problems

4.4.1 Introduction

Kinetic modeling has gradually become an essential part of combustion research. In past, the majority of combustion chemistry modeling was used to study simplified problems, such as shock tubes, flow reactors, and laminar flames [72]. This work provided a great deal of valuable information [15], particularly in terms of fundamental kinetic rates and reaction mechanisms. In more recent studies, kinetic modeling has provided understanding of more practical problems, especially concerning systems such as internal combustion engines. In some kinetic models of combustion chemistry, spatial transport effects have no influence, while in others transport effects interact closely with kinetics to determine the important results.

4.4.2 Zero–dimensional modeling with kinetics

In many combustion environments transport effects may be safely neglected without losing the essential features of a problem. In the laboratory, the best examples of such systems are shock tubes, flow reactors, and static reactors. There are literally thousands of scientific papers dealing with chemical kinetics in these systems. In shock tubes, the primary variable of interest is the ignition delay, the time interval between the arrival of the shock wave and the ignition of the fuel–oxidizer mixture. The most important reactions in most of these systems include the unimolecular decomposition of the fuel, the abstraction of H atoms from the fuel, and chain branching reaction [16]

$H + O_2 = O + OH$.

In flow reactors and static reactors, initial temperatures are somewhat lower than in shock tubes, and different reactions are important. These zero–dimensional studies have led to a general understanding of the kinetics of hydrocarbon ignition and oxidation over a very wide range of temperatures and pressures. These kinetic models, tested against carefully controlled laboratory experiments, are then used to study more complex combustion problems under practical conditions.

Many practical ignition problems can be considered to be essentially zero–dimensional in nature, following the pattern of the shock tube. For example, in detonation waves, conditions are very similar to those in shock tubes, and kinetic analysis has been of considerable value in understanding the fundamentals of detonation stability and propagation [26,73]. Another practical combustion system in which the kinetics of ignition play an important role is the pulse combustor. In this system, used in many industrial processes and for gas heating in home furnaces, hot residual combustion products mix with fresh reactants and lead to ignition in a periodic manner. Keller *el al* [74,75] used kinetic ignition models to show how the interaction of gas–phase ignition with a resonant pressure wave and mixing processes controls the operation of the pulse combustor. Until this system had been studied with the use of kinetic modeling, the basic processes controlling pulse combustion had never been explained in technical terms.

The practical problem of engine knock in internal combustion engines represents another study of hydrocarbon ignition at somewhat lower temperatures, below 800 K. This problem is also essentially zero–dimensional and emphasizes the kinetics of alkylperoxy radical isomerization, as developed by Benson [74] and summarized by Pollard [75]. The essential features of engine knock involve the thermal autoignition of the engine end–gases, those reactants which are the last to be burned by the flame front in the engine. These end gases are compressed and heated both by the motion of the piston and by the propagation of the flame, and they react at a rather slow rate. They would eventually ignite, but under normal operating conditions these end gases are consumed by the flame front and knocking operation is avoided. However, under extreme conditions of initial temperature, for unusually slow flame propagation, or for extremely reactive fuels, thermal ignition is completed quickly and knocking occurs before the flame can arrive and consume these gases.

Recent studies of engine knock chemistry [76,77] have shown that factors such as fuel molecule size and structure influence knock tendency in several major ways. The most important fact is that internal transfer of H atoms (isomerization) within large alkylperoxy and other related large radical species leads to chain branching and rapid increases in overall reaction rate. These transfers depend sensitively on two main variables, the spatial separation between the initial and final location of the H atoms, and the type of C–H bond that must be broken in order to transfer the H atom. When the initial and final locations for the H atoms are very close together, the "ring strain energy barrier" for the transfer is high and the process in inhibited. Furthermore, when a primary C–H bond must be broken in order to facilitate the H atom transfer, the process is also inhibited, a condition which is observed in highly branched hydrocarbon fuels.

This work has shown how highly branched hydrocarbon fuels resist knock, having many primary C–H bonds and high ring strain energy barriers to internal H atom transfer. In contrast, large extended molecules, such as n-alkanes, have low ring strain energy barriers and many weak secondary C–H bonds. These conclusions were extended to include the influence of pro-knock and anti-knock additives [77] on knocking tendency as well as the influence of using various fuel mixtures.

A final example of the use of zero-dimensional modeling to study combustion problems of considerable importance is the development of two very significant techniques to reduce NOx emissions from practical combustors. These are so-called "thermal de-NOx" [78] and "RAPRe-NOx" [79]. In both systems, additives containing single N atoms, either ammonia or cyanuric acid, are added to the product gases containing NOx, and for certain ranges of temperature, elementary reactions in the H–O–C–N system convert the NOx to N_2. The details of the kinetics of these processes have been explained thoroughly by Miller and Bowman [32].

4.4.3 Spatially–dependent models with kinetics

The most common application of spatially variable combustion with detailed chemical kinetics is the simulation of laminar flame propagation under a variety of conditions. These studies have explained the role of halogenated species in flame inhibition [80–82], of pressure on flame propagation [83], and many other problems in flame propagation [72]. However, in terms of application to practical combustion problems, there have been relatively few studies in which detailed chemical kinetic models, coupled with spatially–variable fluid mechanics, have been used to analyze combustion in internal combustion engines or other realistic systems. Perhaps the most important of these applications have involved the interactions between flame propagation and a cooled wall, with its thermal boundary layer and heat transfer.

For many years, it had been commonly accepted that a major source of unburned hydrocarbon emissions from internal combustion engines was the quench layer, a thin region of fuel near the colder walls of the combustion chamber. Since this region is too cold to allow a flame to propagate, the fuel in this region was assumed to remain unburned and eventually to be exhausted as hydrocarbon emissions. However, careful modeling analysis [84] demonstrated that rapid fuel diffusion out of this quench layer resulted in significant fuel consumption on a time scale which was very short, compared with the residence time available in the engine chamber. As a result, the model results predicted that this mechanism could not be responsible for the majority of the observed amounts of unburned hydrocarbon emissions from engines. Subsequent experimental studies showed that this conclusion was essentially correct and that fuel trapped in piston ring crevices and other small volumes was instead responsible for the majority of unburned hydrocarbon emissions.

Finally, combustion in multidimensional geometries represents the future potential of combustion modeling. Since most practical combustion problems are in fact three–dimensional and involve the interaction between chemical kinetics, fluid mechanics and other processes, it will be essential to be able to describe such systems in computer models. At the present time, given the capabilities of current computer hardware, it is common to describe complex chemical kinetics in simple geometries or complex three–

dimensional geometrical systems with simplified treatments of combustion chemistry. An excellent example is the work of Butler *et al* [85], showing how numerical modeling has been able to explain many of the performance characteristics of practical engines. Future models will be able to combine both features in a single system, detailed kinetics, detailed multidimensional fluid dynamics, and complex treatments of the other important physical systems in the combustor. The rapidly advancing capabilities of massively parallel computing provide these possibilities, and we expect that the next few years will show an impressive increase in the powers and productive applications of computer models.   (Charles K. Westbrook and William J. Pitz)

4.5 Applications of Modeling Study to Combustible Gas Flow

4.5.1 Introduction

Conventional combustors are associated with complex and turbulent flows of combustion which can not be always numerically modeled with detailed chemical reactions due to lack of computer capability. For the estimation of the heat release rate in such complex flows, combustion processes are usually assumed with the overall one–step irreversible reaction.

For numerically predicting unsteady combustion processes such as flame propagation, flame quenching, ignition and formation of pollutant substances, detailed chemical reactions of combustion should be taken into account as well as the complicated transport processes in the fluid–dynamic and chemical kinetic equations. The chemical kinetic equations have different time constants which vary widely with species and stages of combustion. This leads to so–called stiffness problems in numerical computations. The numerical integration of a stiff system of differential equations is always associated with serious problems of stability. It comes from integrating the equations of fast reactions at a slower rate which is favorable for other reactions close to equilibrium. Several sophisticated methods such as high order Runge–Kutta method, high order predictor–corrector method etc. have been proposed for the integration of these stiff kinetic equations with satisfactory results. Although these equations of combustion can be integrated numerically with desired accuracy and tolerable stability with these methods, the techniques are prohibitively expensive for practical applications to computing combustion problems because of the severe time–step requirements. Thus, in the numerical analysis of combustion with the elementary reactions, the selection of the time–step of combustion is of a serious problem in addition to the selection of elementary reactions and their rate constants.

Since an implicit finite difference method for the solution of steady–state problems was introduced by Spalding [86], it has been used to resolve the stiffness and instability problems.

The modified Newton's method on a continuously adapted grid [87–89] is an approach to the stiffness solution, which is introduced on the base of the idea of boundary value method. Further, the idea of coarse–to–fine grid refinement to enhance the convergence to the steady–state approach as well as to provide optimal mesh sizes has been introduced. In the method, new mesh points are added in regions where the solution or its gradients change rapidly, after obtaining a solution with the coarse mesh.

The introduction of quasi steady–state assumption for intermediate species and partial equilibrium assumption for elementary steps is also a solution of the stiffness problem. On the basis of these assumptions, the reduced model of methane–air flame was developed [90,91], in which a 58 steps–mechanism of methane–air flame was reduced to a four–steps mechanism. The model gives fairly good predictions of laminar flame structure.

Both the implicit finite difference method and the modified Newton's method are very useful to deal with the stiffness problem of steady flames such as burner–stabilized laminar flame and steady laminar flame propagation. For solutions of multidimensional unsteady flame, another appropriate method should be introduced to overcome the stiffness and instability problems.

The strongly exothermic process of fuel oxidation may localize the reaction into a very narrow region of the combustor which results in spatially steep distributions of species composition and temperature. In this region the transport properties of species composition and temperature play a predominant role in the reaction process, and the time–step of computation should be chosen to satisfy the requirement of species transport into the reaction zone. All the processes of convection, diffusion and chemical reactions of species should be thus equally decoded with same accuracy in integrating both fluid–dynamic and kinetic equations. In order to properly simulate the features of combustion, it is most efficient to calculate these terms with numerical errors of the same order.

The algorithm developed by Sano and Kotake [92] aims at solving the chemical kinetic equations with accuracy and stability of the same order as those of the flow field of heat and mass transfer. In order to deal with the stiffness problem, the chemical rate terms in the equations of species concentration are expressed linearly with respect to the species, and the time constants of the three processes of convection, diffusion and reaction are kept of the same order in magnitude. Here, the algorithm is employed to analyze numerically the flame ignition phenomena of a combustible methane air flow on a hot plate [93].

4.5.2 Numerical algorithm of chemically reactive flows

Generally, the species differential equations are written as

$$\frac{\partial C_i}{\partial t} + V_k \frac{\partial C_i}{\partial X_k} = D_i \frac{\partial^2 C_i}{\partial X_k^2} + \dot{m}_i \tag{4.1}$$

where C_i and D_i are the mole concentration and diffusion coefficient of species i, V_k the k–direction component of velocity and $\dot{m}_i$ the reaction terms of species i. All chemical reactions can be assumed to be elementary reactions such as

$$[C_k]' + [C_l]' \underset{K_{-j}}{\overset{K_{+j}}{\rightleftharpoons}} [C_m]'' + [C_n]'', \qquad j = 1, 2, 3 \ldots JJ. \tag{4.2}$$

where K_{+j} and K_{-j} are the forward and backward reaction constants of reaction j, respectively, and [ ]' and [ ]'' denote the reactant and product species, respectively. The chemical production term $\dot{m}_i$ then given by

$$\dot{m}_i = \sum_{j=1}^{v} K_{+j}(i)\, C_m C_n - \sum_{j=1}^{v} K_{-j}(i)\, C_k C_l \tag{4.3}$$

where $K_{+j}(i)$ and $K_{-j}(i)$ are K_{+j} and K_{-j} of reaction j which concerns production and destruction of species i, respectively. Factorization of C_i from the destruction rate in Eq. (4.3) is introduced as

$$C_i \sum_{j=1}^{v} K_{-j}(i)\, C_l \tag{4.4}$$

For the second order truncation error for the terms of convection and diffusion, Eq. (4.1) is then written in the finite difference form as

$$\left(1 + \alpha \tilde{B}_{k,i} \Delta t\right) C_{k,i} = \left\{1 - (1-\alpha)\, \tilde{B}_{k,i}^{P} \Delta t\right\} C_{k,i}^{P} + \left\{\alpha \tilde{A}_{k,i} + (1-\alpha)\, \tilde{A}_{k,i}^{P}\right\} \Delta t$$

$$\tilde{A}_{k,i} = \frac{1}{\Delta X_k} V_k C_{k,i}^{(v)} + \frac{1}{\Delta X_k^2} D_i C_{k,i}^{(d)} + \left(\sum_{j=1}^{v} K_{+j}(i)\, C_m C_n\right)_k$$

$$\tilde{B}_{k,i} = \frac{V_k}{\Delta X_k} + \frac{D_i}{\Delta X_k^2} + \left(\sum_{j=1}^{v} K_{-j}(i)\, C_l\right)_k \tag{4.5}$$

where $C_{k,i}^{(v)}$ and $C_{k,i}^{(d)}$ are a function of concentrations at the nearest grid point in the k-direction, the superscript p means the value at the time $t=t-\Delta t$ and α is the weighting factor for estimating the time derivative from values at $t=t$ and $t=t-\Delta t$ ($\alpha=0$ means the explicit method).

4.5.3 Numerical calculation of flame ignition

The reactive flow system of ignition is modeled as follows: (1) a combustible mixture of methane and air flows over a flat plate, (2) constructing a two-dimensional laminar boundary-layer flow, and (3) being heated by a plate of constant temperature. (4) The plate surface is assumed to be inactive or active chemically and physically. (5) The pressure of the flow is constant. The governing equations of the two-dimensional reactive laminar flow are solved by using the above mentioned algorithm. The most favorable computational stability was found with the value of $\alpha=0.5$ in Eq. (4.5). The transport properties of viscosity, thermal conductivity and diffusion are calculated theoretically. As the composition of combustion gases of methane-air, 18 species are considered with 61 elementary reactions of methane oxidaton.

Figure 4.9 shows the time histories of contours of temperature and heat release rate of the stoichiometric gas mixture near the plate and Fig. 4.10, the time histories of methane consumption rate, methane concentration, heat release rate and temperature at the ignition point in the case of the plate surface temperature of 2000 K, the pressure of 0.1 MPa and the main flow velocity of 5 cm/s.   (Taeko Sano)

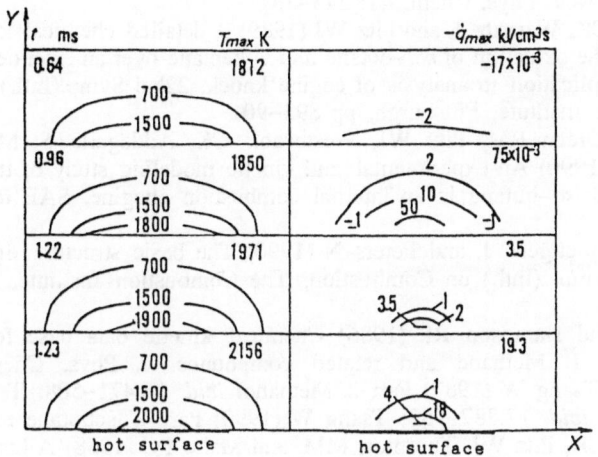

Fig. 4.9   Contours of temperature (T) and heat release rate (-q̇) near plate wall

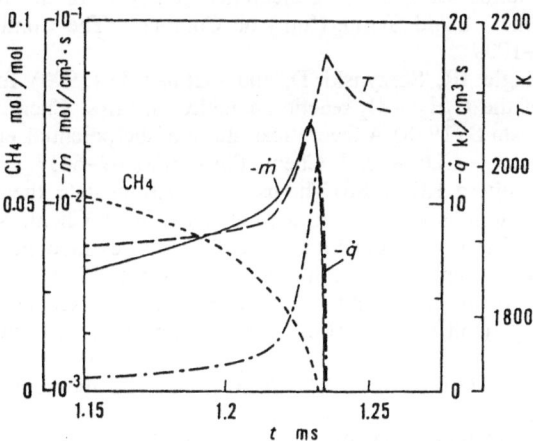

Fig. 4.10   Time histories of methane consumption rate (-ṁ), methane concentration (CH₄), heat release rate (-q̇) and temperature (T) at ignition point

References

[1] Miller JA, Kee RJ, and Westbrook CK (1990) Chemical kinetics and combustion modeling. Ann. Rev. Phys. Chem. 41: 345–387

[2] Westbrook CK, Warnatz J, and Pitz WJ (1989) A detailed chemical kinetic reaction mechanism for the oxidation of iso-octane and n-heptane over an extended temperature range and its application to analysis of engine knock. 22'nd Symp.(Intl.) on Combust., The Combustion Institute, Pittsburgh, pp 893–901

[3] Wilk RD, Green RM, Pitz WJ, Westbrook CK, Addagarla S, Miller DL, and Cernansky NP (1990) An experimental and kinetic modeling study of the combustion of n-butane and iso-butane in an internal combustion enegine. SAE technical paper series 900028

[4] Kennel C, Goettgens J, and Peters N (1990) The basic structure of lean propane flames. 23'rd Symp. (Intl.) on Combustion, The Combustion Institute, Pittsburgh, pp 479–485

[5] Tsang W and Hampson RF (1986) Chemical kinetic data base for combustion chemistry. Part 1. Methane and related compounds. J. Phys. Chem. Ref. Data 15:1087–1279; Tsang W (1987) Part 2. Methanol ibid. 16:471–508; Tsang W (1988) Part 3. Propane. ibid. 17:887–952; Tsang W (1990) Part 4. Isobutane ibid. 19:1–68

[6] Westbrook CK, Pitz WJ, Thornton MM, and Malte PC (1988) A kinetic modeling study of n-pentane oxidation in a well-stirred reactor. Combust. Flame 72:45–62

[7] Westbrook CK, Creighton JC, Lund CM, and Dryer FL (1977) A numerical model of chemical kinetics of combustion in a turbulent flow reactor. J. Phys. Chem. 81:2542–2554

[8] Wilk RD, Pitz WJ, Westbrook CK, and Cernansky NP (1990) Chemical kinetic modeling of ethene oxidation at low and intermediate temperatures. 23'rd Symp. (Intl.) on Combust., The Combustion Institute, Pittsburgh, pp 203–210

[9] Miller JA and Melius CF (1988) A theoretical analysis of the reaction between hydroxyl and acetylene. 22'nd Symp. (Intl.) on Combust., The Combustion Institute, Pittsburgh, pp 1031–1039

[10] Wagner AF, Slagle IR, Sarzynski D, and Gutman D (1990) Experimental and theoretical studies of the $C_2H_5 + O_2$ reaction kinetics. J. Phys. Chem. 94:1853–1868

[11] Aoyagi M and Kato S (1988) A theoretical study of the potential energy surface for the reaction $OH + CO \rightarrow CO_2 + H$. J. Chem. Phys. 88:6409–6418.

[12] Cohen N and Westberg KR (1986) The use of transition-state theory to extrapolate rate coefficients for reactions of O atoms with alkanes. Int. J. Chem. Kinet. 18:99–140

[13] Cohen N (1991) The use of transition-state theory to extrapolate rate coefficients for reactions of H atoms with alkanes. Int. J. Chem. Kinet. 23:683–700

[14] Troe J (1989) Toward a quantitative understanding of elementary combustion reactions. 22'nd Symp. (Intl.) on Combust., The Combustion Institute, Pittsburgh, pp 843–862

[15] Westbrook CK and Dryer FL (1984) Chemical kinetic modeling of hydrocarbon combustion. Prog. Energy Combust. Sci. 10:1–57

[16] Westbrook CK and Dryer FL (1981) Chemical kinetics and modeling of combustion processes. 18'th Symp. (Intl.) on Combust., The Combustion Institute, Pittsburgh, pp 749–767

[17] Westbrook CK and Pitz WJ (1984) A comprehensive chemical kinetic reaction mechanism for oxidation and pyrolysis of propane and propene, Combust. Sci. Technol. 37:117–152

[18] Pitz WJ, Westbrook CK, Proscia WM, and Dryer FL (1985) A comprehensive

chemical kinetic reaction mechanism for the oxidation of *n*-butane. 20'th Symp. (Intl.) on Combust., The Combustion Institute, Pittsburgh, pp 831–843

[19] Coats CM and Williams A (1979) Investigation of the ignition and combustion of *n*-heptane–oxygen mixtures. 17'th Symp. (Intl.) on Combust., The Combustion Institute, Pittsburgh, pp 611–621

[20] Westbrook CK and Pitz WJ (1987) Kinetic modeling of autoignition of higher hydrocarbons: *n*-heptane, *n*-octane, and *iso*-octane. In: Warnatz J and Jager W (ed) Complex chemical reaction systems, mathematical modeling and simulation. Springer–Verlag, Heidelberg

[21] Axelsson EI, Brezinsky K, Dryer FL, Pitz WJ, and Westbrook CK (1987) Chemical kinetic modeling of the oxidation of large alkane fuels: *n*-octane and *iso*-octane. 21'st Symp. (Intl.) on Combust., The Combustion Institute, Pittsburgh, pp 783–793

[22] Kern RD and Xie K (1991) Shock tube studies of gas phase reactions preceeding the soot formation process. Prog. Energy Combust. Sci. 17:191–210

[23] Keller JO and Westbrook CK (1986) Response of a pulse combustor to changes in fuel composition. 21'st Symp. (Intl.) on Combust., The Combustion Institute, Pittsburgh, pp 547–555

[24] Barr PK, Keller JO, Bramlette TT, Westbrook CK, and Dec JE (1990) Pulse combustor modeling: Demonstration of the importance of characteristic times. Combust. Flame 82:252–269

[25] Sloane TM (1984) A computational study of ignition by oxygen dissociation. Combust. Sci. and Technol. 34:317–330

[26] Westbrook CK (1982) Chemical kinetics of hydrocarbon oxidation in gaseous detonations. Combust. Flame 46:191–210

[27] Westbrook CK and Urtiew PA (1983) Chemical kinetic prediction of critical paramaters in gaseous detonations. 19'th Symp. (Intl.) on Combust., The Combustion Institute, Pittsburgh, pp 615–623

[28] Pitz WJ and Westbrook CK (1986) Chemical kinetics of the high pressure oxidation of *n*-butane and its relation to engine knock. Combust. Flame 63:113–133

[29] Westbrook CK, Pitz WJ, and Leppard WR (1991) The autoignition chemistry of paraffinic fuels and pro-knock and anti-knock. Society of Automotive Engineers publication SAE-912314

[30] Cernansky NP, Green RM, Pitz WJ, and Westbrook CK (1986) Chemistry of fuel oxidation preceeding end–gas autoignition. Combust. Sci. Technol. 50:3–25

[31] Griffiths JF, Coppersthwaite D, Phillips CH, Westbrook CK, and Pitz WJ (1990) Autoignition temperatures of binary mixtures of alkanes in a closed vessel:comparisons between experimental measurements and numerical predictions. 23'rd Symp. (Intl.) on Combust., The Combustion Institute, Pittsburgh, pp 1745–1752

[32] Miller JA and Bowman CT (1989) Mechanism and modeling of nitrogen chemistry in combustion. Prog. Energy Combust. Sci. 15:287–338

[33] Jachimowski CJ and McLain AG (1983) A chemical kinetic mechanism for the ignition of silane/hydrogen mixtures. NASA Technical Paper 2129

[34] Britten JA, Tong J, and Westbrook CK (1990) A numerical study of silane combustion. 23'rd Symp. (Intl.) on Combust., The Combustion Insitute, Pittsburgh, pp 195–202

[35] Koda S (1992) Kinetic aspects of oxidation and combustion of silane and related compounds. Prog. Energy Combust. Sci. (in press)

[36] Miller JA, Smooke MD, Green RM, and Kee RJ (1983) Kinetic modeling of the oxidation of ammonia in flames. Combust. Sci. and Technol. 34:149–176

[37] Tieszen SR, Stamps DW, Westbrook CK, and Pitz WJ (1991) Gaseous

hydrocarbon–air detonations. Combust. Flame 84:376–390

[38] Chang W-D and Senkan SM (1989) Detailed chemical kinetic modeling of fuel–rich $C_2HCl_3/O_2//Ar$ flames. Environ. Sci. Technol. 23:442–450

[39] Davidson DF and Hanson RK (1990) High temperature rate coefficients derived from N–atom ARAS measurements and excimer laser photolysis of NO. Int. J. Chem. Kinet. 22:843–861

[40] Koshi M, Yoshimura M, Fukuda K, and Matsui H (1990) Reactions of $N(^4S)$ atoms with NO and H_2. J. Chem. Phys. 93:8703–8708

[41] Yoshimura M, Koshi M, and Matsui H. (1992) Non–Arrhenius temperature dependence of the rate constant for the reaction of $H + H_2S$. Chem. Phys. Lett. 189:199–204

[42] Michael JV and Wagner AF (1990) Rate constants for the reactions $O + C_2H_2$ and $O + C_2D_2$ products, over the temperature range ~ 850–1950 K, by the flash photolysis–shock tube technique. J. Phys. Chem. 94:2353–2464

[43] Fisher JR and Michael JV (1990) Rate constants for the reaction $D + D_2O \rightarrow D_2 + OD$ by the flash photolysis–shock tube technique over the temperature range 1285–2261 K. J. Phys. Chem. 94:2465–2471

[44] Koshi M, Nishida N, and Matsui H. (1992) Kinetics of the reactions of C_2H with C_2H_2, H_2 and D_2. J. Phys. Chem. 97: (in press)

[45] Mahmud K and Fontijn A (1987) A high temperature photochemistry kinetics study of the reaction of $O(^3P)$ atoms with acetylene from 290 to 1510 K. J. Phys. Chem. 91:1918–1921

[46] Shin KS and Michael JV (1991) Rate constants (298–1799 K) for the reactions $C_2H + C_2H_2 \rightarrow C_4H_2 + H$ and $C_2D + C_2D_2 \rightarrow C_4D_2 + D$. J. Phys. Chem. 95:5864–5869

[47] Lange W and Wagner GJ (1975) Massenspektrometrische Untersuchungen uber Erzugung und Reaktionen von C_2H–Radikalen. Ber. Bunsenges. Phys. Chem. 79:165–170

[48] Stephens JW, Hall JL, Solka H, Yan WB, Curl RF, and Glass GP (1987) Rate constant measurements of reactions of C_2H with H_2, O_2, C_2H_2, and NO using color center laser kinetic spectroscopy. J. Phys. Chem. 91:5740–5743

[49] Laufer AH and Bass AM (1979) Photochemistry of acetylene. Bimolecular rate constant for the formation of butadyne and reactions of ethynyl radicals. J. Phys. Chem. 83:310–313

[50] Walker RW (1985) Temperature coefficients for reactions of OH radicals with alkanes between 300 and 1000 K. Int. J. Chem. Kinet. 17:573–582

[51] Atkinson R, Carter WPL, Aschmann SM, Winer AM, and Pitts JN Jr (1984) Kinetics of the reaction of OH radicals with alkanes between 300 and 1000 K. Int. J. Chem. Kinet. 16:469–481

[52] Droege AT and Tully FP (1986) Hydrogen–atom abstraction reaction from alkanes by OH. 3 Propane. J. Phys. Chem. 90:1949–1954

[53] Tully FP, Goldsmith JEM, and Droege AT (1986) Hydrogen–atom abstraction reaction from alkanes by OH. 4 Isobutane. J. Phys. Chem. 90:5932–5937

[54] Ruiz RP and Bays KD (1984) Rates of reaction of propyl radicals with molecular oxygen. J. Phys. Chem. 88:2592–2595

[55] Slagle IR, Balocchi F, and Gutman D (1978) Study of the reactions of oxygen atoms with hydrogen sulfide. J. Phys. Chem. 82:1333–1336

[56] Miyoshi A, Matsui H, and Washida N (1990) Rates of reaction of hydroxyalkyl radicals with molecular oxygen. J. Phys. Chem. 94:3016–3019

[57] Ohmori K (1992) PhD. dissertation, Faculty of Engineering, The University of Tokyo

[58] Asaba T and Fujii N (1971) Shock–tube study of high–temperature pyrolysis of benzene. 13'th Symp. (Intl.) on Combustion, The Combustion Institute, Pittsburgh, pp 155–164
[59] Braun–Unkhoff M, Frank P, and Just Th (1988) A shock tube study on the thermal decomposition of toluene and of the phenyl radical at high temperatures. 22'nd Symp. (Intl.) on Combustion, The Combustion Institute, Pittsburgh, pp 1053–1061
[60] Frenklach M, Clary DW, Gardiner WC Jr, and Stein SE (1984) Detailed kinetic modeling of soot formation in shock–tube pyrolysis of acetylene. 20'th Symp. (Intl.) on Combustion, The Combustion Institute, Pittsburgh, pp 887–901
[61] Fujii N and Asaba T (1973) Shock–tube study of the reaction of rich mixtures of benzene and oxygen. 14'th Symp. (Intl.) on Combustion, The Combustion Institute, Pittsburgh, pp 433–442
[62] Fujii N and Asaba T (1974) Ignition of lean benzene mixtures with oxygen in shock waves. Acta Astronautica 1:417–426
[63] Venkat C, Brezinsky K, and Glassman I (1982) High temperature oxidation of aromatic hydrocarbons. 19'th Symp. (Intl.) on Combustion, The Combustion Institute, Pittsburgh, pp 143–152
[64] Emdee JL, Brezinsky K, and Glassman I (1990) Oxidation of o–xylene. 23'rd Symp. (Intl.) on Combustion, The Combustion Institute, Pittsburgh, pp 77–84
[65] Hippler H, Reihs C, and Troe J (1990) Shock tube UV absorption study of the oxidation of benzyl radicals. 23'rd Symp. (Intl.) on Combustion, The Combustion Institute, Pittsburgh, pp 37–43
[66] Hsu DSY, Lin CY, and Lin MC (1984) CO formation in early stage high temperature benzene oxidation under fuel lean condition: kinetics of the initiation reaction, $C_6H_6 \rightarrow C_6H_5 + H$. 20'th Symp. (Intl.) on Combustion, The Combustion Institute, Pittsburgh, pp 623–630
[67] Thyagarajan K and Bhaskaran KA (1990) High temperature gas phase oxidation kinetics of benzene. Current Topics in Shock Waves, Amer. Inst. Phys., pp 462–467
[68] Kiefer JH, Mizerka LJ, Patel MR, and Wei HC (1985) A shock tube investigation of major pathways in the high–temperature pyrolysis of benzene. J. Phys. Chem. 89:2013–2019
[69] Fujii N, Sakatsume N, and Miyama H (1988) High temperature reaction of the C_6H_6–N_2O system in shock waves. Proc. Nat. Symp. on Shock Wave Phenomena, Shock Wave Research Center, Tohoku University, pp 77–86
[70] Warnatz J (1984) Rate coefficients in the C/H/O system. In: Gardiner WC Jr (ed) Combustion chemistry, Springer–Verlag, New York, pp 197–360
[71] Fujii N (1991) A shock tube study of the oxidation of benzene; effects of H_2 addition. Intl. Seminar on High Temp. Chem. Univ. Tokyo, pp 1–2
[72] Westbrook CK and Miller JA (1983) (ed) Combust. Sci. Techn., Special issue on modeling of laminar flame propagation in premixed gases. vol.34
[73] Atkinson R, Bull DC, and Shuff PJ (1980) Initiation of spherical detonation in hydrogen/air. Combust. Flame 39:287–300
[74] Benson SW (1981) The kinetics and thermochemistry of chemical oxidation with application to combustion and flames. Prog. Energy Combust. Sci. 7:125–134
[75] Pollard RT (1977) Hydrocarbons. In: Bamford CH and Tipper CFH (ed) Comprehensive chemical kinetics vol. 17, Gas–phase combustion. Elsevier, New York, Chapter 2
[76] Westbrook CK and Pitz WJ (1990) Modeling of knock in spark–ignition engines. Intl. Symp. COMODIA 90:11–20
[77] Westbrook CK, Pitz WJ, and Leppard WM (1991) The autoignition chemistry of

paraffinic fuels and pro–knock and anti–knock additives: A detailed chemical kinetic study. Society of Automotive Engineers Report SAE–912314

[78] Lyon RK (1975) Method for the reduction of the concentration of NO in combustion effluents using ammonia. U. S. Patent 3,0900,544

[79] Perry RA and Siebers DL (1986) NO reduction using sublimation of cyanuric acid. Nature 324:657–658

[80] Dixon–Lewis G (1979) Mechanism of inhibition of hydrogen–air flames by hydrogen bromide and its relevance to general problem of flame inhibition. Combust. Flame 36:1–14

[81] Westbrook CK (1980) Inhibition of laminar methane–air and methanol–air flames by hydrogen bromide. Combust. Sci. Techn. 23:191–202

[82] Westbrook CK (1982) Inhibition of hydrocarbon oxidation in laminar flames and detonations by halogenated compounds. 19'th Symp. (Intl.) on Combust., The Combustion Insitute, Pittsburgh, pp 127–141

[83] Westbrook CK and Dryer FL (1980) Prediction of laminar flame properties of methanol–air mixtures. Combust. Flame 37: 171–192

[84] Westbrook CK, Adamczyk AA, and Lavoie GA (1981) A numerical study of laminar flame wall quenching. Combust. Flame 40:81–99

[85] Butler TD, Cloutman LD, Dukowicz JK, and Ramshaw JD (1981) Multidimensional numerical simulation of reactive flow internal combustion engines. Prog. Energy Combust. Sci. 7:293–315

[86] Spalding B (1956) Theory of flame phenomena with a chain reaction. Philos. Trans. Roy. Soc. London 249A:1–25

[87] Smooke MD (1982) Solution of burner–stabilized premixed laminar flames by boundary value methods. J. Comp. Phys. 48:72–105

[88] Smooke MD, Miller JA, and Kee RJ (1982) Numerical solution of burner stabilized pre–mixed laminar flames by an efficient boundary value method, Numerical methods in laminar flame propagation. Friedr. Vieweg & Sohn, Wiesbaden

[89] Smooke MD, Miller JA, and Kee RJ (1983) Determination of adiabatic flame speeds by boundary value methods. Combust. Sci. Techn. 34:79–89

[90] Peters, N. (1985) Numerical simulation of combustion phenomena. Springer, New York, pp 90–109

[91] Peters N and Williams FA (1987) The asymptotic structure of stoichiometric methane–air flames. Combust. Flame 68:185–207

[92] Sano T and Kotake S (1987) A rational algorithm for chemical kinetics; calculation of combustion flow, Numerical methods in thermal problems 5:896–906

[93] Sano T (1991) Flame ignition of premixed methane air mixtures by a high–temperature body. 4th Int. Conf. on Numerical Combustion, pp 186–187

Chapter 5
Soot Formation Fundamentals

5.1 Overview and Characterization of Soot

5.1.1 Introduction

In this chapter topics relevant to soot formation during combustion are given, as well as the results of recent research centered around these. Sooting in flames has long been studied not only from scientific viewpoint but also for practical purposes, such as promotion of radiative heat transfer in boilers and furnaces and reduction of soot emitted as pollutant from combustors and diesel engines. In spite of exhaustive studies, the sooting phenomenon is still unclear and much unknown. This is because the sooting process is not slow enough to observe precisely what takes place in each step of fuel pyrolysis, nucleation, growth, coagulation and oxidation. Thus, difficulties are associated with direct verification of the proposed mechanisms and modeling concepts. To comprehend the essential aspects of the phenomenon, further progress in measuring techniques is necessary. At the same time more precise and advanced theoretical background should be given. Topics compiled in this chapter would respond to these requirements and perspective for understanding soot formation.

Section 5.1 outlines the physicochemical mechanisms of sooting in flames, with the intention of clarifying our current understandings. This follows the topics of morphological characterization of soot particles based on electron–microscopic observations of sampled soot particles.

Section 5.2 gives a detailed chemical mechanism to form polyaromatic hydrocarbons which play a significant role in the nucleation step. Also, a predictive model for soot formation based on this mechanism is described in depth.

Section 5.3 attempts to clarify the likelihood of carbon clusters as being precursors in the nucleation step. Some novel experimental methods for studying the behavior of carbon clusters are also touched on.

Section 5.4 deals with prediction of soot and polyaromatic hydrocarbons using two different models, one being based again on the polyaromatic nucleation mechanism, and the other on a new homogeneous nucleation theory.

Finally, Section 5.5 discusses growth and destruction of soot particles. Recent findings based on various experiments are included to aid further comprehension. Also included are the results of experiments performed on a counter–flow type burner.

5.1.2 Overview

We first look at laminar diffusion flames of gaseous fuel as being the most typical sooting flame. Many minute soot particles are formed in the pyrolysis zone inside the flame boundary. Fig. 5.1 [1] shows measured distributions of soot particle size and number density for acetylene and ethylene flames on a Bunsen burner, where r and x denote radial and axial positions from center axis and burner exit, respectively. The particle size d ranges from 100nm to 300nm and number density C from 10^6 cm^{-3} to 10^{12} cm^{-3}. Some particles will burn out in the presence of oxygen in the outside of the flame, and others are convected well into the flame axis where the number density is the highest. The time elapsed from the flame front to the core is only a few milliseconds during which solid aggregates of 10^6–10^{12} carbon atoms may be formed.

It is accepted that the sooting process from gaseous fuels consists of steps of fuel pyrolysis, nucleation, growth, coagulation and oxidation. We shall briefly look at each step.

1) Thermal decomposition of fuel. In the pyrolysis zone an amount of decomposed species of various kinds may be formed. Analysis of these species has often been carried out, for instance, using mass spectrometry. Prediction of them has become rather successful, partly because the reliable rate constant of each elementary reaction is obtainable and partly because recent progress in computation affords such kinetic calculations for the case when a large number of elementary reactions are included.

2) Nucleation. In condensation of decomposed gaseous species into solid matter, it is natural to assume the presence of nuclei that initiate and afford particle growth, just like those of minute particles and ions in forming rain drops in a subcooled state in the sky. Kelvin's theory tells us that due to surface tension a decrease in the size of nucleus

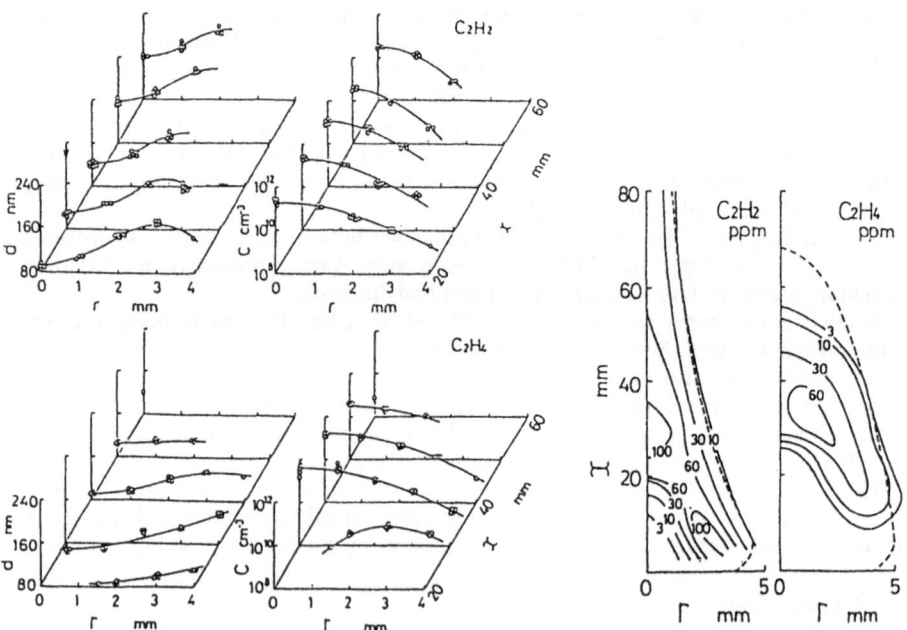

Fig. 5.1   Radial distributions of particle diameters d and number densities C (left) and particle volume fraction (right) measured by a laser homodyne method in acetylene and ethylene flames

particle increases the critical vapor pressure below which condensation may no longer take place. In the case of charged nuclei, the critical vapor pressure is lowered considerably, but still high enough not to permit spontaneous condensation. For this reason, nucleation has been recognized as being a crucial step in soot formation. In flames PAHs (polyaromatic hydrocarbons) are most likely substances of nucleus and have been characterized experimentally [2]. They play a significant role as the nuclei, either in liquid drops or in the solid form after they are subject to dehydrogenation in hotter zones. At much higher temperatures polyacetylenes and carbon vapors are more stable than PAHs and hence they should be taken into consideration in high temperature cases.

3) Growth and coagulation. Some gaseous species will participate on the nuclei which may grow as time goes on. Other than PAHs themselves, saturate and unsaturate carbonaceous gases might be substances for growth. Further clarification of the mechanisms of such solid–gas reactions and their rate constants are required. Collision and attachment between particles are other mechanisms to increase soot particle size. According to electron microscopic observations, the collected soot particles sometimes exhibit chain–like appearance, suggesting that agglomerates may attach side by side at pointed ends due to electrostatic effects.

4) Oxidation. Soot particles may burn up once they enter into oxidative gases. In fuel–rich mixture OH is the most significant oxidizer. Since the rate of soot oxidation decreases with a decrease in temperature, quenching of a soot–containing gas will hinder soot oxidation. Unoxidized soot particles are finally given off outside in a frozen state.

5.1.3 Outline of nucleation theories

What deserves attention is the rapidness of soot formation. It may take only a few milliseconds to form big solid particles having 10^6–10^{12} carbon atoms. Among steps described above, the rate–determining step to reach such big particles is nucleation and growth. For this reason, many studies have been carried out to give a reasonable picture of these steps based on chemical kinetics. However, many proposed theories have their own advantages and disadvantages and are sometimes hindered by difficulties in experimental verification. For this reason, the problem has been left unsolved at present.

Here, nucleation theories proposed previously will be briefly explained. Most theories assume either neutral or ionized heterogeneous nucleation as described earlier. Some other theories rely on the mechanism of homogeneous nucleation.

Some neutral nucleation theories assume a polyacetylene radical mechanism [3] in which carbonaceous nuclei with less hydrogen content are formed as the dehydrogenation of primary molecules proceeds, such as $C_2H_2 \rightarrow C_4H_2 \rightarrow C_6H_2 \rightarrow \cdots \rightarrow$ soot. In fact, concentrations of some polyacetylenes are high in some flames, but this mechanism is not widely accepted because it cannot reasonably explain the rapidness of soot formation. The other theory assumes polyaromatic hydrocarbon mechanism [4], or in short PAH mechanism, in which primary molecules are converted to higher hydrocarbons by condensation and polymerization. This theory reasonably explains the results of shock–tube experiments of soot formation of acetylene decomposition. This theory is also acceptable because it reasonably explains the fact that fuels having stronger bonds such as aromatic fuels and 1, 2–butadiene have greater tendency towards sooting. The details of this theory will be given in later sections. However, there is an argument against to this mechanism; the benzene–ring formation is usually not very fast enough to accelerate soot growth.

While condensing reactions of electrically neutral polyacetylene radicals are usually slow, charged polyacetylenes may react fast enough to proceed sooting. In the theories assuming this mechanism [5], primary ions are formed by chemi–ionization reactions like $CH + O \rightarrow CHO^+ + e^-$ and charged ions are formed by proton–exchange reactions with neutral molecules such as C_2H, CH_2, and CH_3, and larger ions may react fast by some ion–molecule reactions. Charged polyaromatic hydrocarbons may be stable energetically even at higher temperatures, and hence they might be accepted as stable nucleus in flames. However, since hydrocarbons usually have a higher energy for ionization, it is not very certain whether such charged molecules are plenty enough to play an essential role in the nucleation step. Moreover, there is not sufficient experimental verifications for this mechanism because of the difficulty associated with the analysis of positively–charged hydrocarbon molecules.

In addition to the above theories based on heterogeneous nucleation, there are other theories assuming homogeneous nucleation, in which condensation and growth may take place without forming nuclei. These theories [6][7] usually assume carbon vapors such as C_2, C_3 and some sort of polyacetylene radicals, such as C_2H, as primary molecules for forming soot particles. As is widely accepted, the sooting pathway is significantly dependent on temperature; at a very high temperature, sooting proceeds on a more straightforward path, and dehydrogenation proceeds much faster than polymerization to form macromolecules. For this reason, such theories seem to be capable of describing the soot formation at higher temperatures. However, these theories have not been widely accepted because spectra of carbon vapors are little correlated with the soot yield from flames; for instance, a cyanogen–oxygen flame is soot-free in spite of a strong C_2 band emission. This argument against the homogeneous nucleation does not seem to be plausible because the soot yield is much dependent on flame temperature. An extension of the homogeneous nucleation theory might be achieved by including a cluster theory for describing particle growth. In the later section such an extended theory will be given which might be capable of explaining rapid growth of carbon clusters.

Which nucleation mechanism is the best among the ones described above is difficult to answer because each has its own raison d'etre. This seems to be significantly dependent on the conditions to which the mixture is subject, especially on the stability of intermediate substances. According to thermodynamic considerations [7], possible paths would be those via some polynuclear compounds, polyacetylenes, and carbon vapors, each having much higher Gibbs free energy on mass basis than solid carbon. Among them, PAHs are the most stable in the temperature range below 1700K. In the case of fuel–rich premixed laminar flame, PAHs may be temporarily formed in the reaction zone, being held at this temperature range. After conversion to heavier PAHs, they may act as nuclei for sooting in the hotter zone downstream. At the same time, from a thermodynamic point of view, polyacetylenes and carbon clusters may exist in this hot zone. In fact, a mass spectroscopic study [8] suggests the presence of considerable amount of clusters in an acetylene flame. The other extreme case is such that the mixture is suddenly subject to a very high temperature higher than 2000K without passing a middle temperature range. Since polyacetylenes and carbon clusters are more stable than PAHs, clustering would be more significant than others at such a high temperature. This problem will be dealt with in depth in a later section.

(Makoto Ikegami)

5.1.4 Characterization of soot particle

Morphology of soot particles has often been employed to study soot formation during combustion, since the size, shape and internal structure reflect the conditions under which soot is formed during combustion. Soot aggregates and primary particles can be characterized by electron micrography, measurement of light absorption and scattering, X–ray diffraction and other advanced measuring methods. In–situ characterization of aggregates is sometimes carried out by applying laser diagnosing techniques such as laser–light scattering and extinction methods. Furthermore, the internal structure of soot particles has been studied by X–ray diffraction, dark field and phase contrast electron microscopy. Some results of a study performed for sampled soot using electron micrography [9] are given below.

Figure 5.2 shows typical micrograph of soot aggregates which are made of many near–spherical primary particles. In this case of soot particles collected from the exhaust of a diesel engine, the size of each primary particle ranges from 10nm to 50nm, as shown in Fig. 5.2(b). Usually, the size distribution is little affected by the conditions under which soot is formed. The primary particles are connected with each other like a chain, which are sometimes intertwined in a cotton–like configuration. Soot particles rarely exist as individual particles, but as aggregates of up to several tens of individual particles [9]. These may be classified into primary and secondary ones according to their structures. Each primary aggregate is made of carbon layers, the surface of which is continuously connected from one to another. A secondary aggregate is the agglomerate of adhered primary aggregates and is broken up once shear forces are applied.

From the viewpoint of the morphology of internal structures of such aggregates, the use of high–resolution transmission electron microscopy (TEM) is useful. Figure 5.3 shows a TEM image of primary particles of diesel soot together with that of graphite crystal taken on the same magnification. Soot particles exhibit a random crystalline organization, each being oriented in its circumferential direction. This state corresponds to the case when primary particles are joined side by side by adhesion and are fused together in many positions. This suggests that each agglomerate is formed not by collisions of fully carbonized primary particles, but more likely by the growth by condensation of certain gaseous species starting from a minute nucleus.

Unlike the graphite crystal shown in Fig. 5.3, each particle of soot has a closer, but never clearly layered, structure. Electron micrographs at a magnification of up to 10^7, Fig. 5.4, may render visible the layers of hexagonal planes on which carbon atoms are

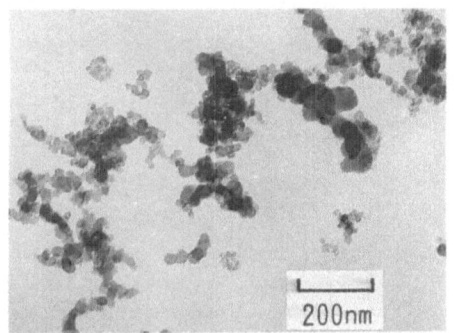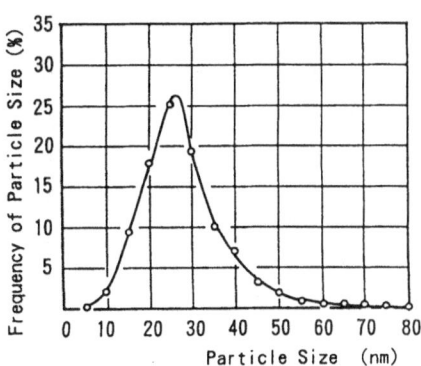

Fig. 5.2   Electron micrograph (left) and size distribution of primary particle
in soot collected from a diesel engine

located as sketched in Fig. 5.5 [11]. It is obvious that interlayers of primary soot look less orderly than those of the crystal and have a spacing of 0.35nm which is greater than 0.335nm for the crystal. This indicates that the soot under consideration is not fully graphitized. In the case when soot particles are aged at 3000°C in an atmospheric inert gas, the interplanar spacing approaches 0.344nm, the planes appearing to be oriented in parallel. As a result, these soot particles become graphitized [12]. The oxidation rate of the graphitized carbon is the slowest among all types from amorphous to single crystal, and therefore the oxidation rate will depend on the degree of crystalline orientation [13].

Figure 5.6 shows electron diffraction patterns for the above two cases, together with that of amorphous carbon. The diffraction pattern of soot particles differs considerably from those of amorphous, having a broad line pattern, and even of a graphite crystal, having bright spots around a hexagonal pattern, and hence we may consider the soot

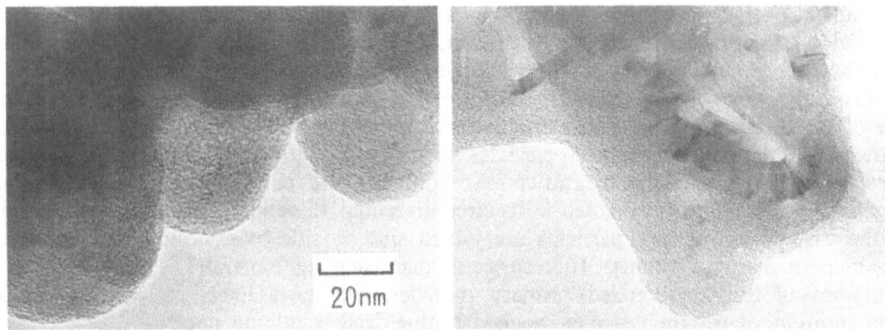

Fig. 5.3   Electron micrographs of primary particle of soot (left) and graphite crystal
(right) at a magnification of 500,000

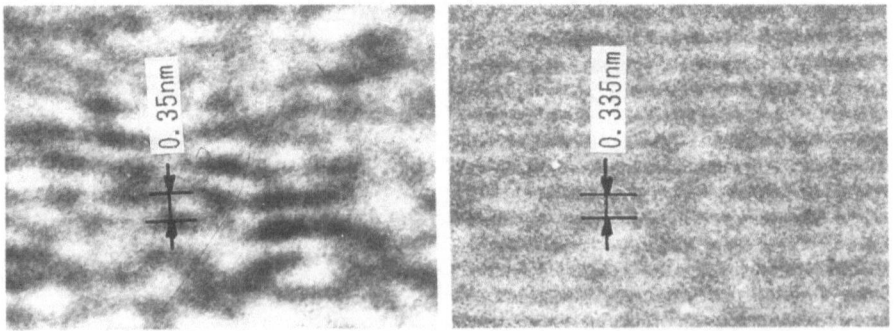

Fig. 5.4   Electron micrographs of layer structures of soot (left) and of graphite crystal
(right) at a magnification of 10,000,000, from [9] with permission

particle to be in between. Generally speaking, lines of diffraction pattern become sharper and more concentrated as the degree of crystallization increases. Thus, the extent of crystallization of the soot particles are somewhere between amorphous and single crystal, exhibiting a turbostratic structure, or a degenerated graphitic structure, in which carbon atoms are located in hexagonal planes, but the layers are randomly distributed in the planes. Such a turbostratic structure has been studied in detail also using an X-ray diffraction technique. According to the results, small crystallites have random orientation being adhered to one another either by amorphous carbon or by hexagonal planes in the manner as shown in Fig. 5.7. This sketch was given by Heckman and Harling [14]. The development of phase contrast electron microscopy has afforded direct imaging of layer planes. The outer planes are oriented as a group in their circumferential directions, but the inner layer has random orientations around the particle center.

(Yasuhiro Fujiwara)

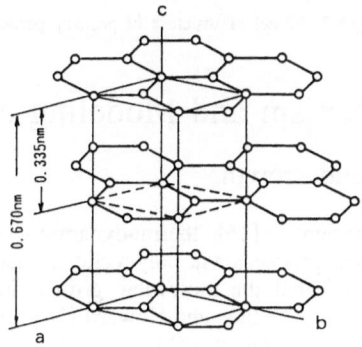

Fig. 5.5 Layer structure of graphite crystal

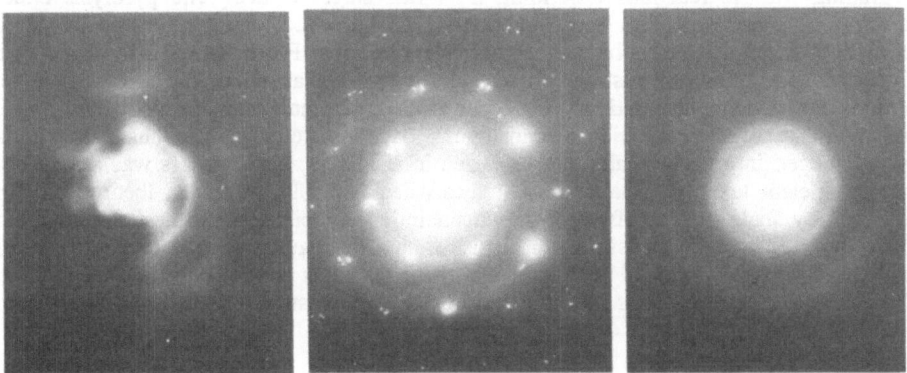

Fig. 5.6   Electron diffraction patterns of soot (left), graphite (center) and amorphous carbon (right), from [9] with permission

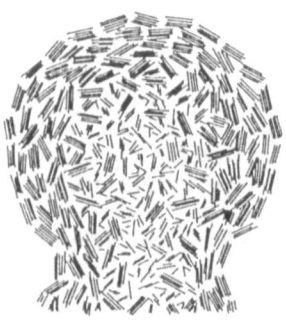

Fig. 5.7 Model of structure of primary particle

5.2 Detailed Mechanism and Modeling of Soot Formation

5.2.1 PAH formation and growth

Growing evidence – experimental [15], thermodynamic [16], and kinetic [4] – has suggested that soot formation proceeds via polycyclic aromatic hydrocarbons (PAHs). Stehling et al. [17] suggested that the molecular growth involves reactions between aromatics and acetylenic species. Bittner and Howard [18] further supported Stehling et al.'s proposal and suggested several possible reaction steps. Analysis of the bell dependence of soot yields on temperature observed in shock–tube experiments indicated that the overall reaction kinetics is consistent with the critical role of aromatic–acetylenic interactions [19,20].

   In order to examine the processes of PAH formation, a chemical reaction mechanism of approximately a thousand elementary reversible chemical reactions was generated [4], beginning with 18 reactions describing the initial stages of acetylene pyrolysis under shock–tube conditions. This initial reaction set was extended, based on physical organic chemistry principles, to include other possible reactions. For instance, knowing that a free hydrogen atom can abstract a hydrogen atom from acetylene, it is reasonable to assume that similar abstractions can occur with larger molecules as well. It was also assumed that the rate coefficients of analogous reactions are the same. Thus all reactions were grouped into classes, about three dozen in all. Each reaction class was assigned a rate coefficient taken from laboratory studies when available, estimated theoretically or assigned by analogy. The effect of the uncertainties associated with these assumptions were then tested using sensitivity analysis. It appears that only three reaction classes have high sensitivity values indicating that these reactions are the rate limiting steps in the reaction sequence. These reaction classes are: (1) abstraction of a hydrogen atom from an aromatic molecule by a hydrogen atom,

$$Ar\text{–}H + H \rightarrow Ar\bullet + H_2$$

where Ar represents an aromatic group; (2) addition of acetylene to the radical formed,

$$Ar\bullet + C_2H_2 \rightarrow Ar-C\equiv CH + H$$

which propagates molecular growth; and (3) cyclization to an aromatic ring, e.g.,

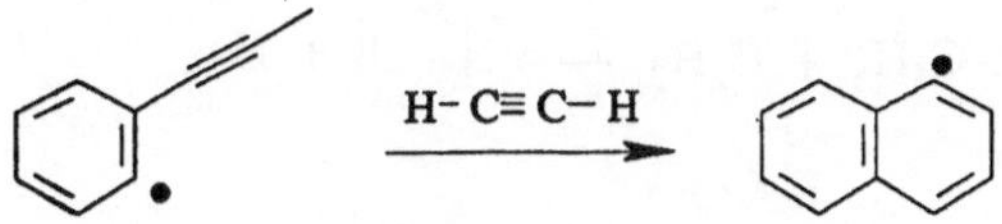

the step in which an additional aromatic ring is formed.

Analysis of the computational results indicated that the principal reaction sequence responsible for PAH growth is in essence a repetition of reactions of the three classes discussed above. This observation allowed us to model the entire reaction sequence to an infinitely–large PAHs using the technique of linear chemical lumping [21]. The essence of this technique is that we do not solve the differential equations for the concentrations of the individual species, but compute moments of the size distribution function of PAHs beginning with a prescribed molecular size. As a result of such computations, we obtain the average size, spread and symmetry of the PAH distribution.

The results of the detailed kinetic modeling study indicated that PAH formation and growth under all conditions proceed via only one or two dominant reaction pathways, even at the initiation stage. The reason for this is a complex network of tightly–balanced chemical reactions. When one reaction flux emerges as the dominant one, it effectively suppresses the fluxes of other competing reactions.

Figure 5.8 presents a reaction sequence identified in the computer simulations to be the dominant pathway under the conditions of shock–tube pyrolysis of acetylene [4].

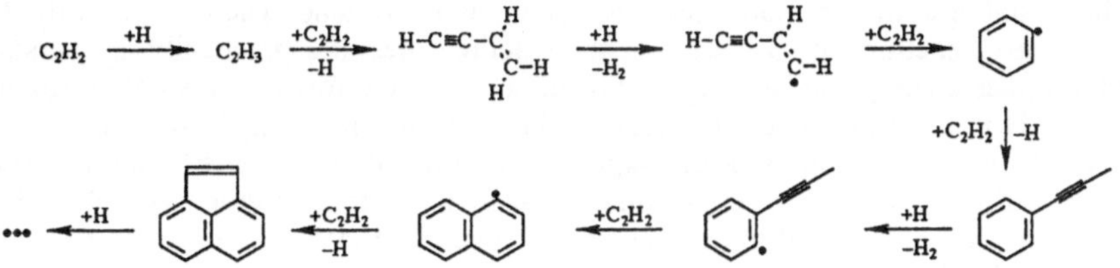

Fig. 5.8 High–temperature acetylene–addition reaction pathway

Initially, it was surprising to find that a single reaction pathway – sequential addition of acetylene as depicted in Fig. 5.8 – dominates the PAH growth. Multiple reaction pathways leading to cyclization and contributing to the growth of aromatics had been anticipated. However, a single dominant reaction pathway was observed in numerical simulations under all conditions, and is in agreement with other modeling studies [22]. It is important to note that this reaction pathway is not the only significant one; on the contrary, under different conditions the PAH production may be initiated by other reactions. But in all cases, as the reaction progresses, the acetylene–addition pathway emerges as the dominant route.

At low temperatures, the formation of the first aromatic ring was found [23,24] to proceed predominantly via another reaction channel,

$$\text{n-C}_4\text{H}_5 + \text{C}_2\text{H}_2 \longrightarrow \bigcirc\!\!| + \text{H}$$

the reaction suggested by Weissman and Benson [25] and Cole et al.[26] Benzene and phenyl are converted to one another via H–abstraction and its reverse reaction. Under some conditions the recombination of propargyl radicals may contribute to the first aromatic ring. Starting with an aromatic fuel, a direct condensation of intact aromatic rings becomes important. For example, in the case of high–temperature pyrolysis of benzene, the following reactions were found to dominate the initial stages of PAH growth [27],

However, as the reaction progresses, the initial benzene molecules decompose forming primarily acetylene. As the concentration of acetylene approaches that of benzene, which occurs shortly after the initial period, the PAH growth switches to the acetylene–addition mechanism discussed above. Thus the reaction system *relaxes* to the acetylene–addition pathway. The relaxation is faster in oxidation [23] as compared to pyrolysis and in mixtures of hydrocarbons [28] as compared to fuel of a single component.

Some of the acetylene addition reactions in the PAH growth sequence form particularly stable aromatic molecules like pyrene and coronene. The change of the free energy in these reactions is so large that the reactions become practically irreversible. This, in turn, has an effect of "pulling" the reaction sequence forward, towards formation of larger PAH molecules. Other acetylene–addition steps are highly reversible, i.e., the rate of the forward reaction is nearly equal to the rate of the reverse reaction. These steps with tightly balanced reaction fluxes create a thermodynamic barrier to PAH growth. A detailed analysis of the computational results [4,29] resulted in the following conclusions. At the very high temperatures, the radicals formed as a result of H–abstraction decompose faster than they grow; and this defines the "ceiling" temperature for PAH formation. At high temperatures, but below the ceiling temperature, the PAH growth is controlled by the superequilibrium of hydrogen atoms. At low temperatures, the growth is determined by the rate of the acetylene–addition step.

Effect of oxygen is twofold [23,24]. Because of the accelerated chain branching it promotes fuel pyrolysis and hence the production of hydrocarbon radicals and H atoms. On the other hand, it removes aromatic radicals and critical aliphatic hydrocarbon radicals, like C_2H_3 and C_4H_3, by reactions with molecular oxygen. The balance of these factors determines the net effect of oxygen addition. In a flame environment, it was found that diffusion of H atoms from the main reaction zone of a laminar premixed acetylene flame to the preheat zone is responsible for the initiation of reactions and the route to PAH formation in particular [24].

5.2.2 Soot particle inception in flames

The computational model used to simulate the inception of soot particle in several laminar premixed flames consists of three logical parts [30]: (*i*) *initial PAH formation*, which includes a detailed chemical kinetic description of acetylene pyrolysis and oxidation, formation of the first aromatic ring, and its subsequent growth to a prescribed size; (*ii*) *"planar" PAH growth*, comprised of replicating–type growth of PAHs beyond the prescribed size; and (*iii*) *spherical particle formation and growth*, consisting of coagulation of PAHs formed in part (*ii*) followed by the growth of the resulting particles by coagulation and surface reactions.

The reaction mechanism for pyrolysis and oxidation of small hydrocarbon molecules responsible for the main flame structures, and reactions describing the formation and growth of PAHs was composed of data taken primarily from several recent sources [31]. The formation and growth of aromatics followed the basic reaction scheme of Frenklach and co–workers [4,24,28]. The chemical reaction mechanism, comprised of a total of 337 reactions and 70 species, was capable of predicting near–quantitatively the measured species profiles including those of aromatics for a number of laminar premixed flames [22,32,33].

The Sandia burner code [34] was used for simulating laminar premixed flame structures and PAH formation and growth up to coronene. The computed profiles of H, H_2, C_2H_2, O_2, OH, H_2O and PAH of a prescribed size (A_ℓ) were then used as an input for the simulation of particle nucleation and growth. Several simultaneous processes were modeled using the method of moments [21], accomplished with an in–house kinetic code. The first part of this model, *nucleation*, describes the planar growth of PAHs via the H–abstraction–C_2H_2–addition reaction sequence. This method provides a mathematically rigorous description of the growth process to an infinite size PAH, which can be schematically represented as $A_\ell \rightarrow A_{\ell+1} \rightarrow A_{\ell+2} \rightarrow \cdots \rightarrow A_\infty$, where A_ℓ represents an aromatic species containing ℓ fused rings.

Examples of the computational results for the concentration profiles of aromatic species are presented in Figs. 5.9 and 5.10. The agreement between the modeling predictions and experimental data was good for the major species [31]. As can be seen in these figures, the agreement for PAHs is not exactly quantitative, but is certainly very close in both shape and absolute values. The important result here is that the near–quantitative accuracy was obtained for PAHs with a single reaction mechanism for all flames simulated.

The PAH species formed in the reaction sequence described earlier were allowed to *coagulate*, that is, all the A_i's $(i=\ell, \ell+1,\cdots,\infty)$ collide with each other forming dimers; the dimers, in turn, collide with A_i forming trimers or with other dimers forming tetramers; and so forth. The coalescence reactions were treated as irreversible having sticking coefficients of unity. As the focus of this work is on very young, small particles, it was assumed that the coagulation dynamics is in the free–molecular regime. Beginning with the dimers, the forming clusters were allowed to undergo surface reactions

$$C_{soot}-H + H = C_{soot}\bullet + H_2$$
$$C_{soot}\bullet + H \rightarrow C_{soot}-H$$
$$C_{soot}\bullet + C_2H_2 \rightarrow C_{soot}-H + H$$
$$C_{soot}\bullet + O_2 \rightarrow products$$
$$C_{soot}-H + OH \rightarrow products$$

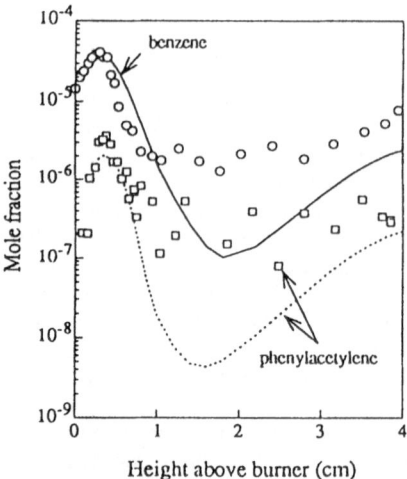

Fig. 5.9   Computed versus experimental aromatic species concentrations in a 46.5%
C$_2$H$_2$-48.5% O$_2$-Ar flame (P=20 torr, cold gas velocity=50cm/s)

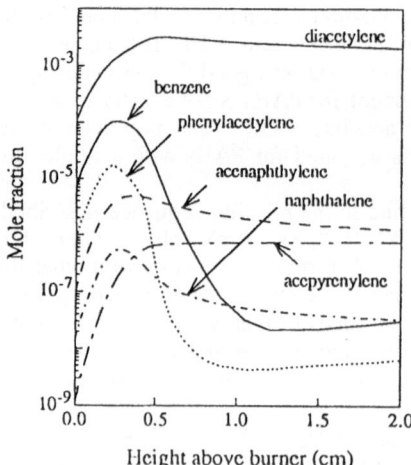

Fig. 5.10 Computed mole fractions of diacetylene and PAHs in a 23.6% C$_2$H$_2$-21.4% O$_2$-Ar flame
(P=90 torr, cold gas velocity=20cm/s).

where C_{soot}–H represents an arm–chair site on the soot particle surface and $C_{soot}\bullet$ the corresponding radical. This mechanism is adopted based on the postulate that the H–abstraction–C_2H_2–addition reaction sequence is responsible for high–temperature growth of all forms of carbonaceous materials. The particle dynamics – the evolution of soot particles undergoing simultaneous nucleation, coagulation and surface reactions described above – was modeled by a method of moments [21]. This method does not require the assumption of a particle size distribution function (PSDF). The closure of the differential equations for the PSDF moments is accomplished by interpolation between the moments.

The computational results of an acetylene flame (Flame B [35] – 25.4%C_2H_2–19.6%O_2–Ar, P = 90 torr and cold gas velocity = 20cm/s) and an ethylene flame (Flame H [36] – 16.5%C_2H_2–17.9%O_2–Ar flame, P = 760 torr and cold gas velocity = 8cm/s) are shown in Figs. 5.11 and 5.12. It can be seen in these figures that the model

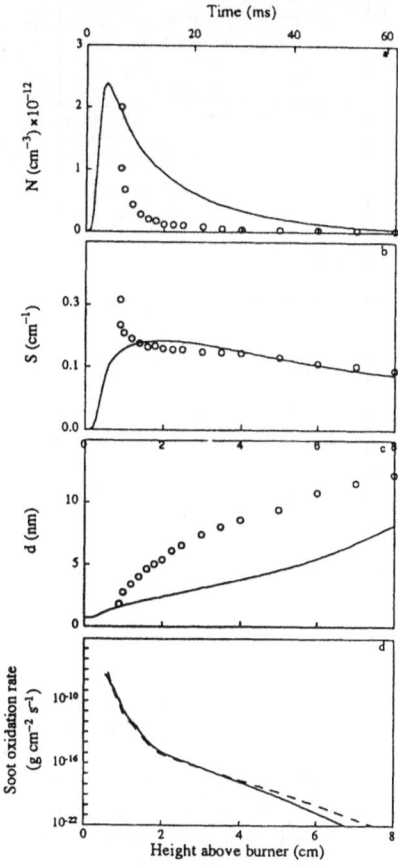

Fig. 5.11 Computed versus experimental particle (a) number density, (b) surface area, (c) diameter; and (d) specific soot oxidation rate by O_2 in Flame B (solid lines: model predictions, dashed line: from the expression of Nagle and Strickland–Constable)

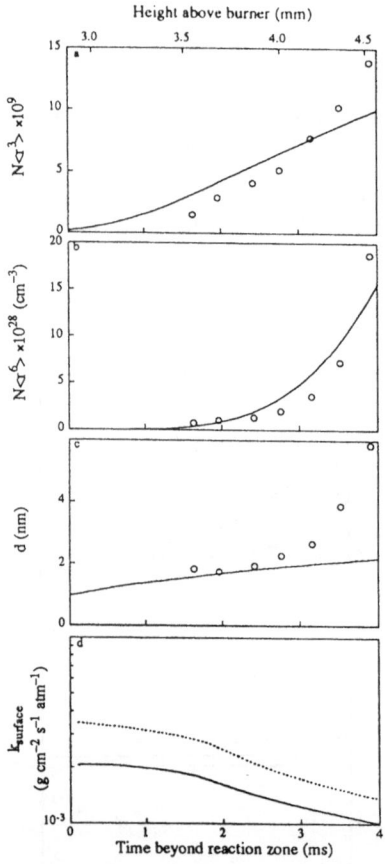

Fig. 5.12 Computed versus experimental soot particle characteristics for Flame H: (a)–(c) –N: particle number density, r: particle radius, $<r^P>$: average r^P, d: average particle diameter; (d) specific surface growth rate coefficient(lines: model prediction, circles: experiments)

predictions are in relatively close agreement with experiment for the particle inception part of the flames. The reliability of the present model is further supported by the facts that the computed net surface growth rate is in close agreement with that determined by Harris and Weiner [36] for Flame H and that the predicted rate of soot oxidation by O_2 for Flame B agrees well with the expression of Nagle and Strickland–Constable [37]

Major results of the present modeling study [30] are summarized as follows:

(*i*) The computed rate of nucleation is balanced by the rate of coagulation throughout the particle inception zone, however, the nucleation rate decays more slowly with the increase in the flame height than is usually deduced from experiment.

(*ii*) Particle inception is primarily determined by PAH coagulation initiated and controlled by PAH coalescence into dimers.

(*iii*) While the average soot particle is computed to contain 10^3–10^5 carbon atoms, the corresponding average PAH size is only 20 to 50 carbon atoms. This indicates that the "crystallites" comprising incipient soot particles should be on the order of 7 to 12Å, in agreement with experiment [38].

(*iv*) The oxidation by OH and O_2 is quite insignificant in the post–flame zone.

(*v*) The surface growth of soot mass is primarily determined by two processes: acetylene addition via the H–abstraction–C_2H_2–addition reaction sequence, and PAH condensation on the particle surface. The relative contribution of each of these processes appears to change with experimental conditions. Thus, while the acetylene addition dominates surface growth in Flame H, PAH condensation prevails in Flame B. The main contribution of the PAH condensation occurs at the early stages of PAH coagulation.

(*vi*) The model predicts the classical structure of soot particles; a less dense core composed of randomly oriented PAH oligomers and a more dense concentrically–arranged particle shell.

(*vii*) Surface processes can be understood in terms of elementary chemical reactions of surface active sites. The number density of these sites is determined by the chemical environment.   (Michael Frenklach and Hai Wang)

5.3 Nucleation and Carbon Clustering

5.3.1 Chemical inception

The soot particles are unfavorable products in flames considering from the chemical equilibrium. Stein and Fair [39] performed a chemical thermodynamic analysis of hydrocarbon molecules at higher temperatures and examined a thermodynamic barrier toward polymerization, taking into account all possible $C_{2n}H_{2m}$ species. According to their analysis, a free energy barrier appears in the range 1400–1800K in the case of the mixture of H_2 and C_2H_2 ($[H_2]/[C_2H_2]>1$). A certain mechanism must exist which allows the rapid passage through this thermodynamical bottle–neck and then generates soot precursors. Thus the understanding of the reactions of hydrocarbon species which lead to the precursors is very important.

The problem can be addressed via two different ways. One way is to identify the species involved in actual flames, to pursue their time evolutions, and then to describe the chemical change of the individual species. The other is to simulate the reaction progress on the basis of chemical kinetic data which are obtained independently. Kern and Xie [40] recently published a review work which deals with the mechanisms of hydrocarbon pyrolysis and their role in the preparticle soot formation process, mainly taking from shock tube kinetic literatures. According to them, the chemical kinetics of small hydrocarbon species containing 1 or 2 carbon atoms are well known, while those of higher hydrocarbons are not understood with comparable reliability. In order to establish a soot kinetic model, both of detailed knowledge of the chemical species in actual flames and the relevant kinetic data are strongly required.

5.3.2 Species in flames

The analysis of flame species have been carried out for a long time by use of mass–spectrometric measurements. The species of higher mass numbers ($>10^3$) are becoming to be identified. In recent measurements of ionic species in a low–pressure acetylene/O_2 flame [41], aliphatic $C_nH_5^+$ and substituted oxyomethylium ions, ions of polyaromatic hydrocarbons (PAH), polyynic ions, and also ions of fullerenes (fullerene is a name of the family of spheroidal carbon clusters) have been identified. Similar neutral C_nH_m species are expected to exist simultaneously. Among them the fullerene compounds are very recently found. Howard et al.[43] succeeded to isolate C_{60} and C_{70} fullerenes from the soot of benzene flames. Now we are required to understand the formation and destruction relationships among the above mentioned species and answer the questions such as; which compounds are precursors of other compounds, which compounds are by-products and so forth.

The plausible route to soot in the ordinary flames, which is suggested on the basis of the time evolutions of individual species along the stream line, may be following. At first fuel hydrocarbons decompose to small hydrocarbon species, including radicals. Subsequent reactions of the decomposed species, however, form some species of higher carbon numbers, which are then aromatized and condensation-polymerized to yield PAH species. Some researchers believe that PAH species then condense to each other to make primary soot nuclei. Though ionic species exist simultaneously with neutral species, they are not considered to be the main precursor species to lead to soot formation.

The pioneer researchers of fullerene chemistry, Zhang and co-workers, mentioned the importance of the large carbon clusters, fullerenes, in considering the formation and morphology of soot. Fullerenes might be involved in a certain stage of the soot formation [43], though such ideas have been criticized by ordinary soot researchers [44]. The fullerenes are, according to some researchers, formed as by-products during the soot formation, or products from the young soot particles. In any case, the mechanism of fullerene formation and its kinetics are very deeply related to the chemistry of soot formation.

5.3.3 Mechanistic study of carbon clusters

In order to study the kinetics of fullerenes, it is desirable to have a clean source of fullerenes. It was realized by adopting the laser ablation-jet expansion method, where a Nd:Yag laser is focused onto a carbon rod and then the ablated materials are seeded into a He carrier gas flow which is then expanded into vacuum. The ablation plasma contains atomic carbon vapors, small carbon molecules and radicals, and probably some ionic species. Such small species react with each other to produce carbon clusters, which are usually identified by use of time-of-flight (TOF) type mass-spectroscopic method. The typical TOF mass-spectra of clusters are shown in Fig. 5.13.

Koda et al.[45] found that the yield of clusters increased when some amounts of hydrocarbon gases such as ethane were added to the carrier gas. Based on this finding, they discussed that the carbon clusters are produced from the small carbonaceous species generated in the laser ablation plasma, and that the clusters can grow very rapidly even in the presence of hydrocarbon gases as well as hydrogen.

The reaction between carbon clusters and molecular hydrogen proceeds, though not so rapidly, to produce a certain kind of C_nH_m compounds as exemplified in Fig. 5.14. Similar reactions have been also studied by Rohlfing [46]. The structure of the products are not fully identified, but PAH species are among the candidates of the products C_nH_m. Considering the reversibility of the reactions, the fullerenes are, to the contrary, expected to be produced from some kind of PAH and/or related compounds. Baum et al.[47] proposed that $C_{48}H_{18}$ which is the main C_{48}-PAH and contains the structural elements of acenaphtylene and benzo[ghi]perylene, which are quite common PAH species, may combine with two benzene molecules to produce C_{60} under flame conditions. In spite of the very rapid development of the study of fullerene reactions, however, their formation mechanism in actual flames and their role in the soot formation are not yet clear.

The possibility of growth of carbon clusters up to solid particles has been discussed by Rohlfing [48] and also by Koda et al. The former researcher reported the observation of black body radiation from the flow of carbon clusters which are irradiated by a laser light, while the latter researchers have tried to determine the particle size distribution and number density by applying the novel technique of laser induced breakdown. The principle of the method (see Fig. 5.15) is that the breakdown event occurs under the irradiation of a laser light which has a stronger fluence over a certain threshold value. The threshold fluence is expected to become larger for smaller particles which are present in the laser focal volume. Thus by analyzing the intensity of the light emitted

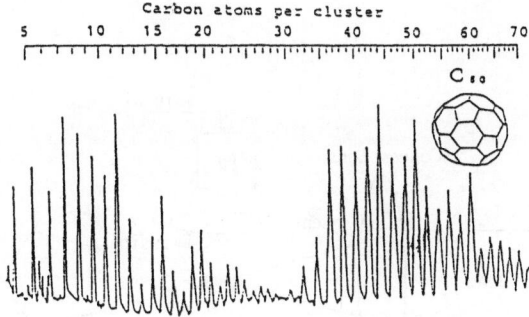

Fig. 5.13   Typical TOF mass spectrum of carbon clusters obtained by
the method of Nd:Yag laser ablation onto a carbon rod

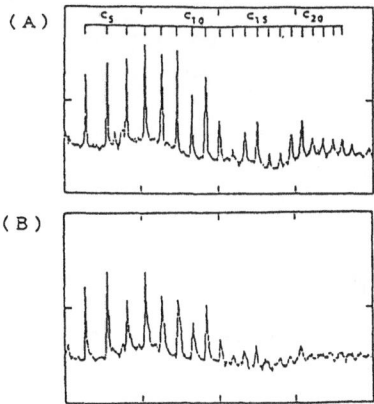

Fig. 5.14   Comparison between the TOF spectra obtained in He (A)and in H_2 (B) carrier gas (The broadness
of the latter spectra indicates the existence of C_nH_m species)

from the breakdown and determining the threshold fluence by changing the intensity of the breakdown laser, it is possible to estimate the size of the particles. A typical histogram of the breakdown light intensity is shown in Fig. 5.16, where the breakdown laser irradiated the He flow of clusters which were produced by laser ablation onto a carbon rod. The peaks in the histogram at higher S values (S means the time–integrated light intensity emitted from the breakdown) imply the presence of particles in the focal volume of the breakdown laser. The number density of the particles was estimated from the probability of the occurrence of the breakdown event. It was shown that a large amount of particles whose diameters are less than 100 nm are present in the cluster flow. Thus it is probable that the carbon clusters can grow up to solid particles under appropriate conditions.

The above mentioned experimental studies indicate that the formation of solid carbon particles from the carbonaceous small species are indeed possible, though it is not clear whether the environmental conditions in actual flames are favorable for the formation

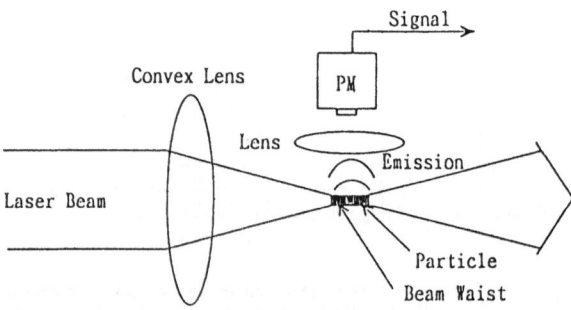

Fig. 5.15 Illustration to show the principle of the laser breakdown technique to obtain the particle size (The emitted light from the breakdown is observed by use of a photomultiplier tube)

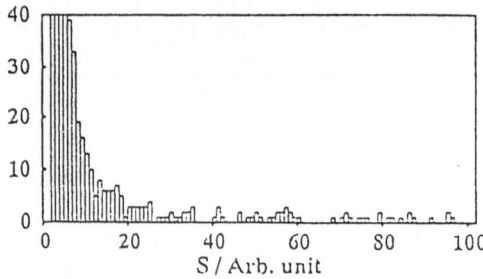

Fig. 5.16 A histogram of the breakdown light intensity S (The total laser shot, 5000. Peaks at S of less than 10 are mainly due to the background light noise)

mechanism or not. The soot formation mechanism under flame conditions should be established on the basis of reliable thermochemical and kinetic data. The data are still of severe shortage, and the analyses of flame species are not yet enough. It is desirable to increase the size of the relevant data base, particularly for the newly found species such as fullerenes.   (Seiichiro Koda)

5.4 Prediction of Soot and Soot Precursors

5.4.1 Predictions based on PAH model

The PAH kinetic scheme proposed by Frenklach et al. [4] was successful in explaining formation of soot precursors during decomposition of acetylene in their own shock–tube experiments. They predicted the bell–shaped profile in the soot yield versus temperature relations. Since then, the model based on this scheme has often been adopted to study other sooting cases. Frenklach and Warnatz [24] and Bastin et al. [49] applied this to the case of low–pressure acetylene flames and showed that predicted results using this model are sufficiently consistent with the experimental results to explain soot precursors sooting at least qualitatively. Furthermore, they estimated the pathway to reach benzene rings based on measured intermediate products. Harris et al.[22] predicted the formation of polycyclic species in an ethylene–air premixed laminar flame and compared them with experimental results. According to their results, the PAH model may predict not only benzene formed, but also phenylacetylene and styrene. Sano [50] also included the PAH model in numerical calculations of atmospheric acetylene–air premixed flat flame to study PAHs such as benzene, naphthalene and their derivatives.

Thus, the PAH model may be helpful in giving qualitative explanations to the polycyclic substances that are recognized as being precursors of soot formation. However, one should not overestimate its predictability, partly because this model is not fully matured yet, and partly because difficulties are often encountered with such numerical predictions; there are uncertainties in rate constants of formation reactions of aromatic compounds, ambiguities in determining certain reverse rate constants, ignorance of trace species that may happen to affect PAH formation, uncertainties in thermochemical data, and so forth. Hopefully, these difficulties will be much reduced and improvements be attained in the near future.

As an example of the prediction relying on the PAH model, Sano's approach [50] will be described below, together with some elucidations concerning PAH formation in a flame. In this study, a fully–kinetic one–dimensional numerical calculations are carried out for acetylene–air flat laminar flames by considering 47 species and 168 elementary reactions including oxidation of benzene, naphthalene and their derivatives with O_2 and OH. Among species considered are $C_6H_5(A1X)$, $C_6H_6(A1)$, $C_6H_5-C_2H(A1A)$, $C_6H_4-C_2H(A1AX$ and A1AE), $C_6H_4-C_2H_2$, $C_{10}H_8(A2)$, $C_{10}H_7(A2X)$, $C_{12}H_8(A2R5)$ and $C_{12}H_7$. Equilibrium constants for determining reverse rate constants are obtained from JANAF table, Burcat's table [51], Stein's data [52], Bahn's data [53] and sometimes by applying Benson's rule [54]. Transport coefficients are given by theoretical calculations.

In Fig. 5.17 calculated mole fractions of some inorganic substances X_i and temperature T are shown against distance from the burner exit, d, for two equivalence ratios of $\phi=1$ and $\phi=2.5$ at atmospheric pressure and at a temperature of unburned mixture of 298 K, in an adiabatic case without heat loss from the flame. At $\phi=1$ acetylene is completely consumed early at the beginning of combustion, whereas at $\phi=2.5$ it remains appreciable in the burned gas. In Figs. 5.18 and 5.19 mole fractions of benzene and its

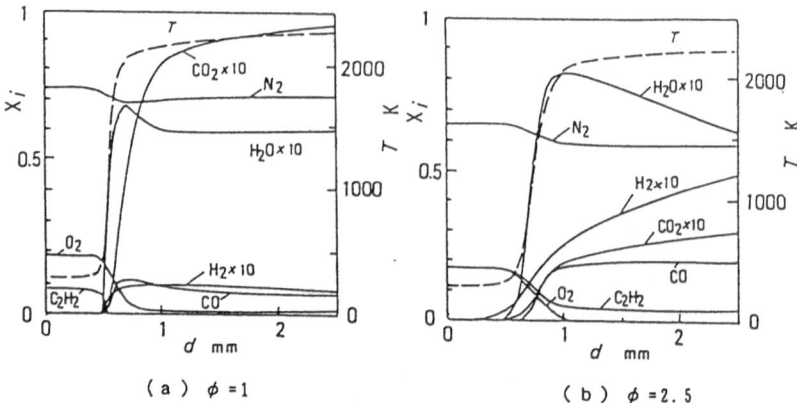

Fig. 5.17   Predicted distributions of temperature and concentrations of major products in one–dimensional atmospheric acetylene–air flame without heat loss, at unburned gas temperature of 298K

derivatives and of radicals are shown, respectively, for the same cases. Much benzene and derivatives are rapidly formed in the visible flame zone where activated species such as hydrogen atom are more abundant and combustion reactions proceed actively. Each yield exhibits wide, but different, spread in the reaction zone depending on species as well as on equivalence ratio. At stoichiometric ratio their concentrations are peaked at much higher levels, but are less spread than the counterpart. In the fuel–overrich case, the mixture is long subject to lower temperature and hence more ring–like substances can be formed.

According to the above calculations, benzene is mainly produced on the following path in a hot reaction zone.

$C_2H_2 + H \rightarrow C_2H_3$
$C_2H_2 + C_2H_3 \rightarrow n-C_4H_5$
$C_2H_2 + n-C_4H_5 \rightarrow n-C_6H_7$
$n-C_6H_7 \rightarrow c-C_6H_7 \rightarrow$ benzene

Also, bigger polyaromatic hydrocarbons are produced via isomers of $C_6H_4-C_2H$ with reactions $n-C_8H_5 \rightarrow C_6H_4-C_2H(A1AE)$ and $i-C_8H_5 \rightarrow C_6H_4-C_2H(A1AX)$.
Bicyclic substances are formed via the former isomer on the steps as follows.

$C_6H_4-C_2H(A1AE) + C_2H_2 \rightarrow C_6H_4-C_4H_3 \rightarrow$ naphthyl
naphthyl $+ H_2 \rightarrow$ naphthalene
naphthyl $+ C_2H_2 \rightarrow$ acenaphthylene

From a series of similar calculations, it has been shown that the more the benzene ring in a molecule the closer the peak of the concentration is located to the low–temperature zone where the acetylene concentration is still high. Also predicted is the fact that heat loss due to radiation causes little change in concentrations of benzene, naphthalene and their derivatives in the flame zone, excluding phenyl, but affords

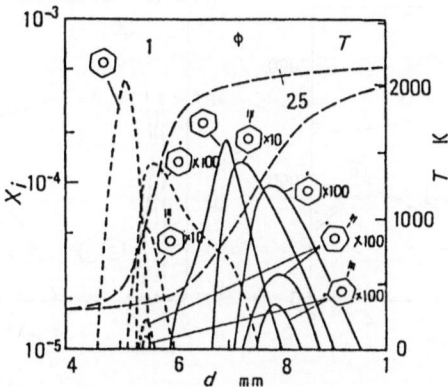

Fig. 5.18 Predicted concentrations of benzene and the derivatives in an adiabatic flame

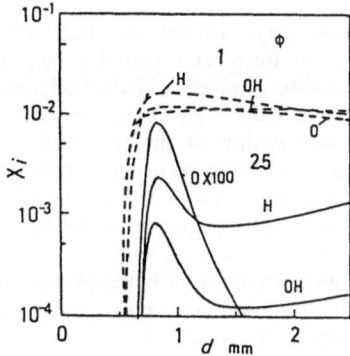

Fig. 5.19 Predicted concentrations of H, O and OH in an adiabatic flame

producing benzene, phenylacetylene and naphthalene in the post–flame zone owing to the decreased temperature, as is shown in Fig. 5.20. The benzene formation in the post–flame zone steps from $n,i-C_8H_5$ to $C_6H_4-C_2H$, phenylacetylene ($C_6H_5-C_2H$), phenyl and final benzene, unlike a more significant step $n-C_6H_7 \rightarrow c-C_6H_7 \rightarrow$ benzene in the flame zone. Since the reverse reaction $C_6H_4-C_2H \rightarrow n,i-C_8H_5$ is slower at lower temperatures, a decrease in temperature leads to an increase in benzene, phenylacetylene and naphthalene.   (Taeko Sano)

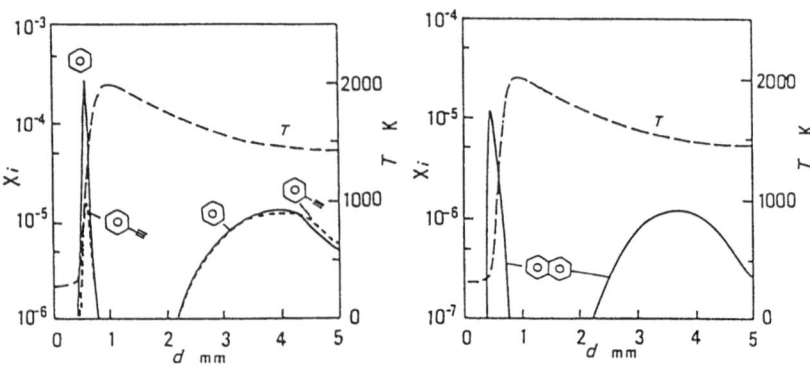

Fig. 5.20 Formation of benzene and phenylacetylene (left) and naphthalene (right)
in the postflame zone in a non–adiabatic flame

5.4.2 Soot prediction based on homogeneous nucleation

At a relatively low temperature, say, around 1500K, polymerization and even
condensation of gaseous species to form soot particles will take place faster than
dehydrogenation, forming intermediate products of polycyclic aromatic compounds that
act as soot precursors, as has been stated earlier. Whereas, at temperatures higher than
2000K, thermal decomposition or splitting of the fuel molecule will be faster than
polymerization. The homogeneous nucleation theories proposed may afford description
of sooting taking this fact into account. Jensen's theory [6] is one such theory on the
assumption that soot is formed via carbon vapors and polyacetylenes like C_2, C_3 and
C_2H.

Yoshihara and Ikegami [7] proposed a different homogeneous nucleation theory based
on cluster kinetics for carbon. Carbon cluster is an aggregate consisting of carbon atoms,
the smallest cluster being carbon monomer C_1 and the large extreme being graphite. The
main points of this theory are as follows. Most gas–phase reactions are dealt with by
applying the rate controlled partial equilibrium method [55] in such a manner that Gibbs
free energy of the system may be minimized under constraints imposed, while slow
reaction steps, such as fuel pyrolysis some sort of oxidation reactions and the clustering
process, are calculated in a kinetic manner. The following reactions are assumed for the
cluster growth and oxidation of cluster of n–mer C_n:

$$C_n + C_i + M = C_{n+i} + M \quad [i=1,..., 5]$$
$$C_n + OH = C_{n-1} + CO + (1/2)H_2$$

where C_i is the colliding partner and M is the third body. The forward step in the former
reaction states that collision between n–mer and i–mer is successful to form $(n+i)$–mer
if the third body M removes the surplus energy at each collision. Oxidation of cluster
only by OH is included, all others being ignored, because the mixture is normally
fuel–rich enough in many sooting cases. Rate constants for cluster reactions are given

by applying Bauer and Frurip's self–consistent kinetic model [56]. This model is based on the following concept: The forward reaction comprises of complex formation and stabilization. The former rate is determined on the assumption of rigid–sphere collision between the stabilizer M and the complex of n–mer and i–mer. The equilibrium concentration of the complex is estimated based on an assumed Lennard–Jones potential which is given by relevant thermodynamic data. The reverse reaction is determined by the equilibrium constant as estimated from thermodynamic considerations.

Figure 5.21 shows the results obtained from this theory and those obtained experimentally by Frenklach and co–workers [4] at the pyrolysis acetylene in shock–tube experiments using a laser–extinction technique. The results given in the figure show that the theory clearly predicts the bell–shape dependence of the soot yield on temperature. What deserves attention is the fact that inclusion of cluster reaction may predict the rapidness of sooting.

According to the present theory, the bell–shaped temperature dependence of soot yield is attributed to the temperature dependence of the rates of the cluster reaction. In Fig. 5.22 the forward and reverse rate constants of reaction of n–mer and dimer, $k_{n,2}^+$ and $k_{n,2}^-$, respectively, are plotted against cluster class n for a few temperatures. Both rate constants quickly increase with n in the range of n greater than 100. $k_{n,2}^+$ is little affected by temperature, but $k_{n,2}^-$ has considerable positive dependence on temperature. Owing to this high rate of reverse reaction high temperatures, clustering no longer proceeds, little soot being formed. On the contrary, at very low temperatures, soot growth is again slow because the concentrations of colliding partners C_1 through C_5 are too low to maintain the cluster growth. Therefore, it is expected that there is a temperature range in which sooting proceeds most rapidly. Similar tendency holds for other colliding partners. In the same figure we may notice the presence of minimum $k_{n,2}^-$ in the cluster

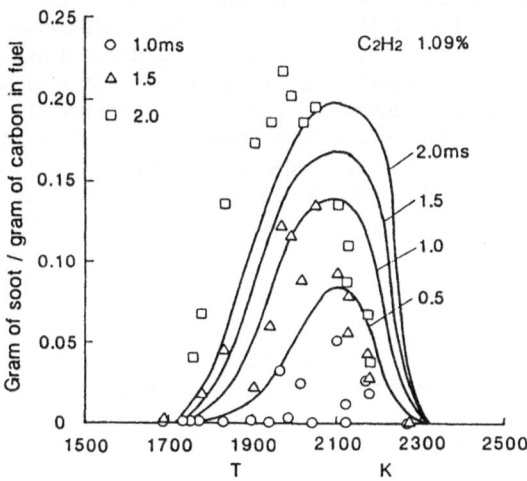

Fig. 5.21 Conversion ratio of carbon of fuel to soot obtained theoretically and experimentally at acetylene pyrolysis

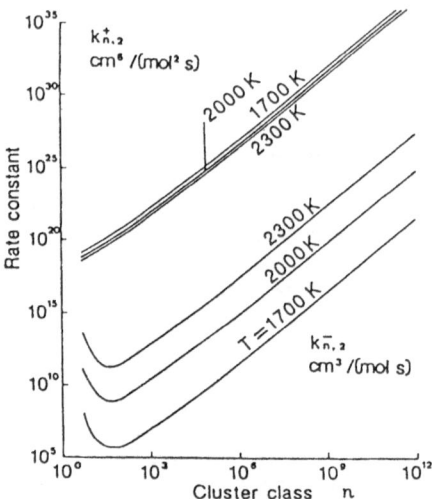

Fig. 5.22 Forward and reverse rate constants of cluster reaction of dimer and n-mer

class n between 10 and 100. This corresponds to critical diameter for nucleation in the droplet model.

The above cluster theory clearly describes rapid growth of soot particles but has two disadvantages; one is that elementary reactions are not dealt with fully kinetically, instead relying upon a simpler partial equilibrium method, and the other is that polyacetylenes, such as C_2H and C_3H, are ignored in the cluster reactions. Such species are likely condensable ones that may act as colliding partners at higher temperatures.

In an improved homogeneous nucleation theory by Yoshihara et al. [57], calculations for all gaseous reactions of hydrocarbons having two carbons in a molecule were performed in a fully kinetic manner and supplementary cluster reactions were included. In cluster reactions with polyacetylenes, radicals are more likely than neutral ones. On the assumption that C_2H is more stable than else among polyacetylene radicals, they included only C_2H as the most likely to be condensable and supplemented cluster–radical reaction of

$$C_n + C_2H = C_{n+2} + H$$

The rate constant was estimated from similar radical reactions. H formed in this reaction may promote formation of C_2H from acetylene in the following way.

$$C_2H_2 + H = C_2H + H_2$$

The above two reactions may proceed in a chain–like manner and accelerate production of C_2H and cluster growth. In Figs. 5.23 and 5.24 the calculated mass fractions of clusters, of some gaseous species and of soot yield are given against time for

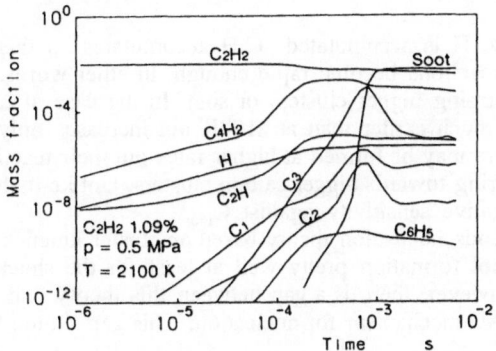

Fig. 5.23 Predicted histories of mass fractions of some gaseous species, clusters and soot at 2100K, at acetylene pyrolysis

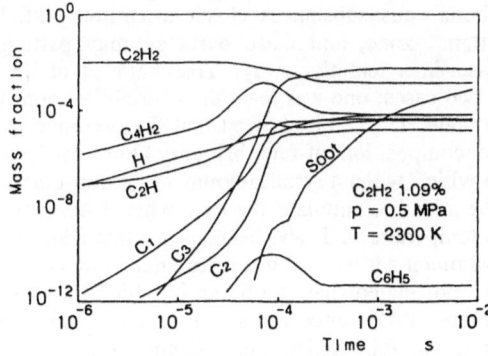

Fig. 5.24 Predicted histories of mass fractions of some gaseous species, clusters and soot at 2300K, at acetylene pyrolysis

temperatures of 2100K and 2300K, respectively, again for the case of shock–heated acetylene pyrolysis at a pressure of 0.5 MPa. In the case of 2100K, C_2H shows the highest concentration among condensables in earlier stages, while yields of carbon vapors C_1–C_3 begin to increase only lately and decrease with onset of soot. According to sensitivity analysis, the main path to form C_2H is

$$C_2H_2 + M = C_2H + H + M$$

in earlier stages, but the above reaction between acetylene and hydrogen atom and

$$C_4H_2 + H = C_2H_2 + C_2H$$

become significant once H is accumulated. C_2H accumulates in this way but is not consumed until cluster reactions become rapid enough. In other words, clustering itself is the bottle-neck in forming bigger clusters or soot. In the case at 2300K, Fig. 5.24, soot begins to increase much earlier than at 2100K but increases only gradually. This implies that small clusters may be formed at higher rates but their reactions with carbon vapors suppresses clustering towards bigger carbon clusters. Unlike in the case at 2100K, carbon vapors have negative sensitivity against C_{100}.

The above homogeneous nucleation theory based on cluster kinetics based on cluster kinetics may explain soot formation pretty well at least for the shock-heated case at higher temperatures. However, there is a gap between this theory and widely accepted theories based on the PAH mechanism for nucleation. This gap should be bridged in the future.

A sooting laminar premixed flame differs from the case when a uniform mixture is changed from a cold state to a hot constant state as in a shock-heated situation, at least in points as follows: One is that H radical preferentially diffused to the colder zones upstream may trigger production of polyacetylene radicals that will activate cluster growth to soot, as suggested by the above homogeneous nucleation theory. The other is that PAHs once formed in colder zones where burning are not completed may be decomposed to more carbonaceous substances closer to carbon particles in the hotter reaction zone and post-flame zone, and these particles may participate the cluster growth. To see if the case in a tentative way, Yoshihara et al. [57] made further calculations as before for two cases; one was that when certain amount of H radical was assumed to exist in initial state. The result showed that the presence of H in very early stages promotes thermal decomposition of fuel, bringing about earlier onset of sooting. The other was the case in which when a small amount of carbon clusters of 6-mer and 12-mer was assumed to be added to simulate the case when PAHs are dehydrogenated into clusters at high flame temperatures. It was been shown that addition of such clusters makes the onset of sooting much earlier and increases final soot yield depending on its amount. Hence, it might be postulated that in a flame in which the mixture is subject to slower change of temperature, PAHs once formed in less hot zones may promote the cluster growth in a hotter flame zone and the post-flame zone.

(Makoto Ikegami and Yoshinobu Yoshihara)

5.5 Growth and Destruction of Soot Particles

5.5.1 Growth of soot particles

The size of soot particles is mostly determined by the soot growth process after nucleation [15]. The chemical substances covering the soot particle tends to cause serious sickness such as lung cancer, bronchitis or asthma. Hence, in order to predict and control the size and surface substance of soot particles, one needs to know the soot growth mechanism. The growth of soot particles proceeds through two different processes, one being the coagulation and the other being surface growth.

In this subsection, soot growth by coagulation is first summarized briefly at first and then, the surface growth mechanism of primary soot particles is discussed on the basis of past studies.

A typical example of soot growth by coagulation is shown in Fig. 5.25 which

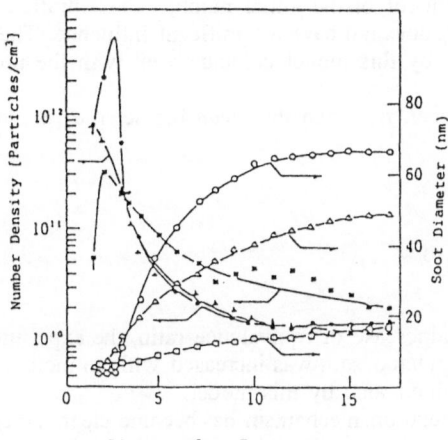

Distance from Burner port

Fig. 5.25 Change of the number density and soot diameter with distance from burner port,
Equivalence ratio: □ 2.2, ▵ 2.5, ○ 2.9

indicates the variation of the soot number density and the soot diameter in the post flame region of propane/O_2 flame for various equivalence ratios [58]. It is noted that the soot diameter increases with a decrease of the number density and also increases with an increase of an equivalence ratio. It is also noted that the number density approaches constant value around 10^{10}/cm³ in the downstream region even if an initial concentration is different for each equivalence ratio. These experimental results can be explained by the following coagulation model.

When the soot particle is smaller than the mean free path, the coagulation rate of soot particles can be expressed by Hidy's free molecular coagulation model [59] as

$$dn/dt = -(6/5)k_{th}f_V^{1/6}n^{11/6}$$ (5.1)

where

$$k_{th} = (5/12)(3/4\pi)^{1/6}(6kT/\rho)^{1/2}\gamma\alpha$$ (5.2)

where n denotes the number density, t the time, f_V the volume fraction of soot particles, ρ the particle density, γ the correction factor of collision cross section and α the factor of influence of the size distribution
Eq.(5.1) can be solved for n as

$$n = n_0(1 + 9.03 \times 10^{-13}n_0^{5/6}T^{1/2}f_V^{1/6}t)^{-6/5}$$ (5.3)

When an initial number concentration n_0 is very large,

$$n=2.84\times10^{14}(T^{1/2}f_V^{1/6}t)^{-6/5} \qquad (5.4)$$

From this equation, the number density approaches the constant value of around $10^{10}/cm^3$ which is independent of the initial number concentration n_0 with time elapse and the volume fraction f_V does not have a significant influence. These calculated results and tendencies predicted by this model coincide well with the previous experimental results.

The relationship between f_V, n and the mean diameter d_p is

$$f_V=\pi n d_p^3/6 \qquad (5.5)$$

Hence,

$$d_p=((6/\pi)f_V/n)^{1/3} \qquad (5.6)$$

As f_V is increased with an increase of equivalence ratio, the experimental results that the final diameter of the coagulated soot was increased with an increase of the equivalence ratio could also be explained also by this model.

In summary, soot coagulation mechanism has become clear and consequently the soot growth by coagulation could be well explained by the mathematical model.

The formation of single soot particles consists of nucleation and surface growth. Tesner indicated that there are clear distinction between nucleation and surface growth by showing that surface growth can occur at temperatures below the nucleation temperature [60]. Narasimhan and Foster showed that an activation energy of the soot surface growth is 244 kJ/mol which is lower than that of the nucleation rate [61]. Since then, a lot of work has been carried out on the surface growth of soot particles. Amongst these a great number of studies focused on an identification of the growth species.

Tesner suggested that simple hydrocarbons such as acetylene and methane were responsible for the surface growth reaction because of their high concentration [60]. Homann and Wagner concluded that acetylene and polyacetylene were the main growth species because the concentration of these species was decreasing in the soot growth region and H/C ratio was kept constant even if C_2H_2 or C_4H_2 was added to the post flame region [62]. D'Alessio et al. measured an axial distribution of soot and PAHs in the post flame region of CH_4/O_2 premixed flame and found that PAH did not tend to be dissipated in the soot growth region, but approached constant value [63]. From the result, he concluded that PAHs was not the soot growth components, but by–products during the soot formation. In addition to PAH, he suggested that the polyacetylene may not become the growth species, as an addition of C_2H_2 in the post flame region did not influence the soot growth rate.

Harris et al. reported that the mass of soot could be increased through the reaction with acetylene because only acetylene was present in high enough concentrations to account for the mass increase provided by surface growth and the calculated concentration of C_2H_2 assuming the reaction probability of 2×10^{-3} given by Tesner agreed with experimental results [64].

They rejected the possibility of polyacetylene as the growth species because the measured concentration of polyacetylene was not high enough to explain the measured soot growth rate unless one could assume the rapid equilibrium between C_2H_2 and polyacetylene and a reaction probability near 1.0. They did not support the possibility of PAH either because the growth rate was similar between benzene flames and aliphatic fuel flames although PAH concentration was about 100 times higher in benzene flames than in flames of aliphatic fuels.

Bockhorn et al. measured the various hydrocarbons in the post flame region of flat low–pressure flames of different hydrocarbons and compared them with the collision model. They showed that the small hydrocarbons under C_6 such as C_2H_2, C_4H_2, C_6H_2 could become the growth species as the calculated concentration of hydrocarbon over C_6 on the basis that the collision model exceeded the measured concentration even if the reaction probability was assumed to be 1.0 [65].

As mentioned above, many studies support C_2H_2 as the main growth species. However, there are some studies that support PAH as the growth species instead of acetylene. Prado et al. measured the concentration of soot and PAH along the turbulent diffusion flame and showed that PAH appeared in advance of the generation of soot and reached maxima early in the flame after which PAH decayed rapidly and soot decayed much more slowly [66]. From these results, they suggested that PAH might be an important growth species. Lam, Howard et al. examined the formation and growth of soot in an atmospheric pressure jet–stirred/plug–flow reactor. They showed that the possibility of tar, which was CH_2Cl_2 soluble material with molecular weight greater than naphthalene could not be excluded as a growth species in addition to C_2H_2 [2]. They indicated the possibility of the new growth process, that is, C_2H_2 adds to tar at first and next, the tar adds to soot. Weiner and Harris observed the large soot precursors in the post flame region of a premixed $C_2H_4/O_2/Ar$ flat flame [67]. They found that hydrocarbon molecules ranging from 500 amu to 1000 amu were significantly dissipated in the soot growth region. From this fact, they suggested that PAH of 500 amu to 1000 amu might be important growth species. They also suggested that the large PAH has a high reaction probability with soot because of a strong van der Waals force and the long contact time with the soot particles.

Recently, Frenklach indicated by computer simulation that the surface growth of soot mass was primarily determined by two processes, acetylene addition via the H–abstraction/C_2H_2 addition sequence and PAH condensation on the particle surface [30]. The relative contribution of each of these processes appears to change with experimental conditions.

In summary, C_2H_2 seems to be the growth species from its high concentration in the post flame region whilst PAH seems to be the growth species from its high reaction probability with soot particles. The previous studies could not, however, identify the main growth species conclusively.

Attentions of previous works have mostly focused on the soot growth just behind the flame region. In this region, however, nucleation and coagulation occur simultaneously and many radicals are generated and react with each other. As a result, it is difficult to extract only the surface growth phenomena from such a complicated reaction zone. In order to examine the surface growth in more detail, we need to examine the soot growth in the post flame region some distance behind the flame where only the surface growth can be observed.

Sadakata and co–workers measured the variation of the size distribution of the primary soot particles and the concentration of aliphatic and aromatic hydrocarbons in the post flame region some distance from the flame. A small amount of C_2H_2 and PAH were introduced respectively in the post flame region to examine the effect of these species on growth. A premixed Merker type burner was mainly used in their experiment. The premixed CH_4/O_2 flat flame was formed under atmospheric condition on the honeycomb ceramics screen located at 20 mm above the burner port for flame stabilization. The cold gas velocity at the burner port was 9.5cm/s at equivalence ratio $\phi=2.0$. Temperature in the post flame zone was controlled from 700K to 1600K by an electrical furnace. Flat temperature distribution could be obtained between 100mm and 400mm downstream from the burner port. The maximum flame temperature at $\phi=2.0$ was 2111K.

Figure 5.26 shows a change of the mean particle diameter D_g, the geometric standard deviation of the particle size σ_g and number density of the primary soot particles with residence time from burner port when the post flame zone was not heated. D_g was gradually increased while σ_g was decreased with time elapsed. On the other hand, the number density of particles N was almost constant. From these results, it was ascertained that the growth of particles is not due to the coagulation but due to the surface reaction. Figure 5.27 shows the change of the size distribution of the primary particles with residence time. It should be noted that the proportion of small particles was sharply reduced with time elapsed and consequently this led to the increase of D_g and decrease of σ_g.

From this result, one can assume that small particles are more reactive rather than large particles. This assumption is backed by the fact that the small particles correspond to young age soot which must be more reactive than old age large soot.

The concentration of hydrocarbon with low molecular weight from C_1 to C_6 for different reaction times are shown in Fig. 5.28. All of the measured species were gradually dissipated as the residence time in the post flame zone was increased. However, it should be noted here that concentrations of C_2H_2 and C_6H_6 were of a higher order than that of other species except for CH_4. It should also be noted that dissipation of C_2H_2 for residence time from 25ms to 100ms was significant compared with other hydrocarbons, while the dissipation of C_6H_6 was slight. It seems that C_2H_2 contributes directly or indirectly to the growth of soot particles in the first half of soot growth region as Homann [62], Harris [64], and Bockhorn [65] indicated in their previous studies.

Concentrations of PAHs for different residence times from the flame are shown in Fig. 5.29. The maximum detectable PAHs were those with 6 benzene rings. The concentration of each PAH was represented by the ion intensity since the quantitative

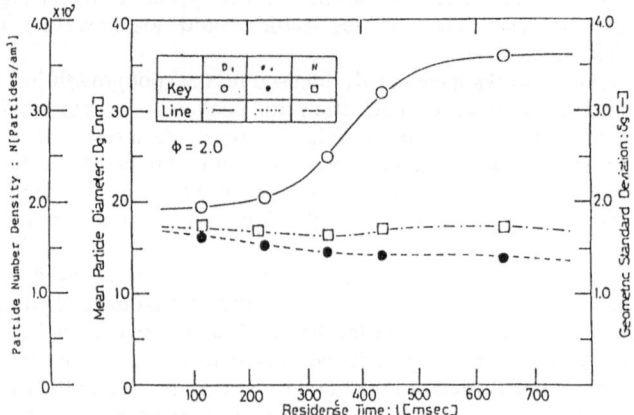

Fig. 5.26 Change of D_g, σ_g and number density of the primary soot particles with residence with residence time from premixed flame burner port

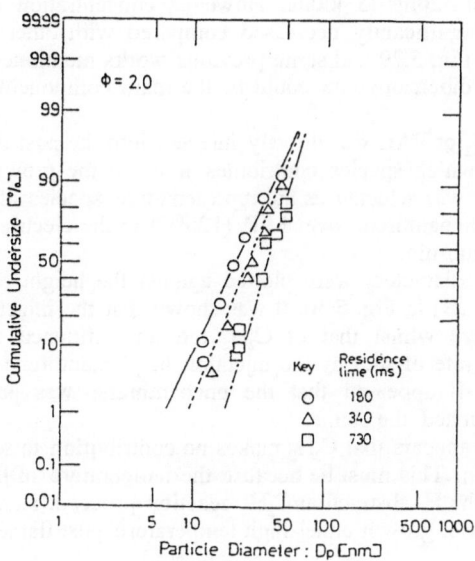

Fig. 5.27 Change of size distribution of the primary particles with residence time

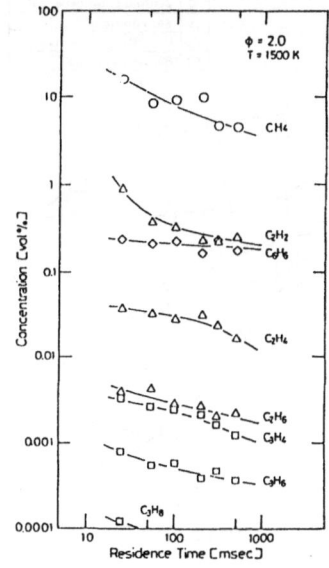

Fig. 5.28 Change of hydrocarbon concentration with residence time in the port flame zone

estimation of concentration for all PAH species was impossible. All species dissipated for residence time from 20ms to 400ms. However, concentration of phenanthrene and dibenzopyrene were significantly decreased compared with other species.

From the result of Fig. 5.29 and some previous works mentioned above, PAHs such as phenanthrene and dibenzopyrene could be the main components for soot growth as well as acetylene.

Finally, either C_2H_2 or PAH was directly injected into the post flame region of 973K in order to ascertain which species contributes more to the growth of soot. Here, the gaseous phenanthrene was selected as the representative species of PAH and generated by heating the solid phenanthrene over M.P. (125°C) in the electrical furnace. The feed rate of C_2H_2 was 53 ml/min.

The mean particle diameters were plotted against the height of the injection point above the burner (H.A.B) in Fig. 5.30. It was shown that the injection of phenanthrene increased the soot size whilst that of C_2H_2 did not influence the soot size at all. However, the growth rate of soot by the injection of phenanthrene was reduced as the H.A.B. became low. It appeared that the phenanthrene was partly cracked as the injection point approached the flame.

From this result, it appears that C_2H_2 makes no contribution to soot growth under our experimental condition. This must be because the temperature of the post flame region was too low to make the H–abstraction/C_2H_2 reaction proceed. Hence, the possibility of C_2H_2 contribution to soot growth under high temperature post flame condition could not

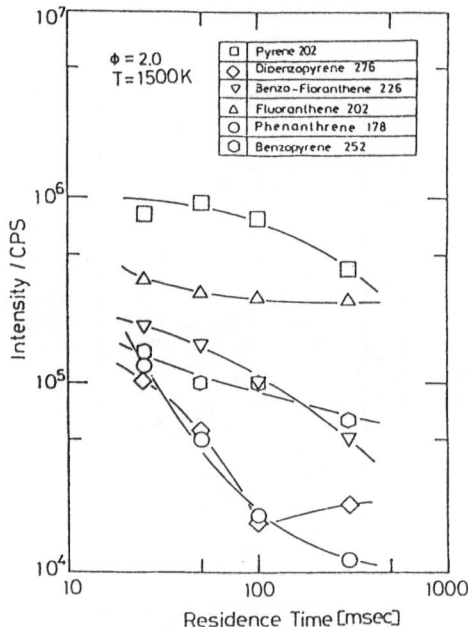

Fig. 5.29 Change of polycyclic aromatic hydrocarbons concentration with residence time in the post flame zone

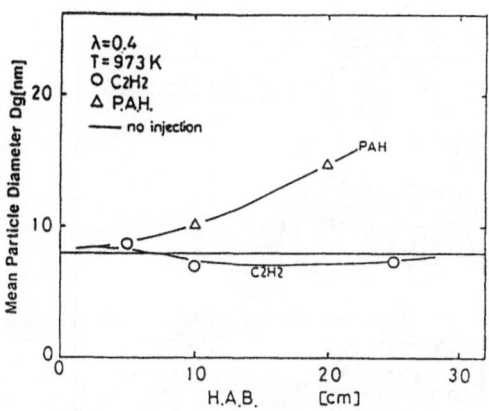

Fig. 5.30 Effect of C_2H_2 and PAH injection

be 'rejected solely on the basis of this experiment.

From these experimental results, the growth mechanism of the primary soot particles in the post flame zone was summarized as PAH such as phenanthrene and dibenzopyrene. Their radicals formed in the flame zone react mainly with the smaller part of primary soot particles in the post flame zone and consequently contribute to the growth of the mean particle size and the reduction of the deviation of the size distribution. (Masayoshi Sadakata)

5.5.2 Soot oxidation

The mechanism and kinetics of the soot oxidation has been one of the most significant areas in the research of soot formation. The main points of previous studies are summarized below.

Experiments of the soot particle oxidation by oxygen have shown that the surface oxidation rate is basically the same as that of graphite. There has been a semi-empirical formula given by Nagle and Strickland-Constable [37] for rate of surface oxidation of pyrolytic graphite. Park and Appleton made shock-tube experiments for soot oxidation [68] and they found that the above formula was true even for soot particles formed in flames, although exceptions were noted in some fuel-rich cases. Figure 5.31 shows the measured specific oxidation rates of soot in terms of unit surface area of soot against reciprocal temperature, together with predictions using the Nagle and Strickland-Constable formula. Also included are other measured rates of soot oxidation obtained at a variety of flames [69-71]. It appears that most data are fairly well correlated with the semi-empirical formula.

The Nagle and Strickland-Constable formula assumes that the surface of graphite consists of reactive sites without coverage of oxides and less reactive sites but capable of desorption. It is also assumed that each site has its own rate dependent on temperature and oxygen partial pressure. Denoting temperature as T and the fraction of surface

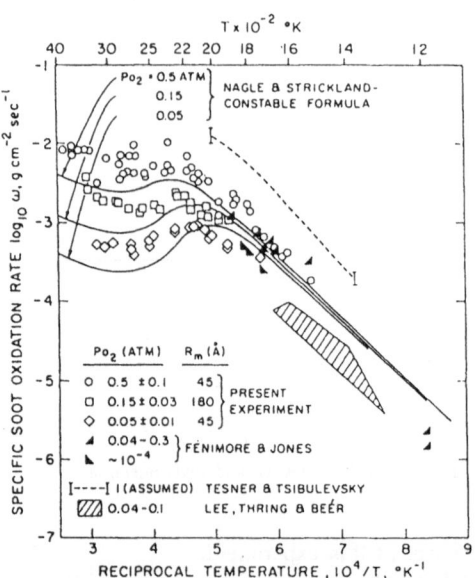

Fig. 5.31   Specific soot oxidation rate versus temperature relations,
from [68] with permission

occupied by the reactive sites as x, the specific oxidation rate in terms of surface area is assumed to be

$$\frac{\omega}{12} = \frac{k_A p_{O2} \chi}{1+k_2 p_{O2}} + k_B\, p_{O2}(1-\chi) \qquad\qquad gcm^{-2}s^{-1} \qquad (5.7)$$

where p_{O2} is the oxygen partial pressure, and

$$\chi = \frac{1}{1 + k_T/k_B p_{O2}} \qquad\qquad\qquad (5.8)$$

$$k_A = 20\exp(-15\ 100/T) \qquad\qquad\qquad (5.9)$$

$$k_B = 4.46\times10^{-3}\exp(-7\ 640/T) \qquad\qquad\qquad (5.10)$$

$$k_T = 1.51\times10^5\exp(-48\ 800/T) \qquad\qquad\qquad (5.11)$$

$$k_Z = 21.3\exp(2\ 060/T) \qquad\qquad\qquad (5.12)$$

In these relationships the cgs unit is used, and pressure and temperature are in atm and in Kelvin, respectively. This oxidation reaction is in the first order at lower oxygen partial pressures and in the zero order at higher oxygen partial pressures. At a given pressure, the surface oxidation rate exponentially increases with temperature until 2000K, but a further increase in temperature slows down and decreases the oxidation rate beyond a certain temperature. In the temperature range of around 2500K, thermal rearrangement from reactive to unreactive sites takes place fast enough on the surface of the soot. At the same time, the reaction rate is dependent on the oxygen partial pressure, unlike at lower temperatures below 2000K. At much higher temperatures the oxidation rate again increases with temperature and becomes to be in the first order.

The above formula seems helpful in giving the soot oxidation rate over a range of conditions. Some other expressions have been proposed for the same purpose of predicting the rate of soot oxidation with oxygen:

$$\omega = 1.085 \times 10^4 p_{O2} T^{-1/2} \exp(-19800/T) \quad \text{by Lee [69]} \quad (5.13)$$

$$\omega = \frac{3.05 \times 10^6 p_{O2}{}^2 \exp(-29\,000/T)}{1 + 3.10 \times 10^{10} p_{O2}{}^2 \exp(-29\,300/T)} \quad \text{by Magnussen [72]} \quad (5.14)$$

In diffusion flames the soot particle oxidation by oxygen proceeds close to the flame boundary. Inside the flame boundary the oxidation of soot may occur in the presence of the penetrated oxygen and other species, and the situation becomes closer to that in a premixed particle flame. In this situation there are many other potential oxidizers other than oxygen in burned gas. It is known that in the case of graphite oxidation, atomic oxygen and hydroxyl radical have a high reaction capability. Fenimore and Jones [70], who made experiments using premixed flames, suggested that the soot reaction rate is only marginally dependent on O_2 and that hydroxyl radical OH is the principal oxidizer. In the burned gas in fuel–rich flames, the OH concentration is much higher than that of O_2, and the O concentration is comparable with that of O_2. Neoh et al. [73] carried out experiments on a fuel–rich flat burner to determine which species was the most likely oxidizer among O_2, O, OH, CO_2 and H_2O that may be formed during combustion. They reached the conclusion that the OH radical is the principal oxidizer of soot particles under fuel–rich conditions.

This conclusion was based on a discussion as follows: From the measured reaction rate and the frequency of collision between a soot particle and an oxidant molecule as determined from kinetic theory, one can obtain reaction efficiency, or reaction probability, defined as the probability at which a carbon atom can be removed from the soot particle at a single collision of the oxidizer molecule with the soot particle. Neoh et al. calculated the collision efficiencies of several oxidizers from their own experimental data of flat–burner experiments. Figure 5.32 shows the results for O_2 and OH against the distance from the exit of the burner, for several cases when an inlet gas produced at an equivalence ratio of 2.10 by a primary burner is burned after dilution, finally reaching an equivalence ratio of $2^\circ\phi$ at the outlet of the test flat burner. The collision efficiency of O_2 exhibits great changes from position to position and has a wide spread depending on burning conditions. A similar tendency was noted for O and CO_2. Unlike these cases, the collision efficiency of OH ranges only from 0.2 to 0.4 almost irrespective of the mixture conditions. Hence, other candidates, such as O_2, O, H_2O and CO_2, were not sufficient to be the main oxidizer. Taking into account experimental ambiguities in determining the collision efficiencies, it can safely be stated from these results that the most likely oxidizer is not O_2 but OH, at least under fuel–rich conditions where O_2 is less abundant. However, the value of the OH collision efficiency is much higher than 0.1 given by Fenimore and Jones [70]. This difference was attributed to the

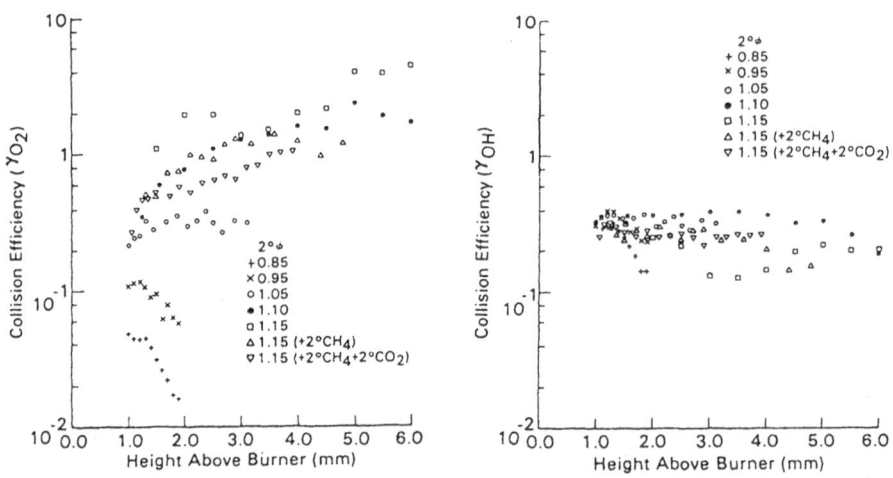

Fig. 5.32   Collision efficiencies of soot oxidation by O_2 and OH,
from [73] with permission

fact that the particle sizing was made optically. Since the soot particles are chain–like agglomerates, the particle size obtained by the optical method underestimated surface area, and the value became much lower once soot particles were sized using electron microscopy. Thus, Noeh et al. reached the conclusion that OH is the oxidizer of soot particle with a collision efficiency having a value between 0.28 and 0.13 in the oxidation of soot in atmospheric pressure flame between 1575K and 1865K and an oxygen mole fraction between 0.05 and 10^{-5}.

There is also a report which shows a discrepancy with the Nagle and Strickland–Constable semi–empirical formula. Yoshizawa and Matuoka [74] measured the oxidation rates of soot particles using a special shock tube. A remarkable feature of their experimental apparatus is a soot generator directly connected with the shock tube for introducing the soot particles into the test section as it is floating in burned gas. The result shows that the oxidation rates are far greater than those by the semi–empirical formula. They also measured the oxidation rates of carbon powder using the same shock tube. The result was correlated with the formula when it was measured in oxygen–argon mixtures. The addition of hydrogen to the mixtures increased the oxidation rates of carbon powder, however the values were yet smaller than those of soot particles. Based on these results, they concluded that the oxidation rates of soot particles in flames may be greater than those of carbon powder and industrially processed carbons such as channel black and furnace black.

Soot oxidation has also been of great concern from a practical point of view, such as staged combustion for low NO_x emission, the use of fuels having higher C/H ratios and regeneration of particulate traps for diesel smoke. In a two–staged combustion, fuel–rich burning takes place in the first stage and a lean burning follows in the subsequent stage. Another example is that of boilers and furnaces in which soot particles should be fully burned out after they act as the promoter of radiative heat transfer. Furthermore, in the combustion chambers of diesel engines, an amount of soot generated during combustion should be burned within a few milliseconds. Otherwise, unburned soot particles are given off outside emitting heavy smoke. In these cases, turbulent combustion is usual and soot oxidation is closely linked with turbulent eddies. It is necessary to clarify this problem in greater detail.   (Yoshio Yoshizawa)

5.5.3 The effect of oxygen addition on sooting in diffusion flames

This subsection deals with the effect of oxygen addition to the fuel side of diffusion flames on soot formation. In the last few years, Glassman et al. [76–79] have studied soot formation for several fuels including ethene and paraffins using a counter–flow type burner [75], and found that the ethene–air diffusion flame is strongly promoted by addition of a small amount of oxygen to the fuel side [78,79]. They concluded that this effect is caused not only by increased temperature, but also by enhanced pyrolysis under the catalyzing effect of oxygen. On the other hand, in diffusion flames of saturate hydrocarbons and air, oxygen addition did not have such a strong promoting effect.

For ethene flames, such a promoting effect of oxygen is also reported by other authors [80–84]. The effects of oxygen addition were also studied for other fuels: Wright [80] and Wey, Powell and Jagoda [85] observed that there is a promoting effect for propane. Dearden and Long [81] and Jones and Rosenfeld [82] found that there is no promoting effect for propane. Saito, Williams and Gordon reported that there is not such a strong effect for methane [86].

More recently, Kono, Sugiyama and Kishi studied the effect of oxygen concentration in the fuel side by changing flame temperature independently; the flame temperature was controlled over a wide range of temperature by changing oxygen concentration in the oxidizer. They also employed a counter–flow diffusion flame, because of its relatively simple structure. Acetylene is used as fuel, because the effect of oxygen addition has not been elucidated fully. They also proposed an experimental formula based on their experiments. This formula includes the time from the onset of soot formation as an independent parameter.

Their experiments were carried out on a porous cylinder burner that was laid under uniform flow of air in a wind tunnel. Fuel emerged through the surface of the cylinder. The diffusion flame was stabilized around the burner. This type of flame is relatively simple and thoroughly studied by Tsuji and Yamaoka [75], and therefore easy to deduce the soot formation rate.

The fuel was acetylene vitiated by nitrogen, the flow rates were $2.5 \times 10^{-3} m^3/s$ and $1.48 \times 10^{-2} m^3/s$. The oxidizer was air in which oxygen concentration was varied from 20.9% to 27.3%. The amount of oxygen addition to the fuel was either 2.2% or 3.3%. All measurements were carried out one–dimensionally along the center stream line which included the front stagnation point. The soot volume fraction ϕ was determined by an optical method based on the Mie light scattering theory. An argon–ion laser was used as a light source. The laser beam was focused at the measuring point. The scattering light intensity was measured by using a photomultiplier which was adapted on a rotating table. The diameter of soot particles was determined from the ratio of intensity at two different angles. In this method, soot was assumed to be a monodisperse sphere. Thus, the number density and volume fraction of soot were determined. Stable species, such as THC, N_2, O_2, H_2, CO, and CO_2 were also measured by gas chromatography. These results were used for the calculations of Farmer's model.

Soot formation rate ϕ was determined by Eq. (5.15) which V_{soot} is the sum of gas

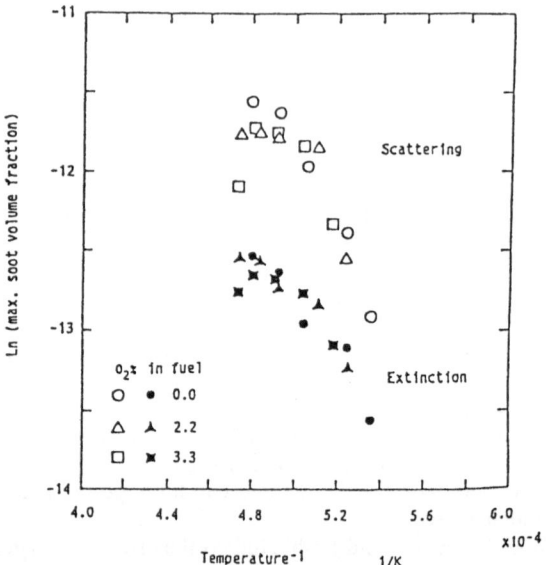

Fig. 5.33 Temperature dependence of soot volume fraction

velocity (u,v) and thermophoresis velocity (u_p,v_t). In the present case, from the symmetry of the flame and the equation of the continuity of fluid, Eq. (5.15) yields Eq. (5.16).

$$\dot{\phi} = \nabla\,(\phi V_{soot}) \tag{5.15}$$

$$\dot{\phi} = (u+u_t)\frac{\partial\phi}{\partial x} + \phi\frac{\partial u_t}{\partial x} + \phi\frac{u}{T}\frac{\partial T}{\partial x} \tag{5.16}$$

The right–hand terms of Eq. (5.16) are obtained from the experiments. LDV measurement gives u, but only a poor LDV signal is obtained at the positions where soot volume fraction is high. Therefore, an assumption was employed to determine the velocity profile, i.e. $u+u_t$ is a linear function of the distance from the burner surface, and equals zero at the position where the soot volume fraction shows maximum value. The term u_t was calculated from the temperature profile.

Figure 5.33 shows the relation between ϕ_{max} and T_{max}. T_{max} is presented in reciprocal value. It is obvious that there is a one to one relation between ϕ_{max} and T_{max}. The oxygen addition to the fuel side does not seem to have a direct effect. Although it raises the flame temperature, most plots shown in this figure still remain on the same line.

Based on these results, an experimental formula was proposed which included the time from the onset of soot formation as an independent parameter. These are given by Eqs. (5.17) and (5.18). The new parameter t is given by Eq. (5.19).

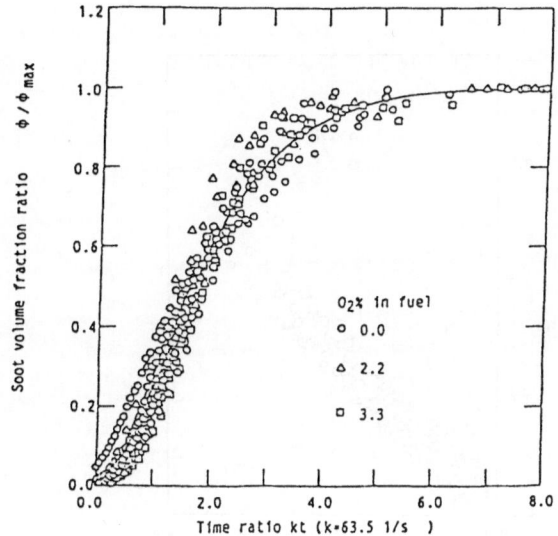

Fig. 5.34 Time dependence of soot volume fraction

$$\phi = \phi_{max}(1 - \exp(-kt) - kt\exp(-kt)) \tag{5.17}$$

$$\dot{\phi} = \phi_{max} k^2 t \exp(-kt) \tag{5.18}$$

$$t = \int_{x(\phi=0)}^{x} \frac{dx}{u + u_t} \tag{5.19}$$

Figure 5.34 shows the results of ϕ. The line in this figure shows Eq. (5.17). This equation formally corresponds to the formation rate in a two stage chemical reaction which has a common rate constant k. The best fit is given when k is 63.5 s^{-1}. As for Eq. (5.18), a similar agreement is given by k=63.5 s^{-1}.

Of course k is merely an apparent rate constant, and therefore the physical meaning of Eqs. (5.17) and (5.18) are not very clear. Nevertheless, these equations do describe the profile of ϕ and $\dot{\phi}$ in the flame under various oxygen conditions in a uniform manner. From the above–mentioned result, the controlling factor of soot formation in diffusion flames is T_{max} and t, because ϕ_{max} is dependent only on T_{max} as shown in Fig. 5.33.

Generally, soot formation may be divided in two physical processes; one being nucleation and the other being surface growth. Figure 5.34 shows the correlation between number density N and ϕ. In the present case, N is lower than 3×10^{15} m^{-3}. Then, the rate of free molecule collision is not so high and N is controlled by the nucleation process only. Hence, nucleation seems to be the most significant process to determine soot formation. Strict discussion needs more information about interaction among soot

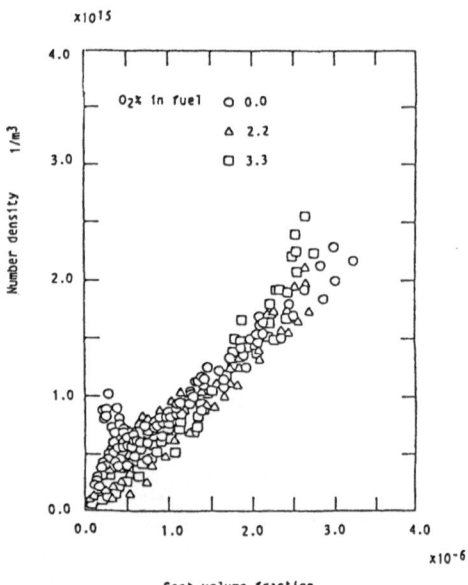

Fig. 5.35 Correlation between soot volume fraction and number density

particles such as electrostatic force. In Fig. 5.35, there are some points which do not fit on the straight line. These points are found nearby the maximum temperature region, and so, there may be some effects of soot ionization.

Unlike the result by Glassman et al., N does not show a constant value in the flame. Therefore, accurate surface growth rate was not calculated. But, apparent surface growth rate (soot formation rate/soot surface area) shows features which are similar to their result.   (Michikata Kono, Gen Sugiyama and Takeyuki Kishi)

References

[1] Ikegami M, Shioji M, and Sakagami Y (1986) Measurement of particles in flames by laser homodyne technique. Bulletin of JSME. 29-247: 156-163

[2] Lam FW, Howard JB, and Longwell JP (1988) The behavior of polycyclic aromatic hydrocarbons during the early stages of soot formation. 22nd Symp.(Intl.) on Combust., The Combustion Institute, Pittsburgh, pp 323-332

[3] Bonne U, Homann KH, and Wagner HGg (1965) Carbon formation in premixed flames. 10th Symp. (Intl.) on Combust., The Combustion Institute, Pittsburgh, pp 503-512

[4] Frenklach M, Clary DW, Gardiner WC Jr, and Stein SE (1985) Detailed kinetic

modeling of soot formation in shock–tube pyrolysis of acetylene. 20th Symp.(Intl.) on Combust., The Combustion Institute, Pittsburgh, pp 887–901

[5] Calcote HF (1981) Mechanisms of soot nucleation in flames – a critical review. Combust. Flame 42: 215–242

[6] Jensen DE (1974) Prediction of soot formation rates: a new approach. Proc. R. Soc. Lond. A.338: 375–396

[7] Yoshihara Y and Ikegami M (1989) Homogeneous nucleation theory for soot formation. JSME International Jour. II–32–2: 273–280

[8] Gerhardt Ph, Löffler S, and Homann (1988) The formation of polyhedral carbon ions in fuel–rich acetylene and benzene flames. 22nd Symp. (Intl.) on Combust., The Combustion Institute, Pittsburgh, pp 395–401

[9] Fujiwara Y, Tosaka S, and Fukazawa S (1990) The microcrystal structure of soot particulates in the combustion chamber of precombustion type diesel engines. SAE Paper 901579: 1–8

[10] Lahaye J and Prado G (1981) Morphology and internal structure of soot and carbon blacks. Particulate carbon formation during combustion. (Ed. Siegel DC and Smith GW), Plenum Press: 33–51

[11] Donnet JB and Voet A (1976) Carbon Black – Physics, Chemistry and Elastomer Reinforcement. Marcel Dekker Inc.

[12] Marsh PA, Voet A, Mullens TJ, and Price LD (1971) Quantitative micrography of carbon black microstructure. Carbon, 9: 797–805

[13] Smith WR and Polley MH (1956) The oxidation of graphitized carbon black. Phys. Chem. 60: 689–691

[14] Heckman FA and Harling DE (1966) Progressive oxidation of selected particles of carbon black: further evidence for a new microstructural model. Rubber Chem. Technol. 39: 1–13

[15] Haynes BS and Wagner HGg (1981) Soot formation. Prog. Energy Combust. Sci. 7: 229–273

[16] Stein SE (1978) J. Phys. Chem. 82: 566

[17] Stehling FC, Frazee JD, and Anderson RC (1962) Mechanism of nucleation in carbon formation. 8th Symp. (Intl.) on Combust., Williams and Wilkins, Baltimore, pp 774–784

[18] Bittner JD and Howard JB (1981) Composition profiles and reaction mechanisms in a near–sooting premixed benzene/oxygen/argon flame. 18th Symp. (Intl.) on Combustion, The Combustion Institute, Pittsburgh, pp 1105–1116

[19] Frenklach M, Taki S, and Matula RA (1983) A conceptual model for soot formation in pyrolysis of aromatic hydrocarbons. Combust. Flame 49: 275–282

[20] Frenklach M, Taki S, Durgaprasad MB, and Matula RA (1983) Combust. Flame 54: 81–101

[21] Frenklach M (1987) In Complex Chemical Reaction Systems, Mathematical Modelling and Simulation (J. Warnatz and W. Jäger, Eds.), Springer–Verlag, Berlin, pp 2–16

[22] Harris SJ, Weiner AM, and Blint RJ (1988) Formation of small aromatic molecules in a sooting ethylene flame. Combust. and Flame 72: 91–109

[23] Frenklach M, Clary DW, Yuan T, Gardiner WC Jr, and Stein SE (1986) Mechanism of soot formation in acetylene–oxygen mixture. Combust. Sci. Technol. 50: 79–115

[24] Frenklach M and Warnatz J (1987) Detailed modeling of PAH profiles in a sooting low–pressure acetylene flame. Combust. Sci. and Tech. 51: 265–283

[25] Weissman M and Benson SW (1984) Int. J. Chem. Kinet. 16: 307–333

[26] Cole JA, Bittner JD, Longwell JP, and Howard JB (1984) Formation mechanism

of aromatic compounds in aliphatic flames. Combust. Flame 56: 51–70

[27] Frenklach M, Clary DW, Gardiner WC Jr, and Stein SE (1986) Effect of fuel structure on pathway to soot. 21st Symp. (Intl.) on Combust., The Combustion Institute, Pittsburgh, pp 1067–1076

[28] Frenklach M, Yuan T, and Ramachandra MK (1988) Energy & Fuels 2: 462–480

[29] Frenklach M (1988) On the driving force of PAH production. 22nd Symp. (Intl.) on Combustion, The Combustion Institute, Pittsburgh, pp 1075–1082

[30] Frenklach M and Wang H (1990) Detailed modeling of soot particle nucleation and growth. 23rd Symp. (Intl.) on Combust., The Combustion Institute, Pittsburgh, pp 1559–1566

[31] Wang H and Frenklach M (1989) In chemical and physical processes in combustion. Fall Technical Meeting of the Eastern States Section of the Combustion Institute, Albany, New York, October 1989, Paper 12

[32] Westmoreland PR (1989) Experimental and theoretical analysis of oxidation and growth chemistry in a fuel–rich acetylene flame, Ph.D. thesis, Massachusetts Institute of Technology; Westmoreland PR, Dean AM, Howard JB, and Longwell JP (1989) J. Phys. Chem. 93: 8171–8180

[33] Bockhorn H, Fetting F, Heddrich A, and Reh Ch (1986) Poster paper P51, presented at the 21st Symp. (Intl.) on Combust., Munich; Bockhorn H, Fetting F Wenz H (1983) Ber. Bunsenges. Phys. Chem. 87: 1067–1073

[34] Kee RJ, Grcar JF, Smooke MD, and Miller JA (1985) Sandia Report No. SAND85–8240, December 1985

[35] Wieschnowsky U, Bockhorn H, and Fetting F (1988) Some new observations concerning the mass growth of soot in premixed hydrocarbon–oxygen flames. 22nd Symp. (Intl.) on Combust., The Combustion Institute, Pittsburgh, pp 343–352

[36] Harris SJ and Weiner AM (1983) Determination of the rate constant for soot surface growth. Combust. Sci. Tech. 32: 267–275

[37] Nagle J and Strickland–Constable RF (1962) Oxidation of carbon between 1000–2000°C, Proceedings of the Fifth Conference on Carbon, Pergamon Press, London: 154–164

[38] Ebert LB, Scanlon JC, and Clausen CA (1988) Energy Fuel 2: 438

[39] Stein SE and Fair A (1988) High–temperature stabilities of hydrocarbons. J. Phys. Chem. 89: 3714–3725

[40] Kern RD and Xie K (1991) Shock tube studies of gas phase reactions preceding the soot formation process. Energy Combust. Sci. 17: 191–210

[41] Gerhardt Ph and Homann KM (1990) Ions and charged soot particles in hydrocarbon flames. 2. Positive aliphatic and aromatic–ions in ethylene/oxygen flames. J. Phys. Chem. 94: 5381–5391

[42] Howard JB, McKinnon JT, Makarovsky Y, Lafleur A, and Johnson ME (1991) Fullerenes C_{60} and C_{70} in flames. Nature 352: 139–141

[43] Zhang QL, O'Brien SC, Heath JR, Liu Y, Curl RF, Kroto HW, and Smalley RE (1986) Reactivity of large carbon clusters: Spheroidal carbon shells and their possible relevance to the formation of morphology of soot. J. Phys. Chem. 90: 525–528

[44] Frenklach M and Ebert LB (1988) Comment on the proposed role of spheroidal carbon clusters in soot formation. J. Phys. Chem. 92: 561–563

[45] Harano A, Kinoshita J, and Koda S (1990) Decomposition of gaseous hydrocarbons in a laser–induced plasma as a novel carbonaceous source for cluster formation. Chem. Phys. Lett. 172: 219–223

[46] Rohlfing EA (1990) High resolution time–of–flight mass spectrometry of carbon and carbonaceous clusters. J. Chem. Phys. 93: 7851–7862

[47] Baum Th, Loeffler S, Weilmuenster P, and Homann KH (1991) Polycyclic aromatics, fullerenes and soot particles as charged species in flames. American Chemical Society; Division of fuel Chemistry Symposium Vol. 36: 1533–1538

[48] Rohlfing EA (1988) Optical emission studies of atomic, molecular, and particle carbon produced from a laser vaporization cluster source. J. Chem. Phys. 89: 6103–6112

[49] Bastin E, Delfau J, Reuillon M, Vovelle C, and Warnatz J (1988) Experimental and computational investigation of the structure of a sooting C_2H_2–O_2–Ar flame. 22nd Symp. (Intl.) on Combust., The Combustion Institute, Pittsburgh: 313–322

[50] Sano T (1992) Trans. Jpn. Soc. Mech. Eng., (in Japanese), 58–554 B (in print)

[51] Burcat A (1984) Combustion Chemistry, Springer–Verlag: 455–504

[52] Stein SE and Barton BD(1981) Thermo. Acta 44: 265–281

[53] Bahn GS (1973) NASA CR-2173

[54] Benson SW (1971) Methods for the estimation of thermochemical data and rate parameters. John Wiley & Sons

[55] Keck JC and Gillespie D (1971) Rate–controlled partial equilibrium method for treating reacting gas mixtures. Combust. Flame 17: 237–241

[56] Bauer SH and Frurip DJ (1977) Homogeneous nucleation in metal vapors. 5. a self-consistent kinetic model. Jour. Phys. Chem. 81-10: 1015–1024

[57] Yoshihara Y, Natake S, and Ikegami M (1992) Kinetics of the soot cluster formation at high temperatures. Trans. Jpn. Soc. Mech. Eng. (in Japanese), 58–549 B: 1557–1565

[58] Prado G, Jagoda J, Neoh K, and Lahaye J (1981) A study of soot formation in premixed propane/oxygen flames by in–situ optical techniques and sampling probes. 18th Symp.(Intl.) on Combust., The Combustion Institute, Pittsburgh, pp 1127–1136

[59] Hidy GM and Brock JR (1970) The Dymanics of Aerocolloidal Systems. Pergamon Press, Oxford

[60] Tesner PA (1962) The activation energy of gas reactions with solid carbon. 8th Symp.(Intl.) on Combust., Williams and Wilkins, Baltimore, pp 807–814

[61] Narasimhan KS and Foster PJ (1965) The rate of growth of soot in turbulent flow with combustion products and methane. 10th Symp.(Intl.) on Combust., The Combustion Institute, Pittsburgh, pp 253–257

[62] Homann KH and Wagner HGg (1967) Some new aspects of the mechanism of carbon formation in premixed flames. 11th Symp. (Intl.) on Combust., The Combustion Institute, Pittsburgh, pp 371–379

[63] D'Alessio A, Di Lorenzo A, Sarofim AF, Beretta F, Masi S, and Venitozzi C (1974) Soot formation in methane–oxygen flames. 15th Symp.(Intl.) on Combust., The Combustion Institute, Pittsburgh, pp 1427–1438

[64] Harris SJ and Weiner AM (1983) Surface growth of soot particles in premixed ethylene/air flames. Combust. Sci. Tech., 31: 155–167

[65] Bockhorn H, Fetting F, Heddrich A, and Wannemacher G (1984) Investigation of the surface growth of soot in flat low pressure hydrocarbon oxygen flames. 20th Symp. (Intl.) on Combust., The Combustion Institute, Pittsburgh, pp 979–988

[66] Prado GP, Lee ML, Hites RA, Hoult DP, and Howard JB (1976) Soot and hydrocarbon formation in a turbulent diffusion flame. 16th Symp.(Intl.) on Combust., The Combustion Institute, Pittsburgh, pp 649–661

[67] Weiner AM and Harris SJ (1989) Optical detection of large soot precursors. Combust. Flame, 77: 261–266

[68] Park C and Appleton JP (1973) Shock–tube measurements of soot oxidation rates, Combust. Flame, 20: 369–379

[69] Lee KB, Thring MW, and Beer JM (1962) On the rate of combustion of soot in a laminar soot flame, Combust. Flame, 6: 137–145

[70] Fenimore CP and Jones GW (1967) Oxidation of soot by hydroxyl radicals, J. Phys. Chem. 71: 593–597

[71] Tesner PA and Tsibulevsky AM (1967) Gasification of dispersed carbon hydrocarbon diffusion flames, III. flames of acetylene–hydrogen and acetylene–water vapor mixtures, Combust. Explosion and Shock–Waves, 3: 163–167

[72] Magnussen BF (1970) The rate of combustion of soot in turbulent flames, 13th Symp. (Intl.) Combust., The Combustion Institute: 869–877

[73] Neoh KG, Howard JB and Sarofim, AF (1981) Soot oxidation in flames. Particulate carbon formation during combustion (ed Siegel DC and Smith W), Plenum Press, New York: 261–277

[74] Yoshizawa Y and Matsuoka S (1983) A shock–tube study on the oxidation rate of soot particle. Proceedings of 1983 ASME–JSME Thermal Engineering Joint Conf. Hawaii, The Jpn. Soc. Mech. Eng., 4: 253–259

[75] Tsuji H and Yamaoka I (1967) The counterflow diffusion flame in the forward stagnation region of a porous cylinder. 11th Symp. (Intl.) on Combust., The Combustion Institute, Pittsburgh, : 979–984

[76] Vandsburger U, Kennedy I, and Glassman I (1984) Sooting counterflow diffusion flames with varying oxygen index. Combust. Sci. Tech. 39: 263–285

[77] Vandsburger U, Kennedy I, and Glassman I (1984) Sooting counter–flow diffusion flames with varying velocity gradients. 20th Symp. (Intl.) on Combust., The Combustion, Pittsburgh, : 1105–1112

[78] Hura HS and Glassman I (1987) Fuel oxygen effects on soot formation in counterflow diffusion flames. Combust. Sci. Tech. 53: 1–21

[79] Hura HS and Glassman I (1989) Soot formation in diffusion flames of fuel/oxygen mixtures. 22nd Symp. (Intl.) on Combust., The Combustion Institute, Pittsburgh, : 371–378

[80] Wright FJ (1974) Effect of oxygen on the carbon–forming tendencies of diffusion flames. Fuel 53: 232–235

[81] Dearden P and Long R (1968) Soot formation in ethylene and propane diffusion flames. J. Appl. Chem. 18: 243–251

[82] Jones JM and Rosenfeld JLJ (1972) A model for sooting in diffusion flames. Combust. Flame 19: 427–434

[83] Schg KP, Manheiner–Timnat Y, Yaccarino P, and Glassman I (1980) Sooting behavior of gaseous hydrocarbon diffusion flames and the influence of additives. Combust. Sci. Tech. 22: 235–250

[84] Chakraborty BB and Long R (1968) The formation of soot and polycyclic aromatic hydrocarbons in diffusion flames III – Effect of additions of oxygen to ethylene and ethane respectively as fuels. Combust. Flame 12: 469–476

[85] Wey C, Powell EA, and Jagoda JI (1984) The effect on soot formation of oxygen in the fuel of a diffusion flame. 20th Symp. (Intl.) on Combust., The Combustion Institute, Pittsburgh, pp 1017–1024

[86] Saito K, Williams FA, and Gordon AS (1986) Effects of oxygen on soot formation in methane diffusion flames. Combust. Sci. Tech. 47: 117–138

Chapter 6
Emissions and Heat Transfer in Combustion Systems

6.1 General

A variety of combustion systems that employ turbulent diffusion combustion have been major sources of air pollutants such as NOx, particulates and hydrocarbons in spite of their high thermal efficiency compared to other types of combustion systems. For diffusion combustion systems, including compression–ignition or diesel engines and industrial furnaces, increasingly stringent emission regulations have been imposed to improve air quality in urban areas in the largest cities. In addition, associated with the recent issues of the global warming and energy conservation, a reduction in emissions is urgently required along with minimizing fuel consumption. This necessitates better understanding of the emission formation mechanism in real combustion systems.

As far as fuel economy with the combustion engine is concerned, reducing heat loss from the in–cylinder charge is desirable to improve thermal efficiency, whereas for the furnace, heat transfer from the burned gas must be enhanced to utilize the heat energy of the fuel. In any combustion system, therefore, the mechanism of convective and radiative heat transfer should be explored in more detail to eventually allow the prediction of the heat flux between the burned gas and the surrounding wall. This will help conserve energy and protect the combustion chamber wall from excessive thermal load at high temperatures as well as encourage the utilization of low heat rejection wall materials typified by ceramics.

Thus, this chapter deals with recent topics related to the formation mechanism of emissions and their control as well as heat transfer, and the use of various experimental combustion apparatuses intended to reproduce combustion and heat transfer in real combustion systems.

6.1.1 Emissions in combustion systems

In order to control exhaust emissions from diffusion combustion systems, particularly from diesel engines, numerous engineering–oriented methods have been developed or proposed. To reduce NOx, for example, decreasing combustion gas temperatures and/or increasing equivalence ratios or inert gas ratios in localized fuel–air mixtures in the combustion chamber are among the most effective methods. For this specific purpose, the most realistic approaches are the control of fuel injection variables and charge dilution typified by exhaust gas recirculation. On the other hand, to improve smoke and particulates as well as fuel economy, enhancing fuel–air mixing is favorable. To do this in engineering design, increased swirling air, modified combustion chamber geometry, and high pressure fuel-injection are utilized. Especially, the beneficial effects of extremely high pressure injection have been demonstrated in Chapter 3.

Reducing NOx, however, inevitably results in a deterioration in fuel economy and causes other emissions such as particulates and unburned hydrocarbons due to incomplete combustion in locally fuel–rich regions. Improving such emissions, however, tends to lead to an increase in NOx. Thus, engineering optimization, based on detailed understanding of the emission formation mechanism, is necessary to reduce all of the emissions while simultaneously assuring fuel economy in real combustion systems. Diesel combustion is, however, inherently complicated what with its high heterogeneity and transience, involving fuel atomization and evaporation, ignition lag, mixing, chemical reactions including fuel pyrolysis and the resulting heat release. Subsequently, the combustion–generated emissions are influenced by such processes in a very complicated manner.

In order to elucidate the effects of these processes and improve emissions, experimental studies have been carried out in combustion apparatuses intended to reproduce diesel–like combustion. The results are described in the following four sections, Sections 6.2 through 6.5 in this chapter. In these studies, rapid compression machines, a modified compression–ignition engine and a constant–volume combustion vessel have been devised and used to reproduce or idealize some aspects of diesel combustion. These combustion apparatuses make it possible to precisely measure and visualize fuel sprays or the subsequent combustion and emission formation processes using several optical techniques and gas sampling and analysis.

In Section 6.2, special emphasis is placed on the fuel pyrolysis which is influenced by gas temperatures and swirling air in diesel sprays, leading to the generation of different types of hydrocarbons as an intermediate matter. Decomposed hydrocarbons in the cylinder were extracted and analyzed over time and their formation mechanism is discussed.

Sections 6.3 and 6.4 discuss particulate formation in compression–ignited fuel–rich mixtures using a modified engine and in diesel sprays injected into a rapid compression machine. Combustion tests were conducted for important parameters such as equivalence ratios, oxygen and inert gas concentrations and fuel types in the combustion apparatuses. Temperatures and particulate concentrations in the flame were measured. In Chapter 5, more fundamental aspects of the soot formation mechanism in simpler combustion systems are reviewed and discussed in terms of chemistry and physics.

In Section 6.5, the effects of opposed spray injection are demonstrated. This unique injection concept utilizes the dynamic interaction of the two impinging fuel sprays and can be an alternative approach, which is quite different from the high pressure injection concept, to reductions in both smoke and NOx.   (Yasuhiro Daisho)

6.1.2 The role of heat transfer in combustion systems

Wall heat transfer in combustion chambers does not always have a primary influence on combustion processes, but it is the impact of wall heat transfer on the quantitative analyses of combustion processes that motivates the study of heat transfer. Convective and radiant components of heat transfer are separate subjects of research because of their completely different natures of physical processes.

Convective heat transfer on combustion chamber walls is a predominant factor in total heat transfer. The numerical analyses of combustion processes based on computer fluid dynamics is evolving a new method for the quantitative analyses and predictions of convective heat transfer in combustion chambers. There have been a great number of papers published on this subject, including both fundamental and practical areas of research. Fundamental research covers the direct simulations and the detailed mathematical models for time-averaged nature of turbulence which describe near wall characteristics of turbulence. The application of this kind of numerical analysis will be shown in Section 6.6, integrating combustion reaction. Numerical computations for practical combustion systems can not yet incorporate these results of fundamental studies because a tremendous number of computational grids are required for resolving thermal boundary layers and this is beyond the capability of computing speed of most available computers in a practical sense. Therefore, most computer predictions of convective heat transfer in practical combustion chambers adopt the law of the wall as a bridging function between the first adjacent node to the wall and the wall surface. Though the law of the wall model is reasonable for modeling steady flow combustion processes, it is different for reciprocating engine processes. This is because the compression/expansion work exerts on thermal boundary layer and its influence is considerable [1]. Heat transfer models for in-cylinder processes in reciprocating engines are being investigated to replace the law of the wall [2]. Convective heat transfer models either stand on their own as subjects worthy of study or are not unique to combustion systems. Thus they will not be discussed further and readers may find a great deal of literature in the CFD research fields including steady flow combustors and reciprocating engines.

It should be noted that the above discussion about convective heat transfer is concerned with a non-reactive case. Exothermic reaction taking place in combustion chambers adds another factor to convection heat transfer analyses. This problem is unique to combustion systems. It is treated differently depending on the reference gas temperature which is taken for the temperature difference term to define a heat transfer coefficient. If there is no exothermic reaction between the wall and the reference temperature location, the problem is reduced to the non-reactive convection heat transfer. Otherwise, heat release must be taken into account. This situation is seen in the case where fuel-air mixture exists near the wall and exothermic reaction takes place with wall quenching. Such a situation does not seem to be common in practical steady flow combustors, but it is often seen in reciprocating engine combustion chambers. In spark ignition engine combustion chambers, all the near wall regions must be in this situation, because premixed combustion takes place. In diesel engine combustion chambers, the fuel is injected in spray, impinging on the combustion chamber walls, and combustion begins with premixed combustion, followed by diffusion combustion. Thus, it is quite plausible that the heat transfer takes place in a combination of convection and exothermic reaction or quenching locally and time-dependently. As the practical CFD computation of in-cylinder processes can not use grid sizes less than a quench distance which is an order of 100 microns in reciprocating engines, recent research has focused on trying to evolve a heat transfer model for near wall regions which includes exothermic reaction as well as compression/expansion work. There has been only one model presented [3] for this kind of heat transfer and a great deal of further research is still needed.

Investigations in this area are not only interesting from a viewpoint of fundamental research, but necessary from a practical point of view, because much more knowledge is required to analyze heat transfer in insulated engines or so-called adiabatic engines. The engines use ceramics as their combustion chamber wall materials in expectation of much lower heat loss. Due to heat insulation, the wall temperatures reach as high as 1000 K which is around 500 K higher than those in conventional engines. Section 6.6 deals with the combined convective/exothermic heat transfer accompanied by pressure change at such high wall temperatures and show some new experimental and theoretical findings.

Due to much higher wall temperatures in insulated engines, combustion also takes place at much higher temperature levels than in conventional engines. It appears that in many cases the high temperature combustion resulted in lower engine performances such as higher NOx emission and smoke, and worse fuel consumption. The mechanism of the high temperature combustion processes has become another important research topic to solve this problem. In Section 6.7, some experimental analyses are shown of the high temperature diesel combustion.

Radiant heat transfer plays a considerable role in a diffusion flame which is seen in most practical combustion systems including both steady and unsteady combustion apparatuses except spark ignition engines. This is, of course, mainly due to particle cloud radiation which exchanges thermal energy between combustion gas products and walls. For example, measurements of time–averaged heat transfer indicate that 20–40% of the total heat flux is due to radiation in diesel engines [1]. On the other hand, heat transfer computations for spark ignition engines based on gas radiation primarily from the CO_2 and H_2O indicate that the radiation portion is around 10%. Another important aspect of the role of radiation in a diffusion flame is that the thermal energy exchange between particle clouds and walls considerably affects gas temperatures in numerical computations. This subject is discussed in detail in Section 6.8 as an important area of recent research in combustion.

As stated above, most research topics of heat transfer during combustion are found in reciprocating engines. Because of the fast changing nature and complexity of their combustion processes, there are many aspects in heat transfer research in reciprocating engines. Readers interested in this subject should consult the review paper by Borman and the present author [1].   (Kazuie Nishiwaki)

6.2   Fuel Pyrolysis and Fuel–Air Mixing in the Diesel Combustion Process

6.2.1 Introduction

The fuel–air mixture formed in an ignition delay period plays on important roles for ignition and combustion in diesel engines and it has great effects on harmful emissions such as NOx, hydrocarbons and particulates. However, the complex heterogeneous nature of the diesel ignition and combustion process hinders the understanding of events occurring in the combustion chamber.

Many studies have been made to provide information on the relative roles of soot, light and heavy hydrocarbons formation in the combustion chamber using an in–cylinder gas sampling technique. Barbella et al.[4] showed that polycyclic aromatic hydrocarbons were formed during the ignition delay and soot was rapidly produced in the early stage of diesel combustion process. Kittelson et al.[5] measured the soot concentration in a combustion chamber of a direct–injection diesel engine by means of the total in–cylinder sampling system. In this work, soot mass first appears shortly after TDC and reaches a peak between 15 and 30 degrees after TDC. They pointed out that the initial rate of soot formation was essentially independent of engine speed and load and that oxygen availability late in the cycle was a critical factor in determining exhaust soot concentrations. Ikegami et al.[6] analyzed the chemical species, such as soot, light and heavy hydrocarbons in an experimental two–cycle diesel engine by means of a fast gas sampling valve. They proposed that the fuel injected in sufficiently hot compressed air underwent rapid evaporation, thermal decomposition and exothermic reactions near the injection nozzle and that the spray developed as a gaseous jet including light hydrocarbons.

6.2.2 Fuel pyrolysis

To elucidate fuel pyrolysis and mixture formation in diesel combustion, with special regard to thermal decomposition of the fuel injected at an elevated air temperature, in–cylinder hydrocarbons analysis has been made on a total sampling system [7].

This system allows dumping, diluting, quenching and collecting the entire contents of the combustion chamber on a time scale of about 0.7 ms, using the rapid compression machine [8], which can realize diesel combustion in a constant volume. The quantitative analysis of single components of each light hydrocarbon was performed by a FID–gaschromatograph.

First, with the objective of knowing how the ambient gas temperature in the combustion chamber affects the pyrolysis reaction of a spray, light hydrocarbons coming from the thermal cracking of a gas oil spray were measured in a nitrogen environment at temperatures T_i at an initial pressure of p_i=4 MPa. Figure 6.1 shows that a slight amount of light hydrocarbons is detected at a low temperature of 700 K. With an increase in temperature, however, a considerable amount of light hydrocarbons is produced in the hot surroundings without the combustion. Furthermore at 900 K, a lot of light unsaturated hydrocarbons such as C_2H_4 and C_3H_6 are detected. This clearly indicates that at sufficiently high temperature, fuel decomposition exists without

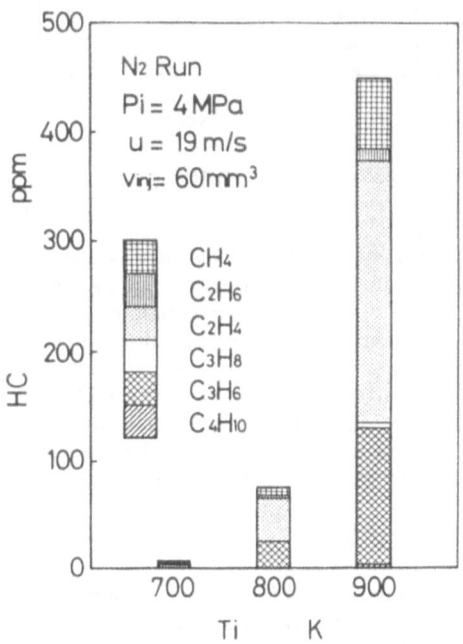

Fig. 6.1 Effect of temperature on thermal decomposition

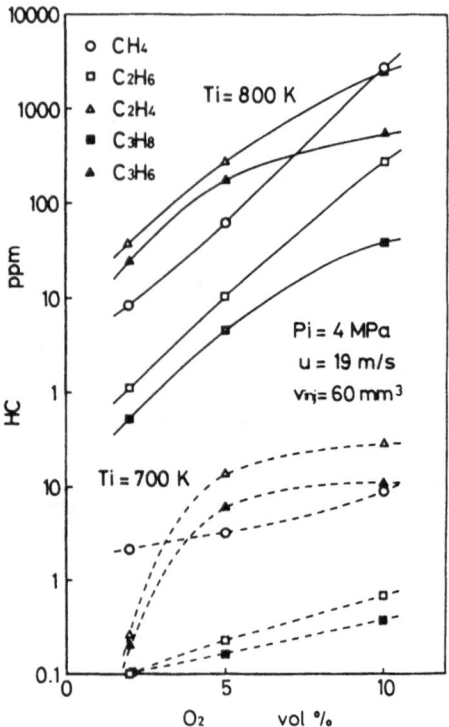

Fig. 6.2 Effect of oxygen on thermal decomposition

temperature elevation caused by oxidation. Next, to elucidate the effect of oxygen on thermal decomposition of this fuel, tests were conducted by changing the oxygen content in the combustion chamber. In Fig. 6.2, the concentrations of each light hydrocarbon are plotted against the oxygen content for initial temperatures of 700 K and 800 K respectively. Concentrations of hydrocarbon coming from pyrolysis and oxidation increase exponentially as the oxygen fraction becomes high. In the case of 800 K, the total hydrocarbon concentration in the 10% oxygen environment reaches 6,000 ppm and the concentrations of saturates such as CH_4 and C_2H_6 become as large as the unsaturates due to the temperature rise by the oxidation.

6.2.3 Diesel combustion process

In this section, in order to know the effects of such decomposition on combustion, gas sampling was carried out as a function of time for the usual combustion tests. Figure 6.3 shows time trends of light hydrocarbons during combustion at 800 K. In this figure, the horizontal axis represents the time elapsed from the injection start and the

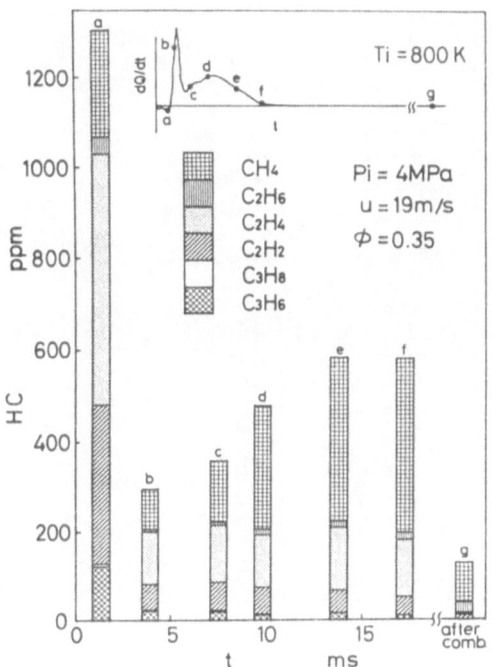

Fig. 6.3 Trend of light hydrocarbons during combustion at the initial temperature Ti=800 K

correspondence between the heat release pattern and the sampling timing is shown in the upper part. As can be seen on the heat release curve, large concentrations of unsaturates due to the effect of the regular oxidation appear before ignition from the timing "a" on. At "b" in the initial stage of combustion, in which the heat release rate is high, the decomposed concentration decreases owing to the combustion of the premixed mixture formed during the ignition delay period. As the combustion goes on to the diffusion burning stage, the decomposed products begin to increase again and keep going up even after the heat release rate begins to decrease. This tendency stops at "f", when the heat release almost ceases and the concentration approaches the exhaust levels at "g". As the combustion proceeds, CH_4 increases while the unsaturates decrease. In the case of higher temperature Ti=900 K as shown in Fig. 6.4, it is noted that light hydrocarbons decomposed in the ignition delay period decrease compared to the case of Ti=800 K. Hence, the peak of the heat release rate in the premixed burning stage lowers. This probably shows that at high temperatures, fuel decomposition does not progress in the combustion chamber due to a short ignition delay and due to a short mixing period, while hydrocarbons formed during the diffusion burning stage increase with time. Especially, the CH_4 concentration in this stage is higher than that at 800 K and remains in a high concentration for a long period of time. This clearly indicates that an extremely

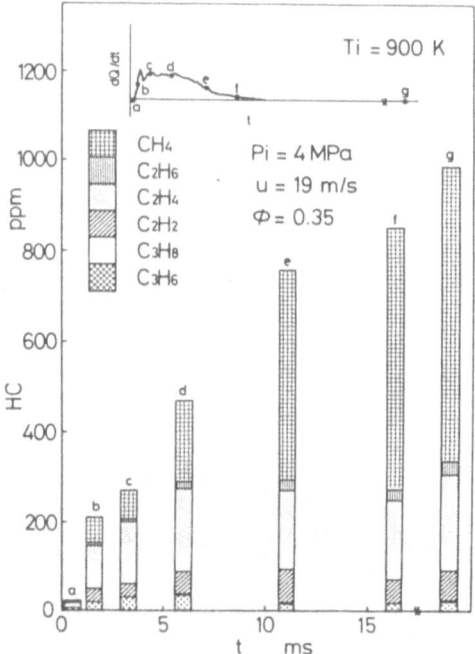

Fig. 6.4 Trend of light hydrocarbons during combustion at the initial temperature Ti=900 K

short ignition delay causes poor air–fuel mixing so that most of the fuel injected is decomposed in the rich diffusion flame.

6.2.4 Conclusions

Diesel ignition and combustion tests were carried out with special regard to fuel pyrolysis and the generation of hydrocarbons by using a rapid compression machine. The results are summarized as follows.
(1) Fuel injected in sufficiently hot compressed gases rapidly undergoes thermal decomposition during the ignition delay period and the gaseous jet containing unsaturated light hydrocarbons like C_2H_4 and C_3H_6 develops as the spray.
(2) Unsaturated products coming from pyrolysis reactions are formed at the start of heat release and quickly consumed during the initial burning stage.
(3) As the ignition delay becomes extremely short, a great amount of saturates such as CH_4 remain in the later stage of combustion due to the poor mixing in the initial and the main combustion stages.   (Kei Miwa)

6.3 Particulate Formation in Compression–Ignited Mixtures

6.3.1 Introduction

Particulates consisting of soot and SOF are among the most undesirable pollutants emitted from the diesel engine. As reviewed by Glassman [9], numerous investigations have been made to establish soot formation fundamentals using open flames and continuous flow reactors. However, little detail is known about the mechanism of particulate formation in the diesel engine due to the highly transient and complicated combustion process taking place at such high pressures and temperatures. Generally, in diesel engines, the spray of fuel injected into the cylinder is heated by the surrounding hot air. It is then evaporated and mixed with the hot air forming heterogeneous mixtures. Because particulate formation and oxidation take place almost simultaneously during the combustion of such mixtures, distinguishing between these two processes is difficult.

In this work, a combustion system has been developed for the specific purpose of reproducing the pressures and temperatures observed in diesel combustion, which are much higher than those used in open flames and reactors. This system uses homogeneous premixed mixtures to avoid the fuel–air mixing process which inevitably involves the oxidation of particulates. Thus, the effects of equivalence ratios and fuel types on particulate formation were investigated.

6.3.2 Experimental apparatus

The combustion system developed includes an inlet system, a single cylinder compression–ignition engine, and measuring instruments [10,11]. In this work, homogeneous fuel–rich mixtures were ignited by compression in the engine to avoid complications associated with atomization, heating, evaporation, and air–mixing of the injected fuel spray and to reproduce local fuel–rich regions where diesel particulates were formed. The engine has a flat piston with a 104 mm bore, a 118 mm stroke, and a 7.8 compression ratio. The engine speed was kept constant at 1,000 rpm with a 50% volumetric efficiency throughout the tests.

To identify the effects of fuel types, n–heptane was chosen as the simplest reference fuel. This fuel has a cetane number of 56, which yields almost the same compression–ignitability as that of ordinary diesel fuel. N–heptane–benzene and n–heptane–cetane blends, n–dodecane, and JIS No. 2 diesel fuel were also used for comparison. The fuel was metered and injected into the inlet pipe by means of an electronically controlled fuel injection system. The injected fuel was heated and evaporated by preheated air and then fully mixed with the air in the mixing chamber.

The inlet temperature of the fuel–air mixture was adjusted to within the range of 433 to 573 K. Such high inlet temperatures allowed the mixture to be ignited by compression even at the low compression ratio. Thus, stable and repeatable compression ignition and combustion were achieved by using fully premixed homogeneous mixtures at high temperatures [11]. A piezoelectric pressure transducer was used to determine the ignition timing, the mean temperature, and the rate of heat release. A mini–dilution tunnel with filters was employed to sample and to weigh the particulates. SOF and solid soot were then separated using dichloromethane as a solvent.

In order to measure soot concentrations in burned gases in the cylinder, the laser extinction method [12] was adopted. A photo–diode with a quartz window was used to detect the light intensities from a He–Ne laser beam introduced into the cylinder through a quartz window opposite. An optical filter was placed before the diode to reduce the radiative emission from luminous soot particles to levels sufficiently lower than the attenuated laser levels. The intensities detected were analyzed to yield soot volume fractions as a function of crank angle, based on Lambert–Beer's law and the Rayleigh scattering theory. Further, soot luminosity was also detected to determine the occurrence of sooting using the same photo–diode without the filter.

6.3.3 Results and discussion

N–heptane, n–dodecane, and JIS No. 2 diesel fuel were burned in the combustion system at different equivalence ratios, and exhaust particulates were analyzed. As shown in Fig. 6.5, in n–heptane and n–dodecane mixtures, exhaust soot began to appear at an equivalence ratio of 2.2 and then rapidly increased with increasing ratios. However, SOF emissions were low for all the equivalence ratios tested. By contrast, in the case of the diesel fuel, the SOF level was high and increased with the equivalence ratio whereas the soot level decreased. Exhaust hydrocarbon compositions were then analyzed for the diesel fuel and its exhaust SOF using high temperature gas chromatography [11]. The result indicates that in the case of the diesel fuel, SOF emissions come from heavy hydrocarbons which in spite of being exposed to high temperatures are not burned due to the very short residence time and lack of oxygen in the combustion system. By contrast, in n–heptane combustion, such heavy hydrocarbons were not detected.

Figure 6.6(a) shows the cylinder pressure diagrams, mean gas temperatures, and the rates of heat release for n–heptane mixtures at three equivalence ratios. Peak pressures up to 3.5 MPa, peak temperatures up to 2,100 K, and combustion periods up 2.0 ms. These conditions are almost equivalent to those in fuel–rich regions where particulates are formed in the earlier part of diesel combustion. Since combustion in this system can be regarded to be spatially uniform, the heat release period is equivalent to the chemical reaction time. In combustion in a closed system, reaction heat will raise pressures and temperatures to much higher levels compared to those reached in experiments using open flames and continuous flow reactors. Thus, in this combustion system, rapid combustion can be reproduced at high pressures and temperatures for simulating diesel combustion.

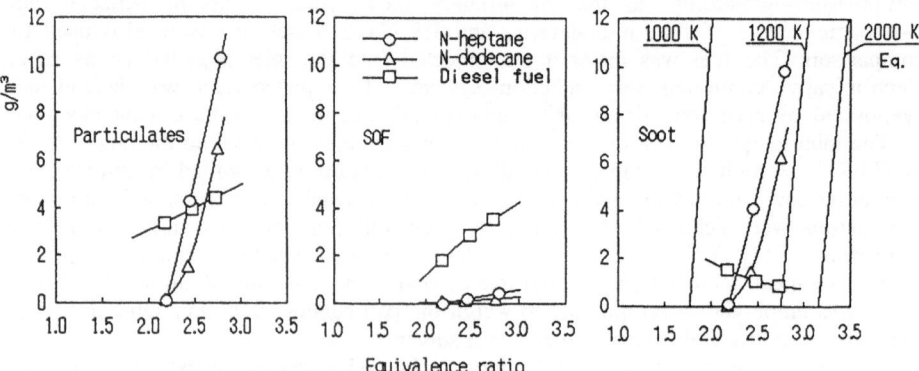

Fig. 6.5 Effects of equivalence ratio on particulates

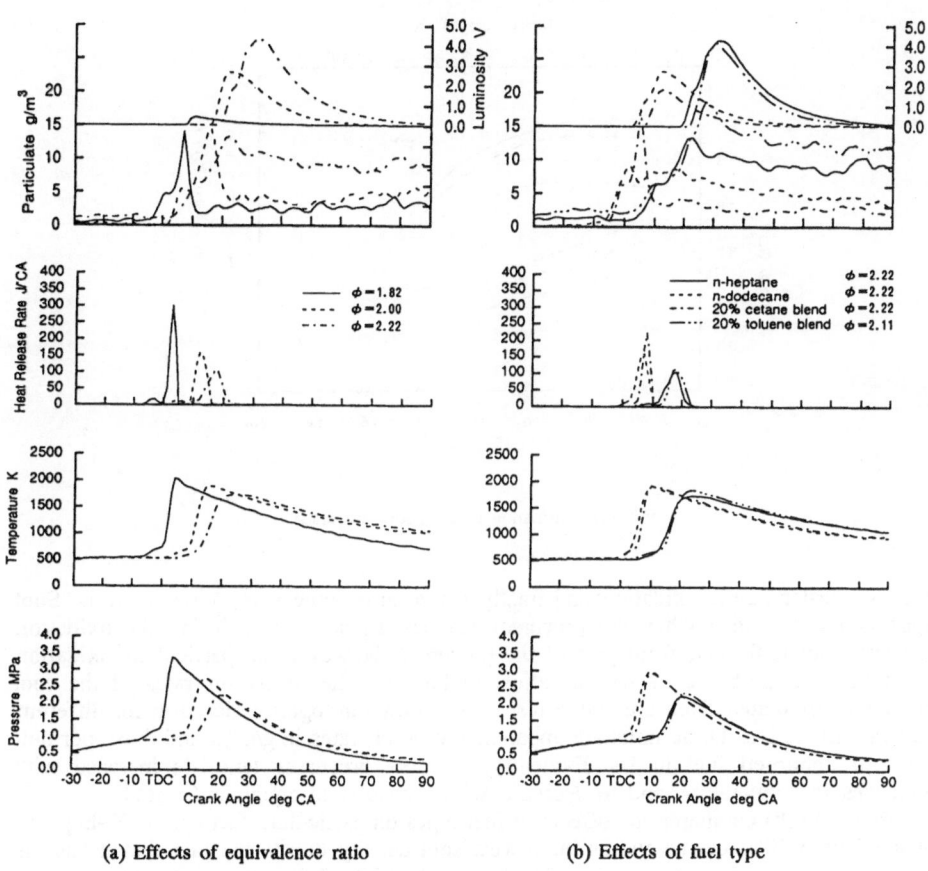

(a) Effects of equivalence ratio   (b) Effects of fuel type

Fig. 6.6 In–cylinder measurements on combustion and particulates

Figures 6.5 and 6.6(a) indicate that soot increases with the equivalence ratio although pressures and temperatures are lowered. This suggests that the equivalence ratio is the most important factor influencing soot formation at the pressures and temperatures reached in this combustion system. Solid carbon densities at the equilibrium state were calculated for n–heptane mixtures at a pressure of 3 MPa and different temperatures. Obviously, real particulates are formed at a lower equivalence ratio than that for equilibrium carbon at 2,000 K. However, in both cases, the dependance of soot formation on the equivalence ratio has some similarity.

As can be seen from Fig. 6.6(a), using the laser extinction method, the rapid occurrence of particulates is observed almost at the beginning of the heat release or ignition. The particulates observed can be regarded as a precursor of soot, which consists of high–carbon–number hydrocarbons and is defined as matter optically equivalent to solid soot, since at this time no luminosity from soot was detected. For mixtures at lower equivalence ratios, the soot precursor tends to increase faster with higher temperatures,

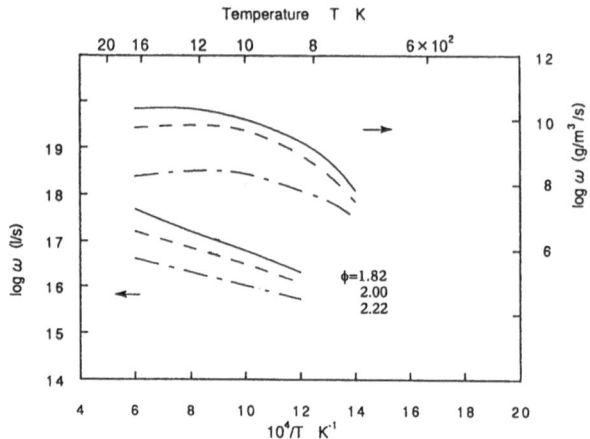

Fig. 6.7 Formation rates of soot precursor

decrease faster due to oxidation and finally freeze to become soot at lower levels. Soot appears no later than when the precursor reaches a peak. Thus, during the oxidation process, soot is formed from part of the precursor. However, no particulate oxidation takes place at higher equivalence ratios. In Fig. 6.7, the formation rates of the soot precursor calculated from the above results are shown in logarithmic form for different equivalence ratios. These units are apparent formation rates in $g/m^3/s$ and those per unit mass of unburned fuel in $1/s$. Evidently, the soot precursor tends to increase with increases in temperature and to decrease with increases in equivalence ratio.

Figure 6.6(b) compares the effects of fuel types on particulate formation. N–heptane blended with 20% cetane yields the lowest soot density due to the fact that it has the highest oxidation rate and in spite of its having the highest formation rate. As far as the paraffinic hydrocarbons are concerned, heavier fuels will form less soot and SOF despite the increase in carbon mass fraction. This tendency can also be seen in the comparison between n-heptane and n-dodecane in Fig. 6.5. This implies that for soot formation, heavier hydrocarbons must be fragmented into lighter hydrocarbons by some chemical course.

By contrast, in case of combustion with 20% toluene blended heptane, more particulates were formed though the equivalence ratio was higher than that of pure n–heptane and the temperature history was almost the same. This supports that aromatic fuels have significant chemical effects on particulate formation.

From these results, it is seen that higher–cetane–number fuels tend to form less soot. In contrast, as shown in Fig. 6.8, in general, diesel engines tend to emit more soot with higher–cetane–number fuels. This can be explained by the fact that in diesel combustion, the shortening of ignition delay due to higher cetane number reduces the time available for mixing with air, thereby leading to a more fuel–rich combustion. Thus, it is suggested that the higher–cetane–number fuel is not the chemical agent that forms soot. On the other hand, in the case of diesel combustion with aromatics–rich fuels, which

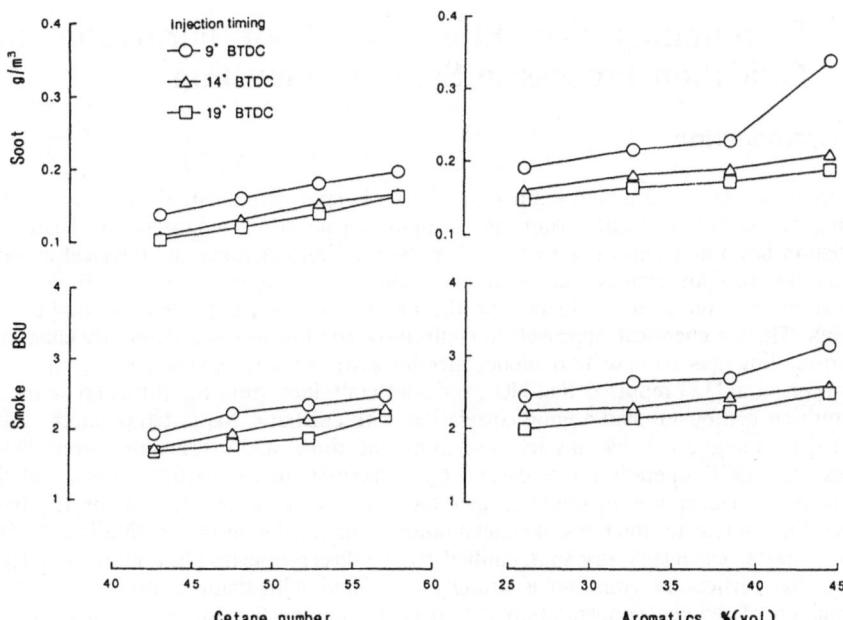

Fig. 6.8 Effects of fuel properties on soot in a diesel engine

have lower ignition quality, the extent of the increase in soot is weakened by the tendency of delayed ignition, providing more time for mixing with air although it has a chemically strong effect on forming soot.

6.3.4 Conclusions

A combustion system has been developed which reproduces the early part of diesel combustion characterized by compression–ignition accompanying fuel–rich premixed burning and sooting. In this combustion system, rapid combustion can be reproduced at pressures up to 3.5 MPa and temperatures up to 2,100 K as observed in diesel combustion.

The combustion test results indicate that the equivalence ratio and the aromatics content in the fuel are dominant factors influencing the particulate formation at the pressures and the temperatures tested. The heavier paraffinic hydrocarbons tend to form less soot than the lighter ones. In diesel fuel combustion, SOF emissions result mainly from some heavier part of the unburned fuel. In combustion of paraffinic hydrocarbons, SOF levels are much lower than soot levels.

Soot precursors are observed using the laser extinction method. It is shown that soot precursors are formed faster at lower equivalence ratios and then oxidized faster yielding less frozen soot. With increasing equivalence ratio, particulate oxidation is reduced and thereby more soot is formed.   (Yasuhiro Daisho)

6.4   Surrounding Gas Effects on Soot Formation and Oxidation Process in Spray Combustion

6.4.1 Introduction

Analysis of the generation and extinction of soot in spray combustion is important for reducing particulate emission from diesel engines, gas turbine engines, etc. Diffusion combustion has both chemical aspects such as combustion reactions and physical aspects such as the mixing process between fuel and air. Changing the in-cylinder gas composition has no direct influence on the mixing process, but does on combustion reactions. Thus, a chemical approach to controlling combustion is possible by changing the surrounding gas composition alone, thereby avoiding any physical changes.

Sawyer, et al.[13] reported that NOx and soot emissions from the diffusion flame of a gas turbine combustor had a good correlation with adiabatic flame temperature. Plee, et al.[14] then suggested that any increase in hydrocarbons and extractables from diesel engines with EGR operation was caused by a decrease in the partial pressure of the oxygen, or by a decrease in flame temperature. Sato and the present author[15] have reported that increasing the oxygen concentration in the intake air is effective in reducing the particulates, especially dry soot, emitted from a direct-injection diesel engine. They verified that particulate emission is strongly correlated with flame temperature, which is a function of oxygen concentration[16]. It is also known that mixing carbon dioxide with intake air results in a reduction in particulate emission.

Thus, a more detailed experimental study has been carried out on soot formation and oxidation in diesel spray combustion with special emphasis on the chemical aspects related to gas temperature and composition.

6.4.2 Experimental procedures and results

(a) Experimental apparatus

To observe the soot formation and oxidation processes of diesel spray combustion, high temperature and high pressure gas was prepared by a rapid compression machine (RCM) developed for this study[16]. The combustion chamber of the RCM is in the shape of pancake, with a diameter of 145 mm and a thickness of more than 48 mm. The shape allows reproduction of a free diesel spray without wall impingement. The piston stroke for compression was 692 mm and the compression ratio 14.6. The piston head and the cylinder head were equipped with windows made of quartz so that observation and optical measurements of the spray flame were possible. High-speed direct photography of flame, analysis of combustion from an indicator diagram, analysis of flame temperature and the KL factor by the two color method applied to a color image of flame were conducted. As fuel, JIS No. 2 diesel oil was used. Fuel was injected by a single-injection system through a single hole nozzle of 0.18 mm diameter. Fuel quantity was 20 mg, valve opening pressure 20 MPa, and the duration of a single injection 3.8 ms.

(b) State of surrounding gas

The main objective of the experiment was to learn the effects of the surrounding gas variables, including concentrations of oxygen and carbon dioxide and temperature, and of the flame temperature on the soot formation and extinction processes in diesel flames. The parameters were defined systematically based on:

(1) composition of charged gas
(2) temperature of charged gas at the end of compression
(3) flame temperature

Flame temperatures were independently controlled by adding argon gas, which has a high specific heat ratio, to a gas consisting of the same oxygen and carbon dioxide concentrations. Figure 6.9 shows the relation between oxygen concentration and temperature of the surrounding gas at the end of compression in the RCM under various set conditions. As the specific heat ratios of oxygen and nitrogen are nearly identical, adding oxygen or nitrogen to the air has very little effect on the pressure and temperature of the surrounding gas at TDC. Adding CO_2, on the other hand, remarkably reduces the temperature of the surrounding gas at TDC. The temperature at TDC can be controlled independently by adding a proper amount of argon gas, for the same combination of oxygen and carbon dioxide concentrations.

(c) Effect of oxygen concentration [Air + N_2 + O_2]

Measured area–averaged flame temperatures for various oxygen concentrations in the surrounding gas are shown in Fig. 6.10. The time of the initial rise in flame temperature corresponds roughly to the ignition time. It is earlier when the oxygen concentration is higher. The maximum flame temperature is higher for a higher oxygen concentration. The time history of an areal integrated KL value is also shown in Fig. 6.10. The initial rise in the integrated KL value occurs earlier for a higher oxygen concentration. The areal integrated KL value rises roughly linearly with time after ignition, reaching its peak value with maximum soot generation.

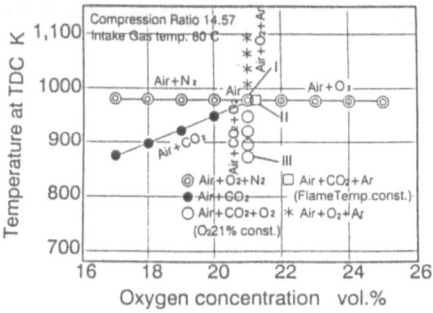 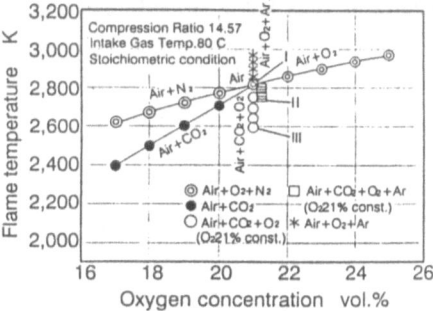

Fig. 6.9   Surrounding gas temperature at the end of compression (left) and estimated stoichiometric adiabatic flame temperature as a function of surrounding gas oxygen concentration for oxygen, nitrogen, and carbon dioxide added to charged air, with initial conditions evaluated at top position of the RCM piston (right)

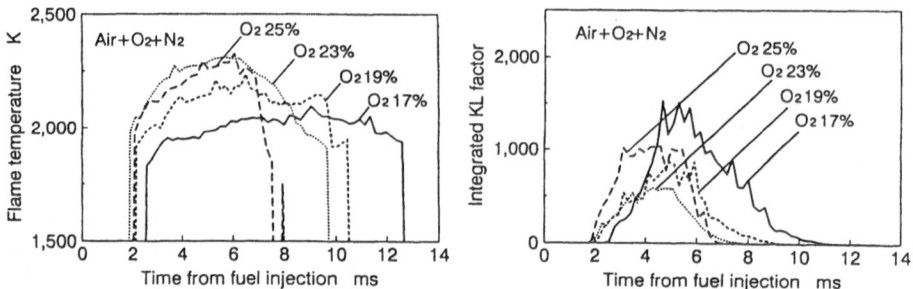

Fig. 6.10 Time history of the measured area-averaged flame temperature (left)
and areal integrated KL value (right)

The peak value is highest for O_2 = 17%, and decreases gradually as the O_2 concentration increases to 19 and 23%. However, it decreases at O_2 = 25%. At O_2 = 25%, a rise in the KL value immediately after ignition is prominent and quick. The time of soot extinction comes earlier with the higher oxygen concentration. Though the peak value of areal integration of KL is high for O_2 = 25%, it decreases sharply at later stages of combustion. Manely, the rate of soot extinction is high, resulting in an earlier extinction time.

The two-dimensional projected area of the flame region was obtained by image processing of the film for flame. The time history of the projected area of the flame is shown in Fig. 6.11. The flame region (bright flame region) is the region where high temperature soot in combustion gas is emitting energy as radiation from a solid surface. Thus, the flame region will serve as an indication of the presence of soot. The initial rise in the projected area of the flame corresponds to ignition delay. It comes earlier for a higher concentration of oxygen. The area of the flame increases rather linearly with time, reaches its peak value around 4 ms, and then decreases gradually. The peak value of the

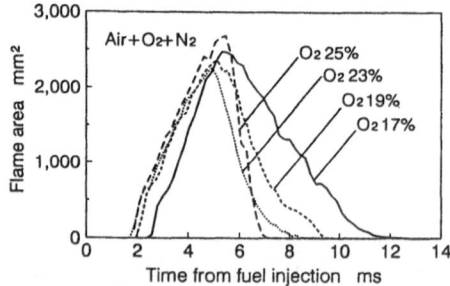

Fig. 6.11 Time history of the projected area of the bright flame region

projected flame area and the time when it appears are nearly identical for various oxygen concentrations. The rate of reduction in the flame area at later stages of combustion is higher for a higher oxygen concentration, or for a higher flame temperature. The extinction time of bright flame is $t_d = 12$ ms for $O_2 = 17\%$, and $t_d = 7$ ms for $O_2 = 25\%$, showing a big difference between these values.

(d) Effect of carbon dioxide [Air + CO_2]

Adding CO_2 to charged air reduced soot generation in flame. This may due to:
(1) Reduction in gas temperature at the end of compression;
(2) Reduction in oxygen concentration;
(3) Increase in ignition delay (Premixed combustion dominates); and
(4) Reduction in flame temperature.
    To clarify the contribution of the above items, the following comparative experiments were conducted:
I   [Air]: Air only
II [Air+CO_2+O_2]: Both CO_2 and O_2 were added to air.
    O_2 was constant at 21% and CO_2 at 19%.
    Condition (2) above was constant to case I.
III [Air+CO_2+O_2+Ar]: CO_2, O_2 and argon were added to air.
    Conditions (1) to (4) above were nearly constant to case I.
    O_2 was constant at 21% and CO_2 at 19%, and argon at 34.1%. The surrounding gas temperature at the end of compression was the same as that in case I.
    The time history of the area–averaged flame temperature and areal integrated KL value are shown in Fig. 6.12. Flame temperatures are nearly identical for case I [Air] and case III [Air+CO_2+O_2+Ar]. I and III have the same ignition delay. The integrated KL value rises initially at the same time for cases I and III. In case III, the rise is rather slow compared to case I [Air] and its peak value is much lower. The decrease rate of the integrated KL value, however, is identical during the late period of combustion. This may reflect the fact that flame temperature and soot oxidation rate are identical under these two conditions. However, in case III, the maximum value of the integrated KL value is small, resulting from the earlier soot extinction time.

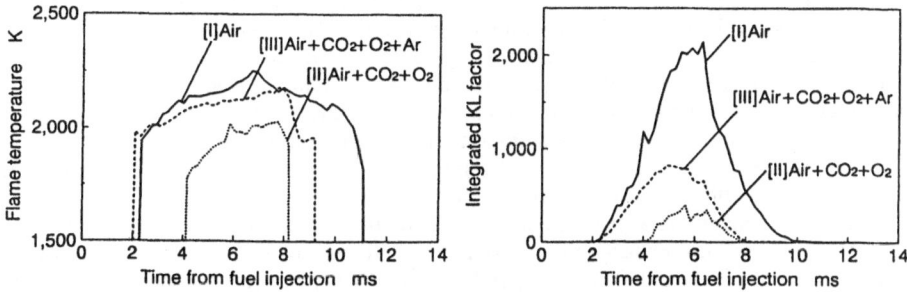

Fig. 6.12 Time history of the measured area–averaged flame temperature (left) and areal integrated KL value (right) for cases I, II and III

(e) Effects of surrounding gas temperature [Air+O_2+Ar]

Comparative tests were conducted by adding oxygen and argon to air. The concentration of oxygen was kept constant at 21%, the concentration of argon was varied to 1, 14.6, and 25.9%, to obtain a higher surrounding gas temperature and a higher flame temperature. When 25.9% argon gas was added, the temperature at the end of compression was higher by 100°C compared to case I (marked * in Fig. 6.9). The rise in flame temperature in this case was 180°C. The time history of the projected area of the bright flame region is shown in Fig. 6.13. Ignition delay is reduced and the peak value of the soot region is increased for the higher argon concentration. The formation rate of the soot region during the early stages of combustion is increased. The reduction rate of the soot region during the later stages of combustion is more prominent for the higher argon concentration. Extinction of the soot region also occurs earlier.

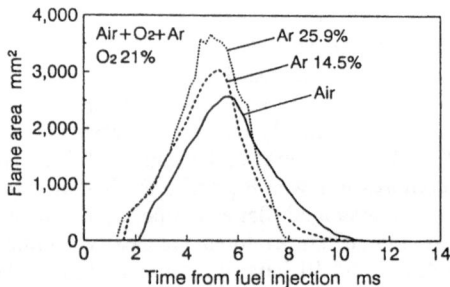

Fig. 6.13 Time history of the projected area of bright flame region

6.4.3 Summary

The formation and extinction processes of soot were observed during the combustion of a free diesel spray injected into gas at a high temperature and pressure using a rapid compression machine. The effects of the surrounding gas composition and flame temperature were investigated. The following results were obtained:

1) Ignition delay is reduced and flame temperature is raised for an increased oxygen concentration in the surrounding gas. The increase rate in the soot region during the early stage of combustion and maximum value of the soot region do not change remarkably, when the oxygen concentration in the surrounding gas is changed from 17 to 25%. The extinction rate of the soot region during the later stage of combustion is enhanced as a result of the increased oxygen concentration.

The above phenomena may be attributed to the flame temperature increase and oxygen partial pressure enhancement. From the results of the comparative tests, in which oxygen concentration and flame temperature were independently changed by adding argon to the surrounding gas, the effect of flame temperature seems to be dominant. These results can be summarized as follows: [a] Formation rate of soot after initiation of combustion tends to be higher for a higher surrounding gas temperature. [b] The oxidation rate and extinction rate of soot in the later stages of combustion tend to be higher for a higher flame temperature.

2) Flame temperature is decreased and ignition delay is increased by the presence of carbon dioxide in the surrounding gas. CO_2 also remarkably decreases soot density in the flame.

In cases when ignition delay and flame temperature were compared by adding carbon dioxide and argon gas so as to keep the oxygen concentration constant, the surrounding gas with carbon dioxide added exhibits a decrease in soot in the flame.

(Norimasa Iida)

6.5 Characteristics of Opposed Spray Combustion

6.5.1 Introduction

One of the major sources of air pollution is internal combustion engines. Especially, in automotive diesel engines, where low–particulate and low–NOx emissions are very much needed with high–load operation, it seems very useful to give a strong turbulence and the so–called Internal EGR(exhaust gas recirculation) effects in the combustion region by the impinging fuel spray injection [17]. The original idea of impinging combustion by Kumagai was first applied to low–NOx and high–load burning of premixed gases such as in spark–ignition engines, by continuously putting the burned gas into the unburned portion just to be burned. Various types of combustion were tried based on this idea. Above all, the impinging combustion has proved to be most effective [18].

As the first step of the present study, experiments have been performed to learn basically how emissions of dry soot and NOx and the maximum burning pressure are influenced by the impinging fuel spray injection in a combustion bomb and the results are compared with those obtained by the single injection.

6.5.2 Experimental apparatus and procedure

As shown in Fig. 6.14, the combustion bomb used is equipped with two pintle–type injection nozzles each on the opposite walls and pair of glass windows along its length for photographic observation. The combustion bomb is of 108 mm diameter, 60 mm length and 700 cc volume. The hole diameter, operating pressure and spray–cone angle of the injection nozzle are 0.8 mm, 14.3 MPa and 15 deg, respectively. Each injection nozzle, protected by a water jacket from preheating of the injectant, is provided individually with a Bosch–type injection pump. The combustion bomb is heated by Siliconit heater elements and the distribution of the air temperature in the bomb is confirmed to be almost uniform except in the vicinity of the wall surface. The fuel quantity per nozzle in the single is twice that in the impinging injection. The soot concentration of the burned gas was measured with a Bosch–type smoke–meter and the NOx concentration by a chemiluminescence method. Experiments were carried out at initial conditions of 0.2 MPa pressure and 623 to 673 K temperature. These initial temperatures are sufficient to establish spontaneous ignition of hexadecane($C_{16}H_{34}$) as fuel. In the present paper, `single injection' means injection with one nozzle and `impinging injection' simultaneous injection with two nozzle facing each other. The nozzle distance for the impinging injection, 95 and 115 mm, is defined as half the distance between two nozzles. The duration of impinging and single injection are 3 and 6 ms, respectively, since the fuel quantity per nozzle in the latter is twice that of the

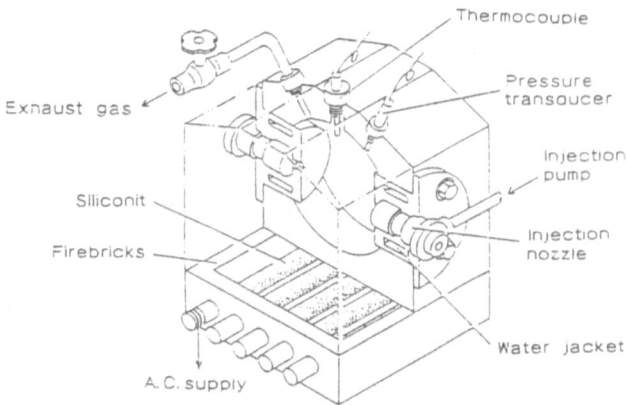

Fig. 6.14 Details of combustion bomb

former. The combustion behavior of impinging and single spray, which was ignited by a small heated nichrome wire located along its outer boundary at the central axis of the combustion bomb, was observed from high speed schlieren photographs taken at an initial condition of 300 K temperature and 0.1 MPa initial pressure, because it was very difficult to observe the combustion behavior through the glass windows at high temperature from 623 to 673 K.

6.5.3 Experimental results and discussion

Figure 6.15 shows typical examples of the pressure diagrams and heat–release rates for spray combustion with the single and impinging injection, respectively. From the estimation of heat–release rate based on the pressure diagram, it can be seen that, in the impinging injection the fuel burns more completely than in the single injection. For the impinging injection, the ignition delay and the burning time are much shorter, and the rate of pressure rise and maximum burning pressure are higher than those for the single injection.

Figure 6.16(a) compares the behavior of fuel sprays without combustion for the single and impinging injection at an initial condition of 300–K temperature and 0.1–MPa pressure. In Fig. 6.16(b), the combustion behaviors are shown for above two types of spray ignited by a small nichrome wire under the same conditions. It can be seen from these figures that combustion by the impinging injection takes place near the central region of the combustion bomb and the flame is strongly stirred due to the impingement of fuel spray, and consequently the burning time becomes much shorter than by the single injection.

Figure 6.17 shows the soot concentration versus excess–air ratio for both types of injection, where initial temperatures of 623 and 673 K were used to achieve spontaneous ignition for the impinging and single injection, respectively. At 623 K, it is impossible for the fuel–air mixture established by the single injection to be ignited. From the result

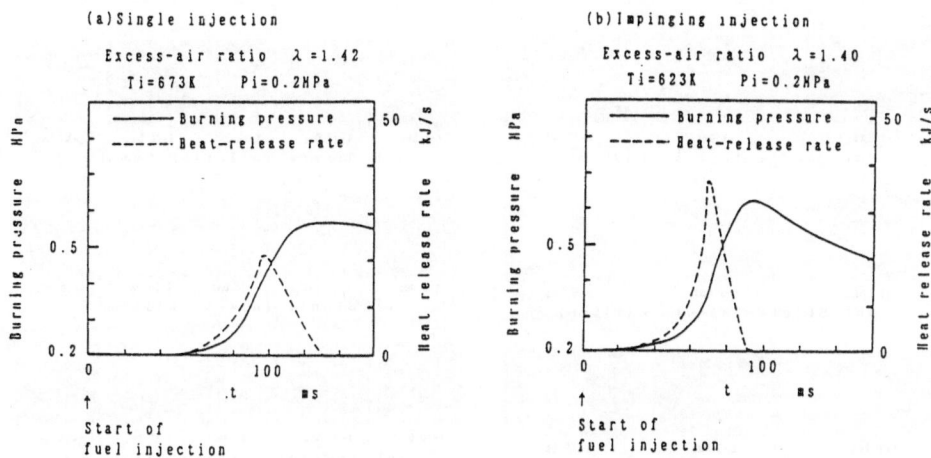

Fig. 6.15 Combustion pressure diagrams and heat release rates for single and impinging injection

it can be seen that at the impinging injection, soot is markedly lower than at the single injection for the excess–air ratios tested. For the impinging injection at 95 and 115 mm nozzle distance, there is not so much difference in the soot concentration between 1.5 and 2.0 in excess–air ratio, but, in the range of 1.0 to 1.5 the soot concentration is lower for 95 mm than for 115 mm nozzle distance. This fact indicates that the effects of fuel–air mixing and Internal EGR in the combustion region are improved when the nozzle distance is reduced particularly at lower excess–air ratios.

The effect of the maximum burning pressure on soot concentration is also shown in Fig. 6.17. From this figure it is found that, though the maximum burning pressure diminishes monotonously with increasing excess–air ratio for both types of injection(not shown), the soot concentration increases with increasing maximum burning pressure. For the impinging injection, however, the soot concentration is remarkably lower than that for the single injection at the same value of maximum burning pressure. Figure 6.17 also suggests that the nozzle distance in the impinging injection is a very important factor to achieve the high–load combustion and low–soot emission. The reduction of soot emission may be due to the internal–recirculation effect, which is realized by impinging spray combustion.

The effects of air ratio and maximum burning pressure on NOx emissions are shown in Fig. 6.18. This figure indicates that NOx emission decreases with increasing excess–air ratio and increases with maximum burning pressure for both types of injection. For the single injection, the emission of NOx is about 210 and 120 ppm when excess–air ratio varies from 1.3 to 2.0, and for the impinging injection, in contrast, it is about 25 ppm at excess–air ratio of 1.0. It is a very interesting fact that the impinging injection is very useful for reducing NOx.

The term EGR refers to providing a portion of combustion products to the burning zone, and it can be performed either externally or internally. The so–called Internal EGR

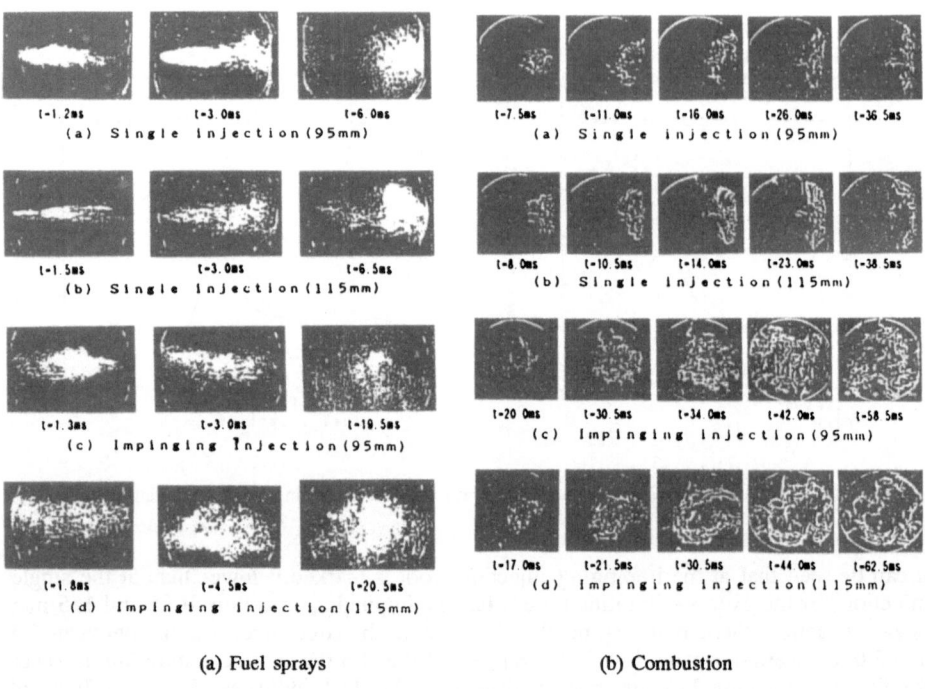

(a) Fuel sprays   (b) Combustion

(initial temperature: 300 K, initial pressure: 0.1 MPa, t: time from injection)

Fig. 6.16 Fuel spray and combustion behaviors for single and impinging injection

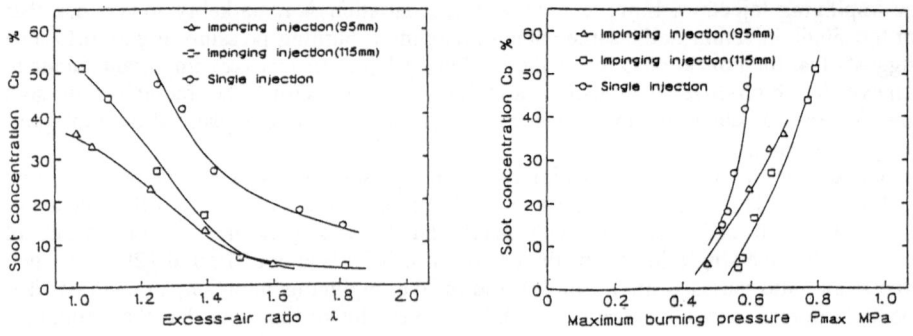

Fig. 6.17 Effects of excess-air ratio and maximum burning pressure on soot

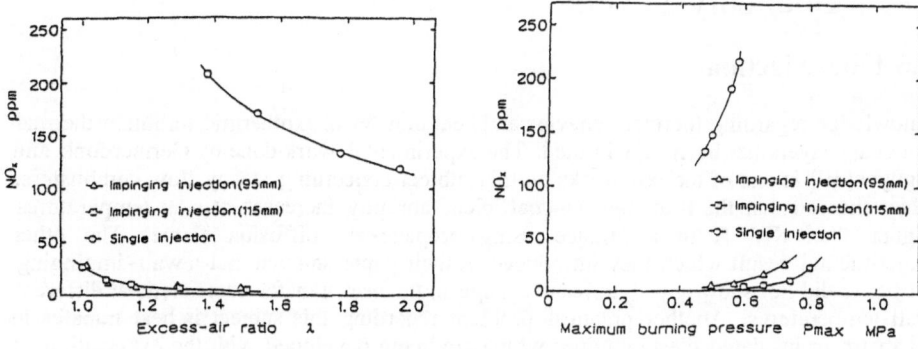

Fig. 6.18 Effects of excess-air ratio and maximum burning pressure on NOX

is the latter case. At the present time it is too early for the author to give an exact explanation of the results. The reduction of NOx emission has two points of view, chemical and physical. The violent mixing and Internal EGR effects, produced by the impinging injection, can reduce the temperatures of the burning zone where NOx is formed. Such effects are partly supported by the high-speed schlieren photographs of the burning zone [18].

6.5.4 Concluding remarks

In order to investigate the compatibility between the low-emissions of particulate and high-load operation in practical diesel engines, experiments have been conducted to examine the effects of intense mixing and the so-called Internal EGR in the combustion region on dry-soot and NOx emissions and the maximum burning pressure, by using the impinging fuel spray injection in a closed vessel.

The results are summarized as follows: 1) It is possible to realize very rapid combustion by this type of injection. 2) This combustion technique makes it possible to predict the conditions under which the low-soot and low-NOx emissions are compatible with high-load operation in diesel engines.

By using the rapid compression machine, further experiments are scheduled at higher initial temperatures and pressures of 800 to 1000 K and 3 to 5 MPa, respectively, including the measurements of HC, CO and CO_2 other than those of soot and NOx in the spray combustion by the single and impinging injection.   (Satoshi Okajima)

6.6 Heat Transfer in Exothermic Turbulent Thermal Boundary Layers

6.6.1 Introduction

Knowledge regarding thermal behavior and heat transfer in exothermic turbulent thermal boundary layers has been very limited. The experimental work done by Germerdonk and Nguyen[19] is one of the few works on this subject concerning steady flow combustors. They showed that the heat transfer coefficient abruptly increases at wall temperatures higher than 873 K in a furnace using propane–air diffusion flame. The other experimental result which they introduced in their paper showed that a wall–impinging jet flame did not show any significant change in the heat transfer coefficient at different wall temperatures. Another practical problem regarding this subject is heat transfer in adiabatic or insulated diesel engines which are being developed with the expectation of reducing wall heat losses by using ceramics. However, several experimental results have posed the problem that higher wall temperatures resulted in higher wall heat fluxes during a combustion period [20,21,22]. On the other hand, there have been nearly the same number of experiments which have shown that the wall heat fluxes during a combustion period decrease in the insulated engines as expected [20,22,23]. There has been no answer given to the contradiction between these results. The present study aims at finding the reason, or in what condition, heat flux increases or decreases as the wall temperature is elevated.

6.6.2 Wall temperature dependance of heat transfer in a steady turbulent diffusion flame – Experimental study [24]

(a) Experimental apparatus

Figure 6.19 shows the test wall and diffusion burner used for the experiment. The test wall was electrically heated from the back to raise the wall temperature to the desired level ranging from around 500 K to 1000 K. The burner was designed so as to produce an approximately two–dimensional flame. It had two slit nozzles; one was for fuel and the other for air. Methane was used for fuel and supplied at a flow rate of 5 l/min with an equivalence ratio of 6.37. The air was fed through the slit at a flow rate of 7 l/min. Schlieren technique was adopted to obtain wall heat fluxes by way of density gradient measurements near the wall. Thermocouples embedded in the wall were not used for the heat flux measurements, because a considerable amount of radiation heat flux from the hot wall to surroundings was estimated and it would give unreliable gas–to–wall heat flux measurements.

(b) Plasma potentials in a flame at test locations

Three different locations were selected along the wall surface (x direction) by checking plasma potentials in the flame. The first location, indicated by x=0 mm in Fig. 6.20, was taken just above the point where the jet flame touched the wall. As seen in the figure, it exhibits the highest plasma potentials reflecting active exothermic reaction. The third location, 40 mm above the first, was selected as the location where almost no plasma potential was detected, suggesting exothermic reaction almost completed. At the second, indicated by x=20 mm, intermediate plasma potentials are seen between those at the first

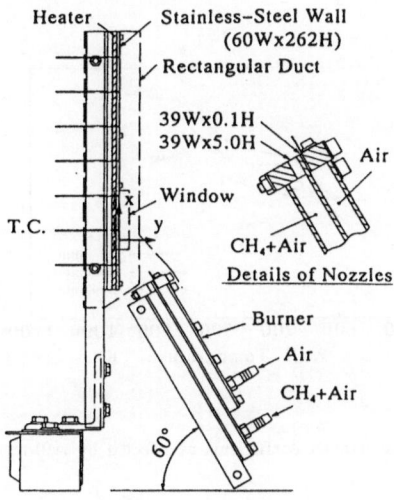

Fig. 6.19 Test wall and burner

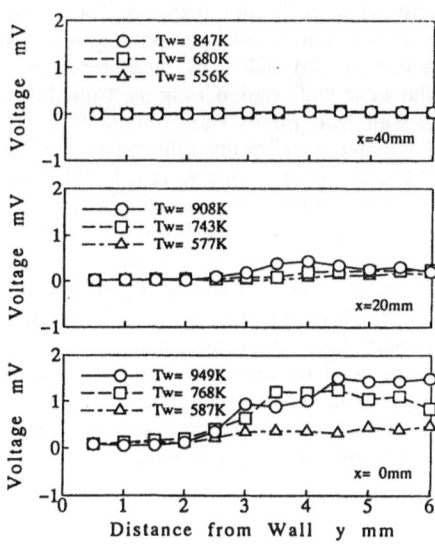

Fig. 6.20 Time-averaged plasma potentials at elevated wall temperatures

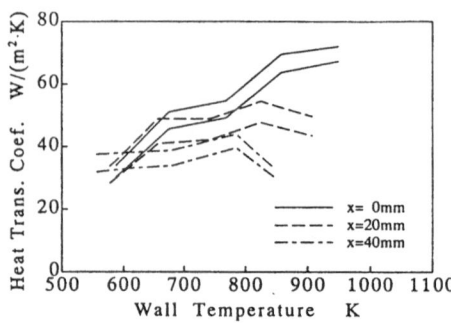

Fig. 6.21 Heat transfer coefficients as affected by wall temperature

and third location. The examination of wall temperature dependence of the plasma potentials at the first and second locations suggests that the exothermic reaction becomes more active with a quench distance becoming shorter as the wall temperature increases.

(c) Heat transfer coefficients at different wall temperatures

A pair of Schlieren photographs were taken with the edge which was set in turn in a direction opposite to one another at each edge position. The crossing point of each pair of brightness distribution curves obtained from the photographs gave the distribution of density gradients perpendicular to the wall. The measurements provided data for the temperature gradients in the near wall region ranging from 0.1 mm to 0.6 mm away from the wall and thus the wall heat fluxes were obtained.

   Figure 6.21 shows heat transfer coefficients affected by the wall temperature at the three test locations; the heat transfer coefficients were calculated by taking the difference between the wall temperature and the maximum gas temperature. Pairs of the same lines shown in the figure refer to the upper and lower limits of errors coming from the errors included in calibration constants. The heat transfer coefficient is seen to considerably increase at x=0 mm and to moderately increase at x=20 mm as the wall temperature increases. These tendencies can be explained by the plasma potential measurements previously shown which imply the exothermic reaction supplies more heat to the boundary layer with increasing wall temperature. The heat transfer coefficient at x=40 mm does not exhibit any significant change. The latter fact is well known as the heat transfer characteristics in a non–reactive thermal boundary layer and thus it gives the validity of the measuring technique used in this experiment.

6.6.3 Wall temperature dependance of heat transfer in an unsteady turbulent flame in a piston–cylinder apparatus–Computational study in one dimension [25]

(a) Outlines of the theoretical model

The system to be solved is a combustion chamber of a piston–cylinder apparatus as shown in Fig. 6.22. The governing differential equations are described in a one-dimensional coordinate perpendicular to the cylinder head surface with the ensemble–averaged dependent variables which include velocity, enthalpy, turbulent kinetic energy, its dissipation rate and mass fractions of fuel, oxygen and combustion products. The continuity equation is described on the assumption of the spatially uniform pressure which varies with time. Submodels introduced in the governing equations are outlined in the following:
(1) The k–ε model combines the model coefficients to take into account the compression–expansion effect, and the model functions to describe the continuously changing nature of turbulence characteristics from a low Reynolds number region (near the wall) to a high Reynolds number region (apart from the wall).
(2) The heat release model is represented by the combination of the Magnussen model and the Arrhenius–type model for a one–step global reaction. The idea of the model is that the time scales characterized by turbulent mixing and chemical reaction are compared and then the longer one is taken as a controlling time scale.

(b) The method of numerical computation

The pressure is calculated by a global thermodynamic model using an assumed global heat release rate. The velocity component parallel to the wall is taken into account to integrate swirling flow effects on turbulence. This velocity is assumed to obey the logarithmic law of the wall. A propane–air mixture is given at the time of ignition (–25 deg. CA) which takes place in the middle between the cylinder head and the piston. The piston–cylinder apparatus has a bore of 80 mm, a stroke of 80 mm and a compression ratio of 9. The crank revolution speed is set to be 1000 rpm. An implicit method

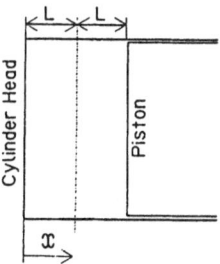

Fig. 6.22 Piston–cylinder apparatus for an unsteady one–dimensional model

numerically solved a set of the governing equations with a grid spacing of 30 microns in the region within 3 mm from the wall and 60 microns beyond it.

(c) Heat fluxes and quench distance

Figure 6.23 shows wall heat fluxes computed for a uniform distribution of fuel–air mixture with an equivalence ratio of 1.0, with several different wall temperatures. It is seen that during compression the higher the wall temperature is, the less the heat flux becomes. This is a well–known tendency seen in non–reacting thermal boundary layers. On the other hand, it is noted that the heat flux peaks higher with increasing wall temperature during combustion. The reason for this is explained by the fact that a quench distance becomes shorter as the wall temperature increases, as is shown in Fig. 6.24; the quench distance is obtained by finding the distance where the 1500 K point of gas temperature comes closest to the wall. It is also seen in the figure that the computed quench distance compares well with the experimental result which was obtained by

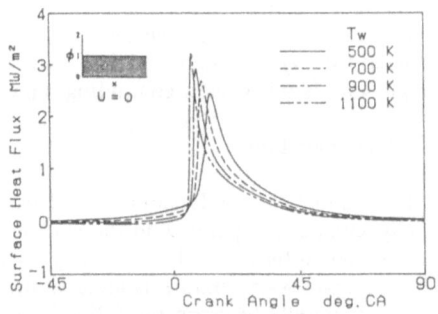

Fig. 6.23 Heat fluxes at elevated wall temperature for stoichiometric uniform fuel distribution

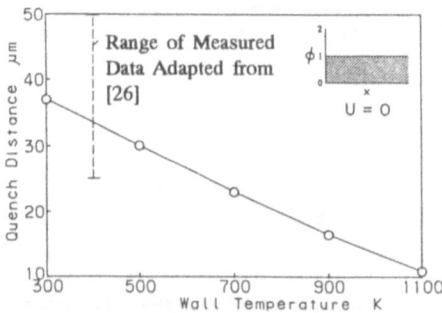

Fig. 6.24 Quench distance versus wall temperature

Lucht et al.[26] at a wall temperature of 400 K under the same conditions as those of the present computation.

(d) Influence of additional flow

Because of the one–dimensional nature of the model, a gas velocity only reflects the compression or expansion with the boundary condition of the velocity=0 at the wall. However, there are some possible cases where fuel–air mixtures are carried by additional gas motion approaching the wall. For example, fuel injection may cause such an additional gas motion; a porous wall such as a deposit layer or a plasma–sprayed– zirconia coating may make gas penetrate into pores. These additional flows are schematically shown in Fig. 6.25. To take these cases into account, an additional flow with a constant velocity, U, is added to the one–dimensional velocity during the combustion period; this means that the velocity is U at the wall.

The computations were performed with the non–uniform fuel distribution, an equivalence ratio being 0.5 at the wall and 1.5 in the middle. Figure 6.26 compares the result computed for U=0 with that for U=0.007Cm (Cm: mean piston speed). It can be seen from this figure that the additional flow makes the heat flux peak lower when the wall temperature is elevated. The examination of burning rate distribution reveals that the additional flow leaves almost no fuel–air mixture to burn when the flame front

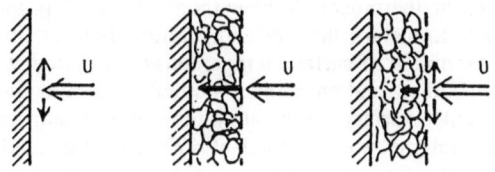

Fig. 6.25 Schematic of models for additional flow

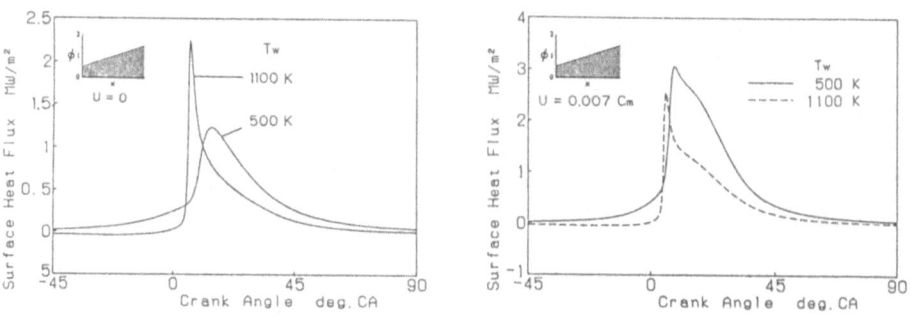

Fig. 6.26 Heat fluxes at different wall temperatures with U=0 and U=0.007Cm
for non–uniform fuel distribution

comes close to the wall and that the hot combustion products, which have completed
exothermic reaction, touch the wall surface. This is the reason why the additional flow
makes the heat flux behavior similar to that seen in a non–reactive case.

(Kazuie Nishiwaki)

6.7 Heat Transfer Measurements in Combustion Systems

6.7.1 Thin–film thermocouple for measuring the instantaneous temperature on surface of combustion chamber wall

The heat flux to the combustion chamber wall surfaces of a reciprocating engine varies
from zero to as high as 10 MW/m^2 and back again to zero in a very short time (msec
order). The instantaneous surface heat flux from a working gas to the wall is obtained
by using measured instantaneous wall surface temperature. In general, the coaxial–type
thin–film thermocouple is used for measurement of the instantaneous wall surface
temperature. The temperature in a wall, produced by the unsteady boundary conditions
at the gas–wall interface, is damped out in quite a small distance (1~2 mm) from the
surface. Hence, to detect the instantaneous wall surface temperature, the size of the
sensing element of the thin–film thermocouple is required to be as small as possible and
the sensing element, as close to the surface as possible. Another requirement is that the
construction materials of the thermocouple have thermophysical properties close to those
of the wall material so as to disturb the wall temperature field as little as possible [27].

Figure 6.27 shows examples of surface temperatures measured with copper–bodied,
Loex (aluminum alloy)–bodied and constantan–bodied thin–film thermocouples which
were embedded in the center surface on the aluminum alloy piston in a 4–stroke spark–
ignited engine when the engine was operated at 4000 rpm full load. Since copper has
a higher thermal diffusivity than the other materials, it is found that the temperature
variation is small. When the thermal diffusivity becomes smaller with that of Loex and
constantan, a bigger difference in temperature variation appears. Figure 6.28 shows the

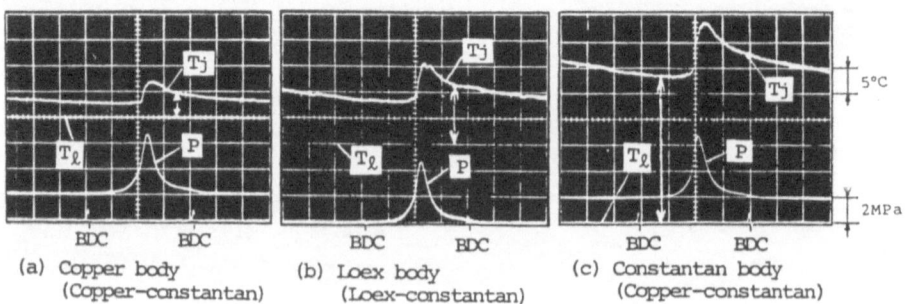

Fig. 6.27 Instantaneous surface temperatures in a cycle of each thin-film thermocouple (4000 rpm, full load)

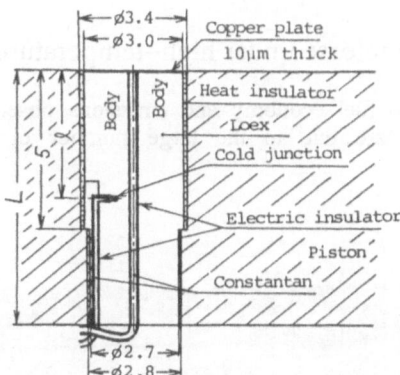

Fig. 6.28 Loex-bodied thin-film thermocouple in piston

Loex–bodied thin–film thermocouple embedded in the piston which was used in the experiment.

6.7.2 Heat transfer in a cycle to the combustion chamber wall surface

As mentioned above, the body material of the thin–film thermocouple should be the same as that of the measuring object, or the thermophysical properties of the body material should be as close as possible to those of the object. To measure heat transfer (heat loss) in a cycle to the combustion chamber walls (piston, cylinder liner, cylinder head, intake and exhaust valves) from combustion gas, an appropriate material was chosen as body material of each thin–film thermocouple. For example, an iron–body iron–constantan thin–film thermocouple was used in the cylinder [28]. Loex–constantan thermocouples with the body material of aluminum alloy Loex were used in the cylinder head and the piston, since the material of the cylinder head was a high–silicon–content aluminum alloy, similar to the material of the piston which was made from Loex. These thin–film thermocouples were embedded in the combustion chamber walls in a water–cooled 4–stroke gasoline engine [28,29].

   Figure 6.29 shows measured examples of piston instantaneous surface temperature at the center (point 1) and the knocking zone (point 16), when the engine was operated at speeds of 2000, 3000 and 5000 rpm, with full load, respectively. Figure 6.30 shows an instantaneous heat flux at measuring point 1 on the crown surface of the piston as a function of the crank angle when the engine is operated at 3000 rpm full load. As shown in the figure, most of the heat flows into the piston during the working stroke. Dotted lines show the mean heat flux throughout the entire stroke.

   Figure 6.31 shows the amount of heat flow into the piston in each stroke at full load. Heat loss in the working stroke increased as the engine speed increased, but there was little change in heat loss in the other strokes. In the case of knocking with a fuel of low octane, only the heat in the working stroke increased by about 70% compared to the operation with regular fuel.

Figure 6.32 shows the heat transfer ratios of the heat losses of the piston, cylinder liner, cylinder head, intake and exhaust valves against respective heat supply values at full load. The total heat loss of all components amounts 7.0 ~ 10.0%.

6.7.3 Real rate of heat release under high–temperature combustion [30,31]

The conflicting results on fuel economy and emissions observed in LHR (low heat rejection) engine tests are due to the large number of possible LHR engine

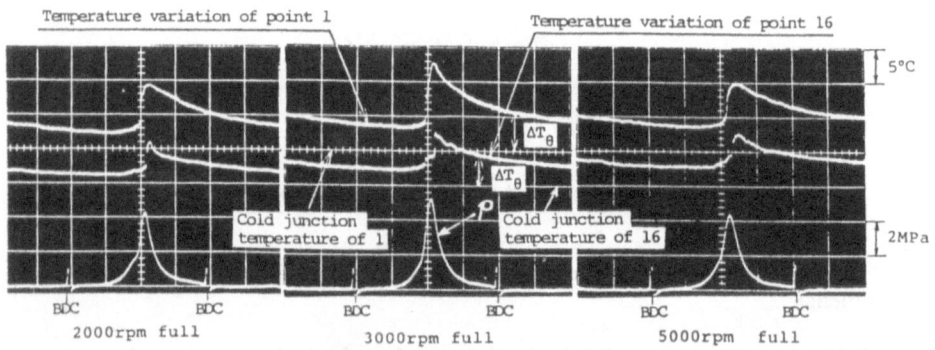

Fig. 6.29 Measuring examples of piston instantaneous surface temperature at center and knocking zone on piston

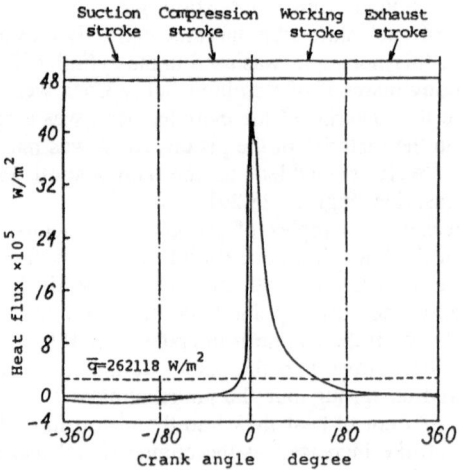

Fig. 6.30 Instantaneous heat flux flowing into a center of piston at 3000 rpm and full load

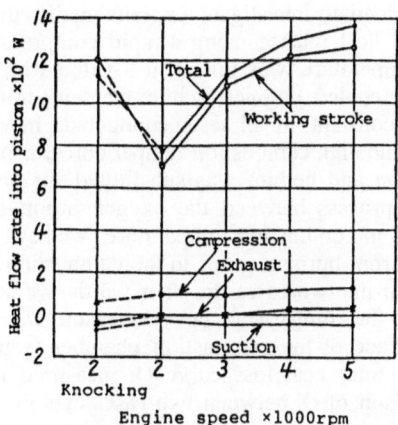

Fig. 6.31 Heat flow rate variation into piston due to engine speed at full load

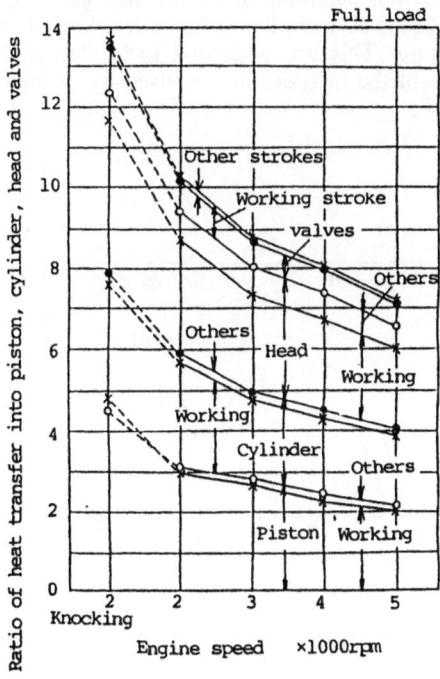

Fig. 6.32 Heat transfer ratio into each combustion chamber wall component

configurations, test conditions, and analysis techniques used. As far as the heat release analysis is concerned, it is difficult with engine experiments to separate the effects of cylinder gas temperatures and combustion chamber wall temperatures on the rate of heat release. This study is intended to investigate experimentally the effect of cylinder gas temperature on the rate of heat release using a rapid compression expansion machine [32]. The cylinder gas temperature was raised to a value which is 300~400 K higher than that in conventional cooled engines, while the temperature of the combustion chamber walls was kept constant in all tests conducted. In order to achieve a high compression temperature and high combustion temperatures, a monatomic molecular gas mixture composed of argon and helium was substituted for nitrogen in air. When we compare the heat release process between the oxygen–argon–helium mixture and air having two different levels of combustion temperature, we have to take into account the precise value of heat loss from burning gases to the combustion chamber walls because the heat loss is different for the two cases. In other words, we have to use the measured real rate of heat release for the comparison. For this reason, we measured local heat flux at 11 locations on the surface of the combustion chamber walls.

Figure 6.33 shows the total heat loss curve Ql measured by iron–body thin–film thermocouples. A comparison of Ql between two cases exhibits a difference amounting to 12% corresponding to the temperature difference in the gas law temperature. The real rate of heat release Qr obtained by adding the total heat loss to the apparent rate of heat release is shown in Fig. 6.34. Now we can compare Qr between two cases having different levels of combustion temperature. The ignition delay is shorter for the high gas temperature case, and this results in the lower rate of initial heat release. In the latter half of the diffusion combustion duration, Qr for the high gas temperature case holds a higher value as compared to the air case, indicating a deteriorated combustion under the high combustion temperature. This was supposed to be due to the decrease in local Reynolds number because of the increase in gas viscosity at high gas temperatures.

(Yoshiteru Enomoto)

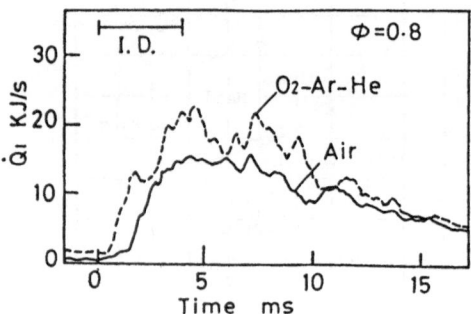

Fig. 6.33   Total heat loss curves, from
[30] with permission

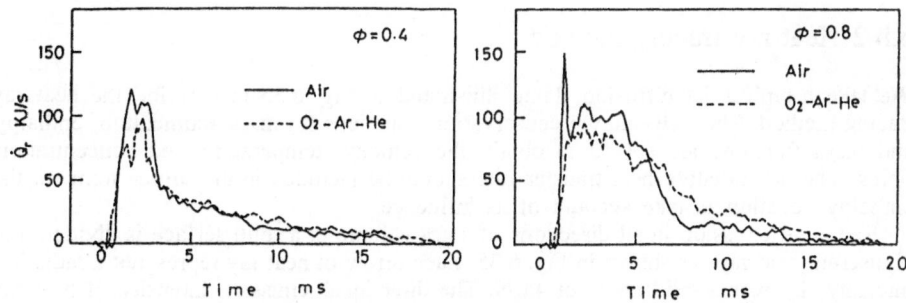

Fig. 6.34   Effect of temperature on real rate of heat release,
from [30] with permission

6.8 Simulation of Radiative Heat Transfer in Flames

6.8.1 Introduction

Radiation is an important way of heat transport in combustion, heating and drying processes in various industries. Nevertheless, it is often ignored in the numerical prediction of combusting flows. On the other hand, when the radiative heat transfer is the subject of interest, the prediction of convection and conduction is eliminated by introducing an approximate evaluation or an assumption of the flow field. It seems that the numerical modeling of radiative heat transfer and that of combusting flow have kept a distance from each other in spite of the significant interaction between them.

The primary reason is that it is difficult to develop a comprehensive numerical method due to the difference in the numerical scheme between radiation and convection/conduction calculations, with the exception of the flux method [33]. Because the radiative heat is transferred over distances, the temperature distribution throughout the whole surrounding system must be known even in evaluating a local temperature, and the phenomenon is described by an integral equation.

On the other hand, the heat transferred by either convection or conduction depends mostly on the local temperature gradient in the flow field, which is expressed by a differential equation. The second is that the mechanism of soot formation and oxidation, which play significant roles in the emissivity of flames, have not been well formulated.

Some analytical or numerical models dealing with the radiative heat transfer are found in literature, such as the zone method [34], Monte Carlo method [35], and flux method [36]. From the computational point of view, however, each of them still has some problems: cumbersome computer programming to calculate complex total exchange areas; requirement of a large computer time to satisfy the statistical accuracy; difficulty in accounting for the distribution of the absorption coefficient in a flame; low capability of dealing with the directionality of radiation.

In order to discuss the soot formation and oxidation in flames it is essential to develop a comprehensive numerical method to deal with combustion and radiative heat transfer simultaneously. The involved "heat ray tracing method" is based on the idea originally proposed by Hayasaka [37]. Its principle is simple as well as facilitating computations, and the distribution of absorption coefficient in a flame can be easily taken into account.

6.8.2 Heat ray tracing method

We take a typical jet diffusion flame illustrated in Fig. 6.35 to describe the heat ray tracing method. The well–known conservation equations for mass, momentum, enthalpy and mass fractions are solved to obtain the velocity, temperature and concentration fields. The net radiative heat transfer rate should be included in the source term for the enthalpy equation to take account of its influence.

Radiative emission in all directions of a gas volume or a solid surface is divided into N discrete heat rays as shown in Fig. 6.35. Each arrow of heat ray represents a radiation intensity, I_0, over a solid angle of $4\pi/N$. The directional emission intensity of heat ray from a wall element follows the Lambert's law. When a heat ray passes through a gas element, a fraction of its radiation energy is absorbed by this element in proportion with the path length, l, and the absorption coefficient, k_a, of the element. The ratio of the heat ray intensity at the exit of the element, I_e, to that at the inlet, I_i, is given by Beer's law.

$$I_e = I_i \exp(-k_a \cdot l) \tag{4.1}$$

Therefore, by tracing all the heat rays emitted from an element, each path length of a heat ray through any element and its local absorption coefficient yields the distribution of the absorbed fraction of the energy emitted from the element. Since the flow field is axisymmetrical two–dimension, the divided gas elements are numbered as (I,J) or (i,j)

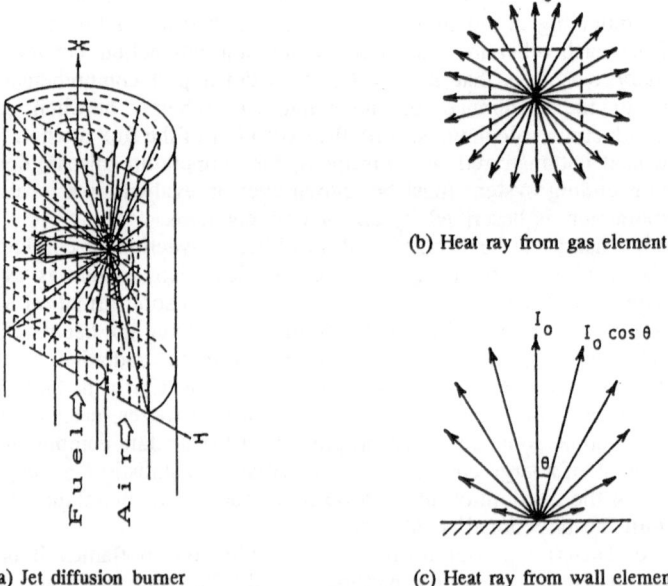

(b) Heat ray from gas element

(c) Heat ray from wall element

(a) Jet diffusion burner

Fig. 6.35 Flow configuration and concept of the radiative heat ray method

to discriminate between emission and absorption of the element, respectively. The inner surface of the combustion duct is also divided into one–dimensional wall elements, and numbered as (I_w) or (i_w) as well. Practically, the gas element is an annular having a rectangular cross section and the wall element is a cylindrical surface in considering the radiative heat exchange between the elements. Each annular gas element or cylindrical wall element includes a number of point sources placed at equal circumferential intervals. Tracing every heat ray from all sources until its intensity becomes lower than the specified value, an absorbed energy distribution RD(IJ– ij) or RD(IJ– i_w), that is the fractions absorbed by the element (i,j) or (i_w) of the unit emission of the element (I,J), is obtained.

The radiation energy emitted by a gas element (I,J) and the wall element (I_w) are given by

$$St_g(IJ)=4\sigma \cdot k_a \cdot T_g^{\,4} \cdot \Delta V(IJ) \tag{4.2}$$

$$St_w(I_w)=\varepsilon_w \cdot \sigma \cdot T_w^{\,4} \cdot \Delta A(I_w) \tag{4.3}$$

where σ is the Stefan–Boltzmann constant, T_g the gas temperature, T_w the wall temperature, k_a the absorption coefficient of the gas element, ε_w the wall emissivity/absorptivity, V(IJ) the volume of the gas element (I,J), and A(I_w) the area of the wall element (I_w). The absorption coefficient, k_a, of a gas element (I,J) is calculated based on the weighted sum of gray gases approximation [38,39].

The radiation energy absorbed by a gas element (i,j) and a wall element (i_w) are given as follows.

$$q(ij)=\underset{I\ J}{\Sigma\Sigma}[RD(IJ\rightarrow ij)\cdot St_g(IJ)]+\underset{I_w}{\Sigma}[RD(I_w\rightarrow ij)\cdot St_w(I_w)] \tag{4.4}$$

$$q(i_w)=\underset{I\ J}{\Sigma\Sigma}[RD(IJ\rightarrow i_w)\cdot St_g(IJ)]+\underset{I_w}{\Sigma}[RD(I_w\rightarrow i_w)\cdot St_w(I_w)] \tag{4.5}$$

Consequently, the net radiative heat transfer to an element (i,j) or (iw) becomes as follows.

$$Q(ij)=\{q(ij)-St_g(IJ)\} \tag{4.6}$$

$$Q(i_w)=\{q(i_w)-St_w(I_w)\} \tag{4.7}$$

The modified formulation with the influence of reflection at walls can be found in Ref. [40].

6.8.3 Results

In numerical calculations, the combustion duct of 70 mm diameter and 400 mm long was divided into seven 5 mm intervals in the radial direction and into ten 40 mm intervals in the axial one. One hundred point emitting sources were equally distributed circumferentially in one element and 5,000 heat rays were emitted from each source. Boundary conditions at the inlet and exit of the duct were open, and the wall was a constant temperature of 300 K.

The predicted profiles of velocity and temperature are compared with the measured values in Fig. 6.36, and the remarkable influence of wall reflectivity on the absorbed

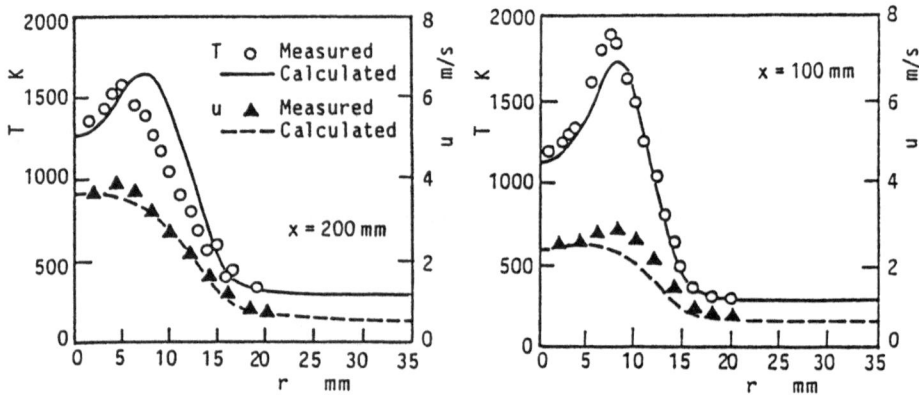

Fig. 6.36 Radial profiles of temperatures and velocities

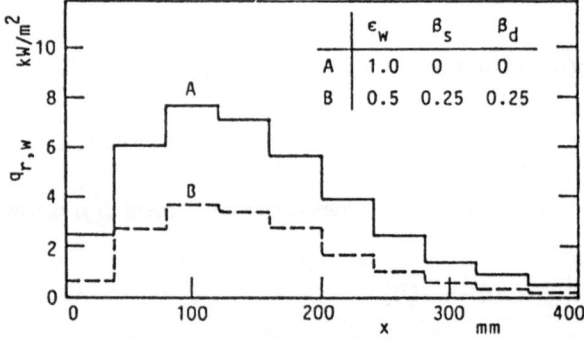

Fig. 6.37 Absorbed radiation energy by the duct wall

energy of the wall is shown in Fig. 6.37. However, only several degrees of gas temperature difference was predicted between the two assumed cases, the black body wall(A) and the gray one(B).

Although the present calculation procedure requires a longer computer time compared with those for non–radiating combusting flows, it is expected to become an effective tool to discuss the combustion processes in which the radiative heat transfer plays an important role, such as the formation and oxidation of soot or the spray combustion.

(Masashi Katsuki)

References

[1]Borman GL and Nishiwaki K (1987) Internal Combustion Engine Heat Transfer, Prog. Energy Combust. Sci., Vol. 13, pp 1–46

[2]Yang J and Martin JK (1989) Approximate Solution–One–Dimensional Energy Equation for Transient, Compressible, Low Mach Number Turbulent Boundary Layer Flows, Trans. ASME, J. of Heat Transfer, Vol. 111, pp 619 – 624

[3]Yang J and Martin JK (1990) Predictions of the Effects of High Temperature Walls, Combustion, and Knock on Heat Transfer in Engine–Type Flows, SAE Paper 900690

[4] Barbella R, Bertoll C, Ciajolo A, D'Anna A, and Masi S (1989) In–Cylinder Sampling of High Molecular Weight Hydrocarbons from a D.I. Light Duty Diesel Engine. SAE Paper 890437

[5] Kittelson DB, Pipho MJ, Ambs JL, and Luo L (1988) In–Cylinder Measurements of Soot Production in a Direct–Injection Diesel Engine. SAE Paper 880344

[6] Ikegami M, Miwa K, and Li XH (1986) Spray Process and Fuel Pyrolysis in the Initial Stage of Diesel Combustion. Bull. of JSME, Vol.29, No.235, pp 2189–2195

[7] Miwa K, Ohmija T, and Nishitani T (1988) A Study of the Ignition Delay of Diesel Fuel Spray Using a Rapid Compression Machine. JSME Int. J., Series II, Vol.31, No.1, pp 166–173

[8] Ishiyama T, Miwa K, and Kanno A (1990) Experimental Study on Fuel Air Mixture Formation and Ignition Process in Diesel Combustion. COMODIA 90, pp 565–570

[9] Glassman I (1988) Soot Formation in Combustion Processes, 22nd Symp. (Intl.) on Combustion, The Combustion Institute, pp 295–311

[10] Kurata O, Daisho Y, Borman GL, Martin JK, and Saito T (1990) Development of a soot generating combustion system using rich premixed mixtures. Proceedings of JSME, Vol. B, No. 900–14 pp 305–307

[11] Ogawa T, Daisho Y, Kurata O, and Saito T (1990) Particulate formation in compression–ignited fuel–rich mixtures. COMODIA 90, pp 633–638

[12] Kamimoto T, Myurng HB, and Kobayasi H (1989) A study on soot formation in premixed constant–volume propane combustion. Combust. Flame 75, pp 221–228

[13] Sawyer RF, Cernansky NP, and Oppenheim AK (1973) Factors controlling pollutant emissions from gas turbine engines. Atmospheric Pollution by Aircraft Engines, AGARD CP–125, Paper No.22

[14] Plee SL, Ahmad T, Myers JP, and Siegla DC (1981) Effects of flame temperature and air–fuel mixing on emissions of particulate carbon from a divided–chamber diesel engine. Particulate carbon – Formation during combustion, Plenum Press, New York, pp 423–487

[15] Iida N and Sato GT (1988) Temperature and mixing effects on NOx and particulate. SAE Trans., Paper 880424

[16] Iida N and Watanabe J (1990) Surrounding gas condition effects on NOx and Particulate. International Symposium COMODIA 90, Kyoto, pp 625–632

[17] Mori M, Okajima S, Iinuma K, and Kumagai S (1987) A study on soot reduction in a constant–volume spray combustion (Report No. 2). 25th Combustion Symposium of Japan, November, pp 85–90

[18] Fujimoto S and Kumagai S (1984) Possibility of low–Nox and high–load combustion in pre–mixed gases. 20th Symp.(Intl.) on Combustion, The Combustion Institute, pp 61–66

[19] Germadonk R and Nguyen N (1985) Increase of the local heat transfer coefficient by convection vive phenomenon. Ger. Chem. Eng., Vol. 8, pp 81–86

[20] Furuhama S and Enomoto Y (1987) Heat transfer into ceramic combustion wall of

internal combustion engines. SAE SP–700, 870153, pp 109–124

[21] Woschni G and Spindler W (1988) Heat transfer with insulated combustion chamber walls and its influence on the performance of diesel engines, Trans. ASME, J. Eng. Gas Turbines and Power, Vol. 110, pp 482–502

[22] Huang JC and Borman GL (1987) Measurements of instantaneous heat flux to metal and ceramic surfaces in a diesel engine, SAE SP–700, No. 870155, pp 137–152

[23] Morel T, Wahiduzzman S, and Fort EF (1989) Heat transfer in a cooled and an insulated diesel engine, SAE SP–785, 890572, pp 185–196

[24] Kojima T, Hagiwara T, and Nishiwaki K (1992) Wall temperature dependence of the heat transfer coefficient in exothermic thermal boundary layers. Trans. JSME, Vol.58, No.553, pp 2812–2818

[25] Nishiwaki K and Hagiwara T (1990) The potential of heat transfer control on insulated combustion chamber walls. Proc. COMODIA '90, pp 315–320

[26] Lucht R P and Maris M A (1987) CARS measurements of temperature profiles near a wall in an internal combustion engine, SAE Paper 870459

[27] Enomoto Y,and Furuhama S (1985) Study on thin film thermocouple measuring instantaneous temperature on surface of combustion chamber wall in internal combustion engine. Bulletin of JSME, Vol.28, No.235, pp 108–116

[28] Enomoto Y, Furuhama S,and Minakami K (1985) Heat loss into combustion chamber wall in 4–stroke gasoline engine (1st report, Heat loss to piston and cylinder) Bulletin of JSME, Vol.28, No.238, pp 647–655

[29] Enomoto Y and Furuhama S (1986) Heat loss into combustion chamber wall in 4–stroke gasoline engine (2nd report, Heat loss into cylinder head, intake and exhaust valves) Bulletin of JSME, Vol.29, No.253, pp 2196–2203

[30] No SH, Kobori S, Kamimoto T,and Enomoto Y (1991) High temperature diesel combustion in a rapid compression–expansion machine SAE Paper 911845

[31] No SH, Kobori S, Kamimoto T,and Enomoto Y (1991) Diesel combustion with a rapid compression–expansion machine (2nd report, Real rate of heat release at high combustion temperature) Trans. of JSME, Vol.57, No.540, pp 2833–2838

[32] Kamimoto T, Kando H, Kobori S, Hatano H, Kobayashi H, Tsuchiya K (1988) Development of a rapid compression–expansion machine to simulate combustion in diesel engines, SAE Paper 881640

[33] Gosman AD and Lockwood FC (1973) Incorporation of a flux model for radiation into a finite–difference procedure for furnace calculation. 14th Symp. (Intl.) on Combustion, The Combustion Institute, pp 661–671

[34] Hottel HC and Cohen ES (1958) Radiant heat exchange in a gas–filled enclosure: Allowance for nonuniformity of gas temperature. AIChE J., 4–1, pp 3–14

[35] Howell JR and Perlmutter M (1984) Monte Calro simulation of thermal transfer through radiant media between gray walls. Trans. ASME, Ser. C, 86–1, pp 116–122

[36] Siddall RG (1974) Flux method for the analysis of radiant heat transfer. J. Inst. Fuel, 47–391, pp 101–109

[37] Hayasaka H (1987) A direct simulation method for the analysis of radiative heat transfer in furnaces. HTD–Vol.74(ASME), pp 59–63

[38] Beer JM (1974) Methods for calculating radiative heat transfer from flames in combustors and furnaces. Heat transfer in flames, John Wiley & Sons

[39] Truelove JS (1976) A mixed grey gas model for flame radiation. Harwell Report, AERE–R8494

[40] Katsuki M, Mizutani Y, Ando A, Hattori Y, Jinja Y, and Lee DH (1990) Numerical prediction of coaxial flow diffusion flames with radiative heat transfer. JSME Int. J., Ser.II, 33–4, pp 772–777

Chapter 7
Effects of Fuel Properties in Combustion Systems

7.1 Introduction

In combustion research, most efforts have been concentrated on an understanding of factors such as chemical kinetics, flame structure, and flame stability, and very little on the effects of fuels. Additionally, research addressed to fuel effects are mostly limited to premixed flames aiming to eliminate the effects of physical mixing. The increasing demand for alternative fuels and environmental concerns require an investigation of the effects of fuels, particularly in diffusion combustion. This chapter deals with two basic studies of fuel effect in diffusion combustion and two studies of application of extremely different fuels, hydrogen and heavy residual fuels, in diesel engines.

Petroleum fuel is a complex mixture of hydrocarbon compounds, which may be roughly categorized into paraffins, olefins, naphthene and aromatic compounds. The paraffin family is chemically stable and does not oxidize in the atmosphere. Paraffins are gaseous with carbon numbers below 4, liquid with 5 to 15, and solid above 16. At constant temperatures, the vapor pressure of hydrocarbons is in the order: olefins > paraffins > naphthene ≒ aromatics. Ignition characteristics generally become better as the carbon number increases, and at the same carbon number the ignitability is in the order: paraffins > olefins > naphthene > aromatics. Particularly aromatics usually have very poor ignition characteristics. Aromatics and olefins generally produce large amounts of soot; but this is also influenced by ignition characteristics and sensitive to combustion system differences.

Investigations of the heat cracking process of hydrocarbons [1] have shown that paraffins are cracked once into low carbon–number hydrocarbons, and then dehydrogenate with the formation of benzene ring compounds through polymerization. This is followed by the formation of multiple ring compounds. Aromatics are only cracked with difficulty, and polymerize directly into larger ring compounds; but at high temperatures some aromatic hydrocarbons also crack into lower carbon–number hydrocarbons. The amount of poly–aromatic hydrocarbons, PAH, produced in the heat cracking process is in decreasing order: aromatics, naphthene, olefins, and paraffins produce the least PAH. Higher carbon number hydrocarbons produce more olefinic compounds during the cracking, resulting in larger amounts of PAH. Start of PAH production occurs at increasing temperatures in the order: olefins, aromatics, paraffins, and naphthene.

Experiments with CI engines have shown that the aromatic and sulfur contents in fuels have very little effect on total unburnt hydrocarbons (UHC) in the exhaust, when ignition lags remain constant [2]. In this case, however, the components of UHC were different, and in general, increasing aromatic content in fuels results in an increased aromatic fraction and decreased C_2–C_4 olefinic fraction in the exhaust [3]. This indicates that even with similar unburned fractions of the fuel, the degree of cracking is different, depending on the fuel properties. Typical additives for improving fuel octane numbers like oxygenate additives have little effect on UHC, while nitrogen–compound additives increase both UHC and NOx considerably [4]. The origin of the UHC in CI engines are the over–lean and over–rich mixture region in the fuel spray, the quenched region close to the wall, and fuel spilled from the suck volume in injectors. Hydrocarbons created in these regions are clearly affected by fuel volatility, viscosity, and the flammable limits of the fuel mixture. Increases in fuel viscosity and fuel amount impinged on the wall increase UHC in the exhaust.

In diffusion combustion, the combustion characteristics of different combustion systems do not always exhibit similar tendencies, and sometimes show conflicting changes, when fuel properties are changed. This happens because fuel properties affect both the physical and chemical characteristics, and physical factors are particularly sensitive in the combustion system. The physical characteristics include the mixing process of fuel and air, and chemical characteristics related to the chemical reaction. Therefore careful investigation must be conducted to separate fuel effects from equipment effects. To ensure this, characteristics which enable generalization of the changes in combustion with fuel properties are very helpful.

As one important investigation on the combustion characteristics helpful for the generalization, Onuma et al. investigated similarity in the flame structure between spray and gas jet diffusion flames [5,6]. They compared droplet and temperature distributions, flow velocity, and gas composition in the flame of an air–atomizing burner. They showed that droplets within the flames do not burn individually with enveloping flames, but that a cloud of fuel vapor generated by them burns as a turbulent gas diffusion flame, and that the spray combustion flame is structurally very similar to the turbulent gas diffusion flame. They confirmed this result with heavy oil as well as kerosene. They also measured NOx distribution in the flames, and concluded that similarity is established even in NOx distribution.

Apart from the commonly used hydrocarbon fuels, hydrogen and heavy residual fuels may be investigated as fuels which have very different characteristics. Hydrogen is considered a promising future fuel, because the combustion product is ideally only water and hydrogen can be produced with, for example, wind power and solar energy. It has the potential of offering clean combustion and is free from resource limitations. Investigations of hydrogen engines may be categorized into two types depending on the fuel supply methods: premixed charge methods and direct injection method. The major problem with the both system is to control combustion with the hydrogen. Investigations of hydrogen engines are currently conducted at the following institutions:

German Aerospace Research Establishment (DLR) and BMW investigate engines aspirating hydrogen from the intake manifold and engines with hydrogen injection into the cylinder. Dimler Benz investigates water injection in hydrogen engines to solve back–fire problem. Other than these institutions, similar research is being performed in Los Alamos National Laboratory, American Hydrogen Association, Piza university in Italy, Merborn University in Australia, and Canterbury University in Newzealand.

In Japan, the Mechanical Engineering Laboratory of MITI, Toyota, and Suzuki investigate premixed charge hydrogen engines. At Mazda, premixed hydrogen is applied to rotary engines utilizing the unique characteristics of these engines. Musashi Institute of Technology has studied direct injection hydrogen engines for 20 years, and tests with engines installed in automobiles are also performed.

The residual compounds of the petroleum refining process, the so called residual fuels, are widely used, but the spray combustion mechanism has not been much investigated. Poor ignitability is considered a major problem of this fuel, and research has concentrated on the improvement of ignition characteristics. However this fuel contains compounds with higher volatility than the compressed air in the engines, and the effect on evaporation and combustion must be investigated and compared with lighter fuels.

For the ignition characteristics of residual fuels, Zeelenberg proposed a CCAI ignition index instead of cetane numbers [7]. The CCAI index is determined easily from fuel density and viscosity, and it corresponds well to the aromatic content of the fuels. Fuels with higher CCAI index generally produce larger amounts of particulate emissions [8]. Nomura et al. proposed an ignition index as a function of the volatility over $300°$ C, and a combustion index as a function of the volatility over $500°$ C, based on the observation that the volatility is closely related to ignition and combustion characteristics [9]. There is also a report that when aromatic content is higher than 30%, ignition and combustion difficulties become significant [10]. It has also been shown that ignition characteristics are strongly correlated with the aromatic content of compounds with more than three benzene rings [11].

As mentioned above, there is only a limited amount of research on the effect of fuel differences on combustion and least research into diffusion combustion. Research on fuel effect is also necessary when considering potential shortages of petroleum and air pollution concerns.   (Tadashi Murayama)

7.2 Basic Studies of Fuel Effects on Combustion

7.2.1 Fuel effects on the turbulent diffusion flame structure

(a) Introduction

When changing fuel properties, flame structures sometimes display different characteristics, depending on the combustion environment. In diesel engines, for example, higher smoke emissions from fuels with higher aromatic hydrocarbon contents depends on the engine. Therefore it is important to know possible combustion characteristics to identify and separate effects of the fuel from effects of the equipment.

It was shown that turbulent diffusion flames have many similar characteristics for a variety of fuels. This section discusses similarities in the flame structure for different fuels and the mechanism why such similarities appear. Details   are described in references [12–14].

(b) Similarity in flame structure

Different temperature distributions in turbulent jet diffusion flames may be measured for various kinds of fuel, nozzle diameter, or jet velocity. However, when the temperature distributions are re-plotted in a non–dimensional manner as shown in Fig. 7.1, all data except for very low volatility fuels arrange on one curve. In Fig. 7.1, the axial temperature is non–dimensionalized by the theoretical adiabatic flame temperature and ambient air   temperature, T_0 and T_∞, and it is plotted for axial distance, non-dimensionalized by the flame length x_r.

Even methanol, where the stoichiometric air–fuel ratio is almost half of LPG, gives the same non–dimensional temperature distribution as LPG. Temperatures in the radial direction are also similar. Generally, fuels of low volatility than methanol show thinner

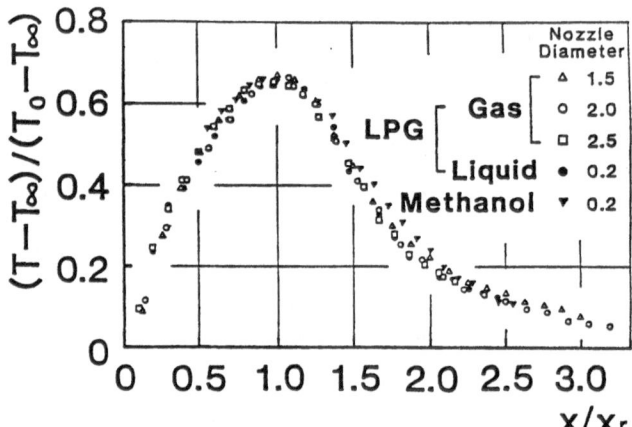

Fig. 7.1 Axial temperature distribution

and longer flames, but the axial temperature is still similar as shown in Fig. 7.1, if the distance from the nozzle is modified to an imaginary downstream nozzle position.

(c) Boundary characteristics and eddy structure

When Schlieren photographs of jet flames are compared for a variety of fuels, the boundary is always straight and the angle of the jet is constant. Fig. 7.2 is an example, which compares non–flame and flame jets of LPG. This boundary behavior is closely related to the mechanism of the temperature similarity as will be discussed later. However it is limited to the steady state condition as seen in Fig. 7.2, where the boundary expands at the moment of extinguishing.

The eddy structure of jets are significantly different with and without flames. Fig. 7.3 was taken by a laser sheet method with titanium oxide tracer particles, T_iO_2, which were mixed in the fuel. The tracer diffuses nearly uniformly in the non–flaming jet, while an apparent eddy structure is observed in the jet flame. This indicates that there are no small scale eddies in the flame jet, due to its higher kinematic viscosity. Fig. 7.3 also indicates that at the micro scale combustion takes place by molecular diffusion even in strong turbulent combustion, because tracers, which are considered to be in the fuel region, do not mix uniformly even far downstream of the flame; this is explained by the reaction of fuel and air taking place by molecular diffusion and the tracer follows the bulk flow, so that tracer particles remain in the fuel region after combustion. Even with this difference in eddy structure, large scale mixing appears to depend only on density as will be discussed later.

(d) Mechanism of similarity

This sub–section explains the mechanism why temperature distributions become similar, first considering the mechanism of straight boundaries. If vortex rings are assumed to

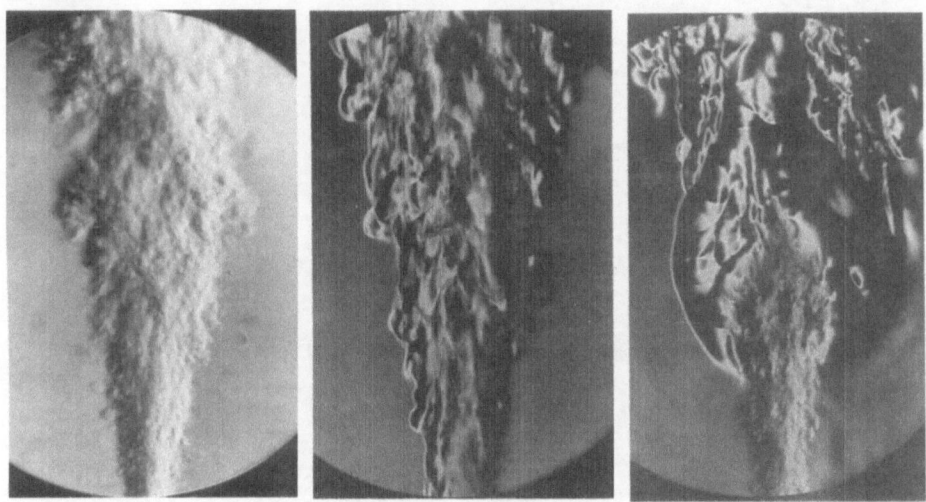

Fig. 7.2 Schlieren photograph of non–flame jet(left), flame jet(center), and extinguishing moment(right)

be emitted continuously from the nozzle, the air entrainment may be considered proportional to the number–density of vortex rings in a unit area and to their intensity, because air is entrained in the jet by turbulent vorticity. From the continuity equation of angular momentum of vortex rings and the equation of number–density of vortex rings, air entrainment through a unit length in the x direction, m_x, can be derived as [14]:

$$m_x \propto \rho \Gamma_e n V_e / u_m b d_0 \propto \rho_0 u_0^2 d_0^2 / u_m b \tag{7.1}$$

Here, ρ is the density, Γ_e is the circulation of the vortex rings, n is the number of rings, Ve is the volume of a ring, b is the diameter of the boundary, d_0 is the nozzle diameter, the suffix 0 indicates the state at the nozzle exit, and m is the state on the axis. For the equations of momentum and continuity, established with this air entrainment, it is possible to show that the only solution is a straight boundary even for non–uniform gas densities along the axis [14].

The physical meaning of the analysis is that the boundary may expand once after ignition, but that the air entrainment decreases inversely proportional to b and u_m until the boundary becomes straight. The decrease in air entrainment is caused by a decreased number–density of the vortex rings due to expansion.

Next, consider why similar temperature distributions appear as shown in Fig. 7.1. One can derive the following analytical temperature and velocity distributions with the straight boundary condition [13] in a similar manner to ref.[15], where it is shown that similarity can be established among enthalpy, velocity, and chemical species in mixing controlled combustion:

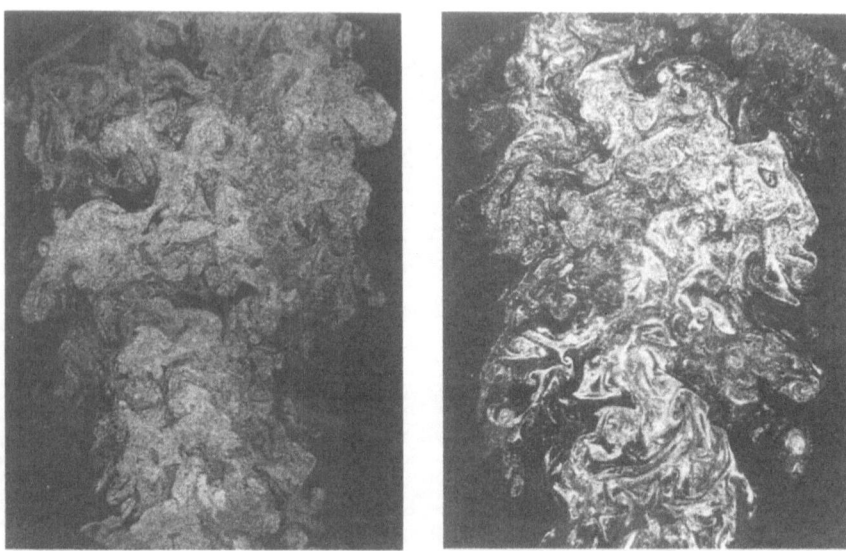

Fig. 7.3 Comparison of eddy structure i non–flame jet(left) and flame(right)

$$u^* \equiv u / u_0 = A \left(\rho_f / \rho_m \right)^{1/2} \left(d_0 / x^* x_r \right) f_{(\eta)}$$

$$T^* = (L + 1)(1 - u^*) / L \quad (\text{inside of flame}) \tag{7.2}$$

$$T^* = (L + 1) u^* \qquad\qquad (\text{outside of flame})$$

where T^* is the dimensionless temperature defined as in Fig. 7.1, A is a constant, L is the stoichiometric air–fuel ratio, ρ_f is the fuel density at the nozzle exit, ρ_m is the density on the center axis, x_r is the flame length, x^* is the dimensionless axial distance defined as x/x_r, and η is the dimensionless radial distance defined as $\eta \equiv r/b$. Similar to ref.[15], the flame length x_r can be derived as,

$$x_r = A (L + 1)(\rho_f / \rho_r)^{1/2} d_0 \tag{7.3}$$

where ρ_r is the density at the flame. Substitution of Eq.(7.3) into Eq.(7.2) shows that T^* is always a function of only x^* in the region outside the flame and a function of the stoichiometric air–fuel ratio and x^* inside the flame.

(e) Flame length and air entrainment

Equation (7.3) is the flame length for a variety of fuels. Table 7.1 compares the measured constant A values in Eq. (7.3) for a variety of fuels and nozzle conditions. The results show that the constant value is approximately 5.0 even for very different fuel

Table 7.1 Comparison of flame lengths

Fuel	Non Flame Jet	LPG				CH$_3$OH	
Phase	Gas	Gas			Liq.	Gas	Liq.
L	–	15.6				6.47	
d$_0$ mm	–	1.5	2.0	2.5	0.2	2.0	0.2
x$_r$ mm	–	360	400	510	510	185	551
$(\rho_f/\rho_r)^{1/2}$	1	3.01	3.01	3.04	47.4	2.19	60.8
A	5.00	4.80	4.00	4.04	3.24	5.65	6.07

densities and stoichiometric air–fuel ratios. The liquid LPG shows smaller A values than the other cases. This appears to be attributed to the fact that the fuel evaporates very rapidly and the actual fuel density is somewhat lower than the pure liquid density. It should be noted that the constant value is same as in a non–flame jet.

The axial temperature change, T_m^*, can be determined from Eqs. (7.2) and (7.3) by considering the density change due to combustion, resulting in $T^* \propto 1/x^{*2}$ outside the flame. Fig. 7.4 shows the analytical temperature curves and experimental results, where $\rho=constant$ corresponds to the theoretical case where combustion takes place at the mixing–rate of the non–flame case; this theoretical condition gives $T^* \propto 1/x^*$. The quick temperature decrease in the variable–density case indicates that air entrainment increases as the temperature decreases in the flow direction, i.e. air entrainment decreases in the flame region. Here it must be noted that the variable–density flame length is longer than in the constant–density case, and the abscissa is non–dimensionalized by these flame lengths. The variable–density line fits well with the experimental result. This indicates that the reduction in air entrainment in the flame region is mainly due to density changes, and the disappearance of small eddies in the flame does not have significant effect on the entrainment, because the analysis does not take the difference in small eddies into account.

(f) Soot and NOx concentration

Emissions such as soot and NOx must be a function of the chemical reaction rate as well as of the mixing rate, even when the combustion rate is simply controlled by mixing. The diffusion equation for the emission concentration, χ_e, is [16],

$$\rho^* (D\chi_e / Dt^*) = \rho^* (1 / R_e S_c)\nabla^{*2}\chi_e + x_f \omega_e / \rho_a u_0 \tag{7.4}$$

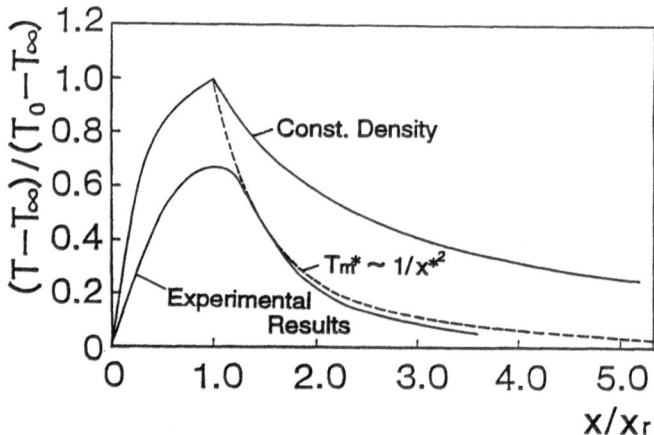

Fig. 7.4 Theoretical temperature distribution and experimental results

where Re and Sc are the Reynolds and Schmit numbers defined by eddy viscosity and eddy diffusivity, ρ^* is density non–dimensionalyzed by the ambient air density ρ_a, and ω_e is the emission formation rate. The effect of the macroscopic diffusion rate, the first term in the right side of Eq. (7.4) can be estimated by considering the case without chemical reaction; from the species continuity and momentum equations χ_e can be shown to be proportional to u^* in this no–chemical reaction case [15]. The second term in Eq. (7.4) is a macroscopic Damköhler number.   Unless micro–scale heterogeneity affects ω_e significantly, it can be further reduced to a time, x_r/u_o, for the same fuel, because ω_e and ρ_a can be treated as constants due to the similar chemical species concentration and temperature, as was seen in the previous sub–section. Thus, for one fuel, the only parameter   affecting emission concentrations is the residence time.

Experimental results showed a reasonable correspondence to the above analytical predictions. However there was scattering in the data, which may be due to differences in soot particle size and micro scale eddy structure. A farther evaluation of this matter is necessary.   (Tadashi Murayama and Takemi Chikahisa)

7.2.2 Liquid fuel properties and combustion characteristics

(a) Introduction

One of the great concerns in combustion systems is the environmental impact of exhaust gas emissions and combustion. Much research on this has been done to improve exhaust gas emissions such as NOx, particulate, soot, and HC. This research has generally been carried out in two ways: combustion system research and fuel composition research.

In combustion system research, there are numerous reports of 2 or 3–stage combustion′ and exhaust aftertreatment in boilers, combustion tuning with EGR as well as exhaust aftertreatment in SI engines, and combustion tuning including chamber configuration and high pressure fuel injection in CI engines [17].

Fuel composition research includes the use of LNG, LPG, or water–emulsions in boilers and gas turbines, aromatics reduction and MTBE addition to gasoline for use in SI engines, and also aromatic reduction and cetane number improvements in diesel fuels for use in CI engines, and so on [18]. Fuel properties influence more or less the combustion and emissions characteristics of combustion systems. Besides combustion research on individual fuels, much remains to be learned about the relationship between fuel properties and emissions as well as combustion.

Of the various combustion systems, the combustion and emissions from CI engines tend to be more affected by liquid fuel properties because their combustion is initiated by self ignition following fuel injection, and proceeds diffusively in the small combustion chamber. This section will deal mainly with effects of middle–distillate diesel fuel properties on combustion and emissions, specially soot and particulate, in CI engines.

(b) Effect of ignitability

Ignitability of liquid diesel fuels is one important factor affecting combustion and emissions in CI engines, and it is commonly evaluated by ignition lag, cetane values, or cetane index.

Lengthening of ignition lags affects exhaust emissions by decreasing dry soot and increasing NOx, unburnt HC, and SOF, as shown in Fig. 7.5 [19]. At longer ignition lags, ignition takes place in a fuel spray zone further away from the fuel injection nozzle where the equivalence ratio may be quite low, causing the decrease in dry soot

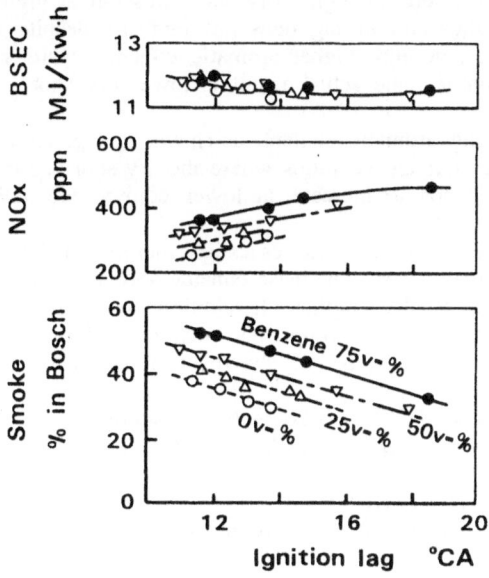

Fig. 7.5 Effects of ignition lags on dry soot and NOx in various fuels, from [19] with permission

formation. Ignition is similar with higher pressure fuel injection, resulting in lower soot emissions [20]. The higher NOx with longer ignition lags is caused by higher in–chamber temperatures due to rapid combustion.

Particulate, which contains dry soot, SOF, and small amounts of other products, does not necessarily show consistent changes with ignition lag changes, because longer ignition lags cause both increased SOF and decreased dry soot. With increasing ignition lags, particulate decreases at higher equivalence ratios where dry soot becomes the major portion of the particulate, and particulate increases at lower equivalence ratios where SOF forms the major portion. The degree of change in emission products such as particulate, dry soot, and SOF caused by changes in ignition lags depends on the engine operating conditions and combustion systems.

On the other hand, the net thermal efficiency in CI engines is affected by ignitability, with a maximum value at a specific ignition lag, which is not defined absolutely but depends on engine and fuel types. The net thermal efficiency tends to decrease for both extremely short and long ignition lags. In diesel combustion, extremely long ignition lags, lower ignitability, brings higher cooling, and mechanical losses as a result of rapid combustion; with extremely short ignition lags, higher ignitability, there is a decrease in the degree of constant–volume combustion mainly due to the longer combustion duration.

(c) Effect of aromatic content

Increasing aromatic content in fuels has been a concern in combustion systems, specially in CI engines, because of the high propensity of aromatics to form pollutants and suppress self–ignition.

Aromatic components act inherently to increase soot and NOx emissions, and increased aromatic content results in higher dry soot emission at higher equivalence ratios, specifically when the ignition lag does not increase despite an increase in aromatic content. However, because higher aromatic content results in lower cetane number and longer ignition lag, the actual result is lower dry soot and higher SOF emissions.

Therefore, with increasing aromatic content, as shown in Fig. 7.6 [21], particulate tends to decrease at higher equivalence ratios where the dry soot is a major portion of particulate, and inversely tends to increase at lower equivalence ratio where SOF dominates.

Increased aromatic content potentially causes a slight increase in NOx and no changes in HC, but NOx and HC actually tend to increase considerably with increasing aromatic content because ignition lags usually lengthen. In addition, soot in Bosch units and NOx in ppm correlate somewhat with fuel properties such as ignition lag, IL, aromatic content, AC, and kinematic viscosity, KV, as shown in the following equations [21]:

$$Soot\ (BU)\ =\ 1.5KV\ -\ 3.0IL\ +\ 0.7AC\ +\ const.$$
$$NOx(ppm)\ =\ -7.0KV\ +\ 4.7IL\ +\ 4.5AC\ +\ const.$$

These equations indicate that soot reduction requires longer ignition lags and lower aromatic content, and that reduced NOx also requires shorter ignition lags and lower aromatic content.

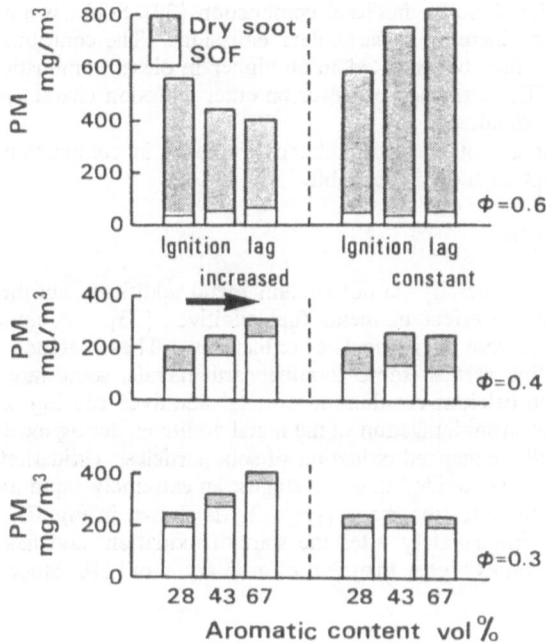

Fig. 7.6 particulate composition for different aromatic contents, from [21] with permission

(d) Effect of viscosity and volatility

In usual fuels, the viscosity correlates almost linearly with density. The NOx, HC, soot, and particulate are slightly affected by fuel viscosity.   When ignition lags are kept constant, a higher viscosity results in a decrease in NOx, and increases in HC, soot, and particulate.   Increased soot and particulate with more viscous fuels is significant at higher equivalence ratios. This suggests that any countermeasure is necessary to obtain cleaner combustion with higher viscosity fuels.

The higher viscosity also results in lower initial burning rates and slightly longer combustion. The lower initial burning rate causes lower NOx and noise, and the longer combustion gives slightly lower net thermal efficiency. The fuel volatility has almost no effect on exhaust gas emissions and combustion, when the mean boiling point remains unchanged.

(e) Effect of sulfur content

Most sulfur in fuels converts to sulfur dioxide, sulfuric anhydride, and sulfate in the flame. The conversion ratio of sulfur to sulfuric anhydride differs among combustion systems, combustion conditions, and the presence of catalysts. The conversion ratio to sulfate is usually 0.5–3.5 % sulfur in diesel combustion [22]. Sulfur is a part of exhaust particulate and causes an increase in particulate emissions. The contribution of sulfate to particulate emissions may be expected to be higher in diesel combustion with lower particulate emissions. The influence of sulfur on other emission characteristics such as NOx and THC is very small.

Corrosion and wear are other sulfur related problems in combustion systems. A decreased sulfur content in fuels is desirable.

(f) Effect of metal content

Commercial diesel fuels generally do not contain metal additives, but there have been many investigations of the effect of metal fuel additives [23]. A number of metal additives strongly affect soot or particulate reductions. These include copper, iron, manganese, some alkaline metals, some alkaline earth metals, some rare earth metals, and others. Barium and calcium are dominant metal additives causing soot reduction.

Soot reductions result from ionization of the metal additives during the soot formation process and a catalytically enhanced oxidation of soot particles. Oxidation processes of metal containing soot can be divided into two stages: an extremely rapid oxidation stage (stage 1) and a moderate oxidation stage (stage 2), as shown in Fig. 7.7.

Stage 1 takes place immediately after the start of oxidation, and has an oxidation velocity more than 100 times higher than that of ordinary soot [24]. Stage 2 follows the

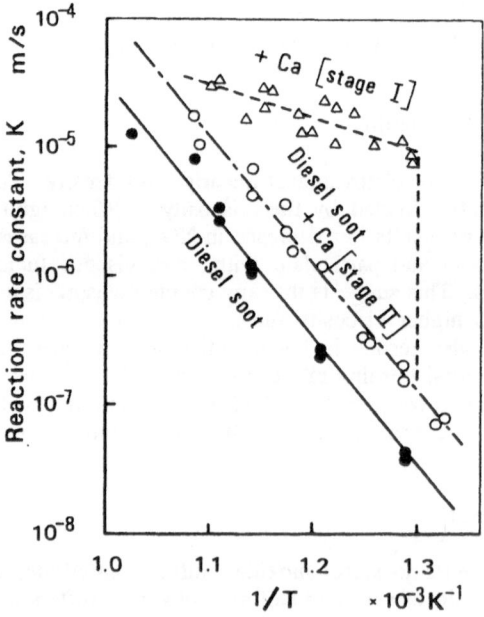

Fig. 7.7 Catalytic oxidation of Ca–containing soot particles, from [24] with permission

rapid oxidation stage and here the oxidation velocity is slightly higher than that of ordinary soot. With increasing metal content of the soot, the oxidation temperature of the soot decreases and the amount of soot oxidizing in the rapid oxidation stage increases. These catalytic oxidation characteristics of soot with metal can apply in many carbon treatment systems such as particulate filter trap systems. The effect of metal additives differs for different fuels.   Soot reduction is more remarkable with lower viscosity fuels, partially   explained by larger amount of soot oxidation in the rapid oxidation stage.

Gaseous additives including SO_2, SO_3, H_2S, NH_3, and others generally cause soot reductions.   Soot reduction with gaseous additives seems to come mainly from the thermal mechanism, a reduction in the rate of pyrolysis of the fuel.

(g) Fuel properties and soot formation characteristics

The sooting limit is defined by the smallest equivalence ratio or the smallest C/O ratio, (C/O)min. Both of these smallest ratios are the ratios where soot first appears, and have been established for a wide range of fuels [25].   The propensity of soot formation is in the following order: aromatics, alkane, alkene, and alkyne. The last has the lowest propensity.   Soot propensity tends to increase with increasing carbon number, independent of fuel type.

The soot yield, S, in premixed flames increases with increases in C/O or equivalence ratios, and it can be described by the relation:

$$S=\{(C/O)-(C/O)min\}^n$$

here n equals about 3.

Many investigations on the velocity as well as the mechanism of soot formation have been carried out, most performed   with a specific fuel. Systematic soot formation velocity expressions for different types of fuel have successfully been established, to some extent, by an extension of Tesner's concept of soot formation [26]. Fig. 7.8 shows

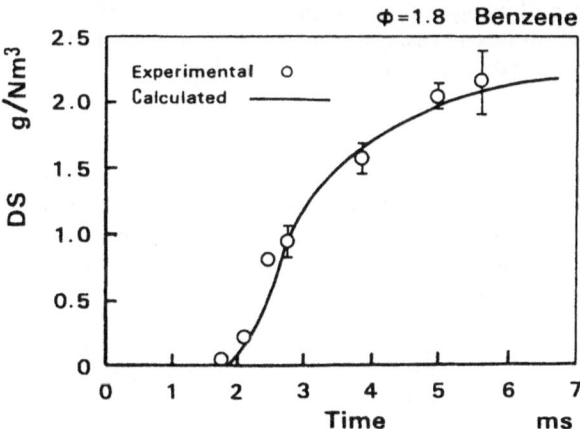

Fig. 7.8 Calculated and experimental soot formation rates, from [26] with permission

a comparison of the calculated and experimental soot formation behavior in premixed flames. All coefficients in Tesner's equations describing the formation velocities of soot precursors and particles are independent of fuel type, while the frequency coefficient A and activation energy E in the equations for the velocity of the soot precursor Np are specific to individual fuels.

Np can be expressed as,

$$Np = AT^{(-1.94)}[HC]exp(-E/RT)$$

where, $[HC]$ is the unburnt fuel concentration on a C_1 basis kg/m^3.

(Noboru Miyamoto and Hideyuki Ogawa)

7.3 Application of Extremely Light and Heavy Fuels for Engines

7.3.1 Hydrogen–injected engines

(a) Introduction

Internal combustion engines now in operation in our life are using fossil fuel. And the air pollution brought about by the exhaust of the engines has become a serious problem of global ecology. In particular, carbon dioxide in the exhaust which plays a great roll in the increase of ambient temperature on earth should be suppressed by all means. From this standpoint, hydrogen has been focused on as one of fuels [27].

Hydrogen is an alternative fuel for fossil fuels such as petroleum and coal which fatally never fail to run short. Hydrogen is also obtained by electrolyzing water and reduce to water again by combustion. When using hydrogen as a fuel, nitrogen oxides (NOx) are only generated in the combustion free from carbon dioxide (CO), hydrocarbon (HC), sulfur oxides (SOx), and particles over which the regulation of the emission of exhaust gas is carried out.

The research work on hydrogen fueled engines began in 1930's but it stopped for about four decades and, in recent years, it has been started again to study the engines using hydrogen as an alternative fuel in respect of ecology.

(b) Problems of hydrogen fueled engines

Hydrogen fuel supply method   Basically, there found are two methods in the supply method. One is pre–mixing method. The other is direct injection method. The former method is such that hydrogen and air are mixed in the intake manifold and the mixed gas is introduced into the engine in suction stroke. The latter method is such that only air is introduced into the engine in suction stroke and hydrogen is injected into the cylinder after the intake valve closes.

In Fig. 7.9, the comparison is made on basis of the calorific value obtained at the stoichiometric mixture ratio with various fuel supply methods in the engine with 1 litter stroke volume. It is found that the calorific value with the intake manifold pre–mixing method is 85% of that of the gasoline engine and the calorific value with the direct injection method is 120%. In practice, the output power with the pre–mixing method of 50% of that of the gasoline engine is attained because, due to backfire to the intake

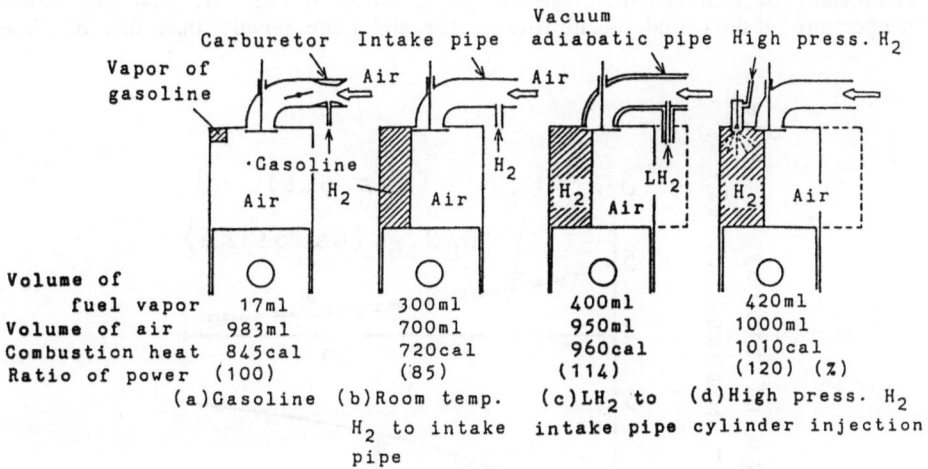

Fig. 7.9 Output power with various fuel supply methods at 1 litter stroke volume

manifold, the engine with the pre−mixing method is forced to run in leaner mixture ratio (about excess air ratio(λ)=2) than the stoichiometric mixture ratio. To avoid this demerit, the direct injection method has been proposed and carried out for hydrogen fueled engines [28].

The effect of decreasing mole number   In contrast with petroleum fuel, the mole number of mixture of hydrogen−air decreases greatly after the combustion. The difference between the mole number reduction of hydrogen−air mixture and that of petroleum−air mixture become larger as the excess air ratio (λ) become 1 (stoichiometric). Figure 7.10 shows the change of the mole number against the excess air ratio. There has appeared the dispute that the mole number reduction may cause output power smaller than that of a petroleum fueled engine. But, the combustion temperature rises according to the amount of the mole number reduction because the molecular specific heat at constant volume Cv of the combustion gas of hydrogen and air is nearly constant. This compensates a drop in the output power of hydrogen engines.

In order to prove this, the combustion pressures both of hydrogen and petroleum fueled engines were calculated out against the amount of heat release per unit mole of air. The results are shown in Fig. 7.11. This figure shows that the combustion pressure of hydrogen−air mixture is approximately same as that of petroleum−air mixture. Namely, the output power of hydrogen−air engine is almost the same as that of petroleum. In conclusion, hydrogen engines do not thermodynamically have any disadvantage in the output power owing to the mole number reduction.

Loss of heat transfer   The thermal conductivity of hydrogen is 6.6 times as great as that of air but, based on the Nusselt's equation, the heat transfer coefficient of the brunt gas of hydrogen−injected engines between the wall and burnt gas is almost the same as that of diesel engines because hydrogen hardly remains in the burnt gas after the completion of combustion.   However, it is found in Fig. 7.12 that the surface temperature at the cylinder head rises higher and more rapidly than that of diesel

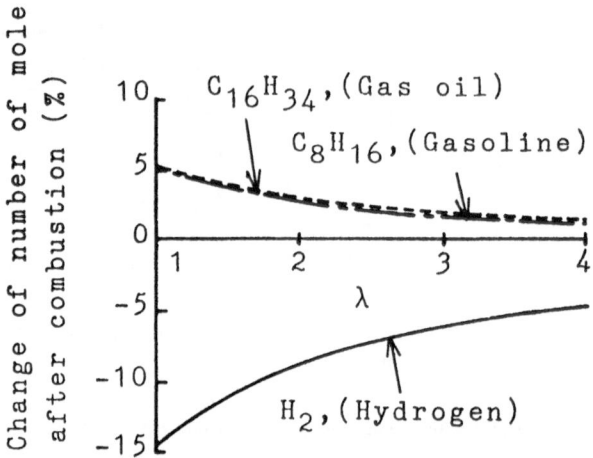

Fig. 7.10 Change of number of mole combustion compared after H$_2$ with gasoline and gas oil, from [34] with permission

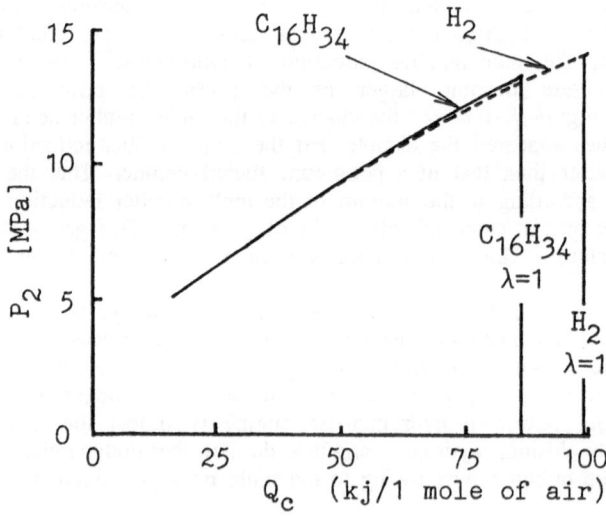

Fig. 7.11 Combustion pressure of H$_2$ compared with C$_{16}$H$_{34}$ against Q$_c$, from [34] with permission

engine's cylinder head. The both temperatures were measured in the same engine operating condition by an instantaneous surface temperature measuring method. Therefore, it can be concluded that hydrogen–injected engines have greater loss of heat transfer than diesel engines.

Ignition method in hydrogen–injected engines In hydrogen–injected engines, it is necessary to install ignition system because of difficulty in compression ignition. As the ignition system, two methods such as hot surface ignition by a diesel glow plug and spark ignition have ever been studied experimentally. As the results, the hot surface of the glow plug is required to keep at the temperature of 1200 K in order to decrease the ignition delay. This makes problems in the life of glow plug and the consumption of electric power. And, once hydrogen leaks into the cylinders at the valve of injector, the hot surface of the glow plug results in the occurrence of the backfire at the start of engine. So, as the ignition system of hydrogen–injected engines, spark ignition method is more advantageous.

(c) Behavior and ignition of hydrogen jet

In order to investigate the mechanism of combustion in hydrogen–injected engines, the behavior and ignition characteristics of hydrogen jet were studied preliminarily using constant volume combustion chamber [29]. Fig. 7.13 shows the pictures taken by a high speed camera by Schlieren method when hydrogen was injected from a single hole nozzle into the constant volume combustion chamber. It is found from the pictures that the hydrogen jet spread out quite slowly in comparison with the injected liquid fuel since the density of hydrogen is very small. Accordingly, the distance from the exit of nozzle to the tip of spark plug is one of the important effects on ignition delay. In addition, the pressure in the combustion chamber changes slightly the behavior of the hydrogen jet.

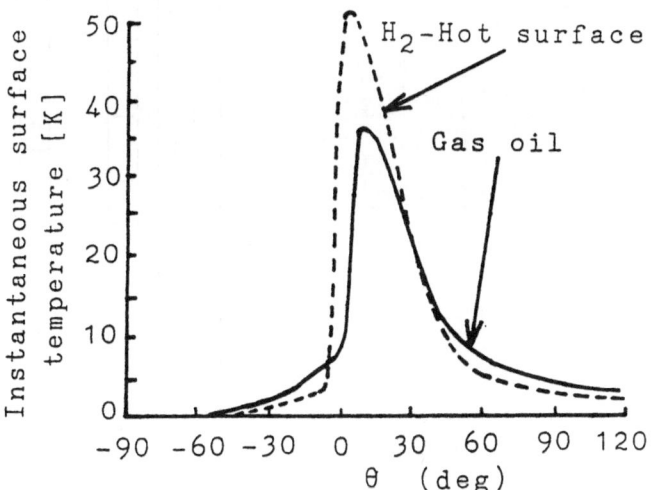

Fig. 7.12 Measured surface temperature of hydrogen engine compared with gas oil engine, from [34] with permission

The selection of the ignition timing of spark plug is another important item to obtain a stable ignition. Figure 7.14 shows the effect of the spark timing and the behavior of hydrogen jet on ignition. As seen in the pictures(a), stable ignition with a short ignition delay was firmly obtained when the spark was made just at the tip of the hydrogen jet. On the other hand, as seen in the pictures(b), when the spark was made after the tip of the hydrogen jet passed through the tip of the spark plug to a large extent, a too rapid combustion was made because the ignition delay became longer and a large amount of flammable mixture was built up. This resulted in the source which generated combustion pressure vibration. More delay of the spark caused an unstable ignition and furthermore delay ignition impossible.

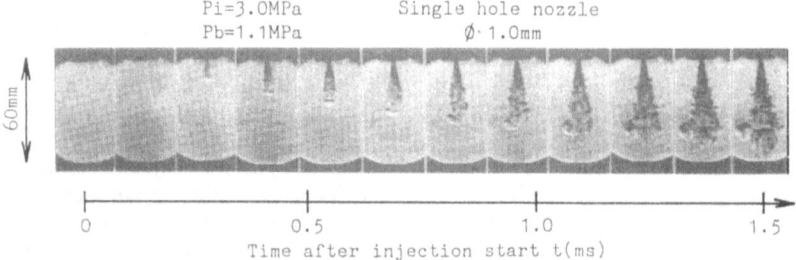

Fig. 7.13 Hydrogen jet (Schlieren method)

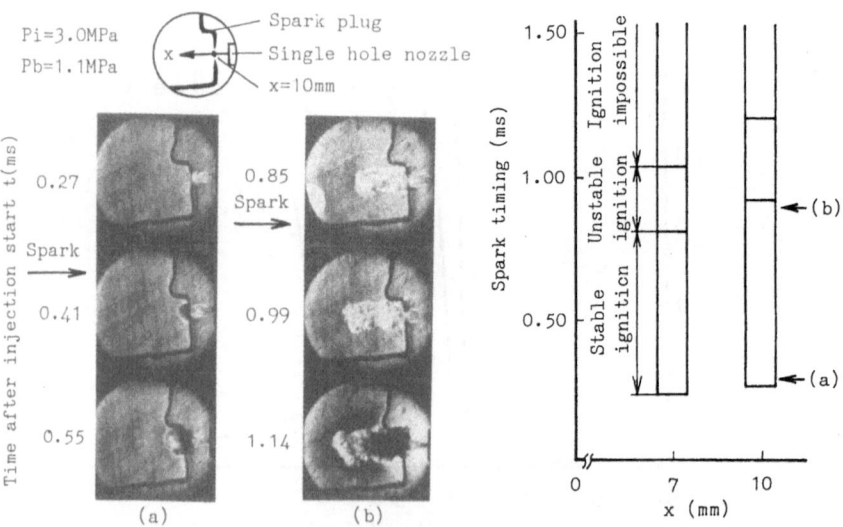

Fig. 7.14 Effect of spark timing, from
[30] with permission

(d) Combustion characteristic of hydrogen-injected engines [30]

As to the hydrogen injected engine with a spark ignition employed, the cylinder head has hydrogen injector and spark plug approximately in the middle part of the cylinder head. Figure 7.15 shows the positions of the injector, the spark plug and the directions of 8 hydrogen jets from nozzles. As described before, it is found also in this case that the distance from the jet nozzle to the spark plug plays an important roll in the ignition delay. Figure 7.16 shows the pictures taken from under the piston through an observation window installed at the bottom of piston cavity when the engine ran actually at the engine speed of 500 rpm. The visual range, in this case, was the inner portion of the cavity of piston. It was observed that the ignition occurred at the tip of the hydrogen jet whose direction faced to the spark plug while the spark timing was the same as

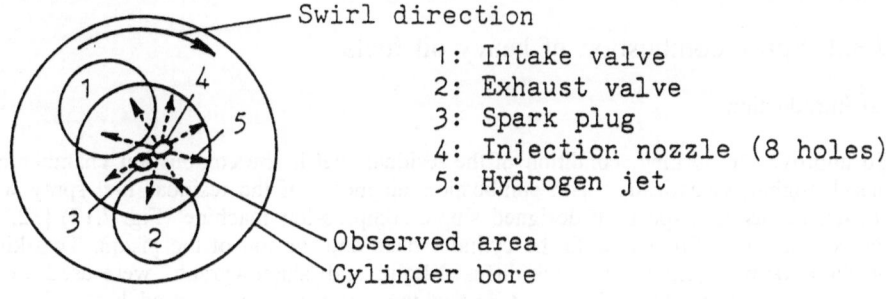

Swirl direction

1: Intake valve
2: Exhaust valve
3: Spark plug
4: Injection nozzle (8 holes)
5: Hydrogen jet

Observed area
Cylinder bore

Fig. 7.15 Combustion chanber

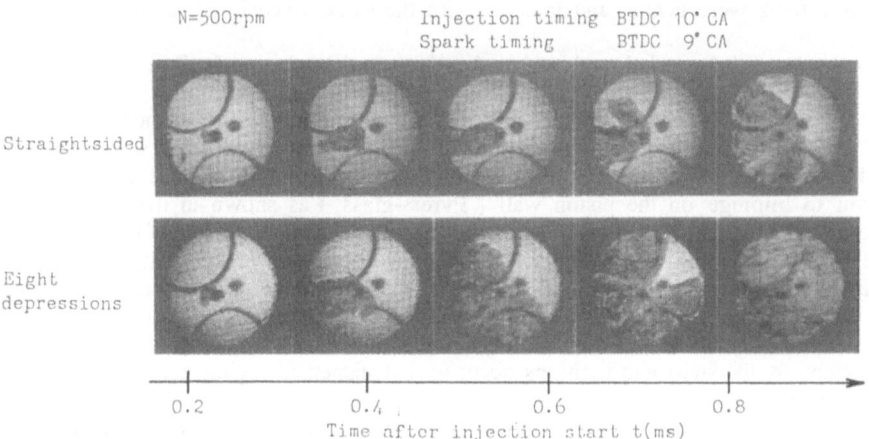

N=500rpm   Injection timing BTDC 10° CA
              Spark timing   BTDC 9° CA

Straightsided

Eight
depressions

0.2   0.4   0.6   0.8

Time after injection start t(ms)

Fig. 7.16 Flame propagation in hydrogen-injected engine

shown in Fig. 7.14 (a), and the flame spread out to the adjacent hydrogen jets while growing. The upper pictures show the flame propagation under the influence of the swirling in cylinder mainly. The part colored gray shows the flame. On the other hand, the lower pictures also show the flame propagation when the turbulence in combustion chamber was strengthened by adding a flower-shaped cavity to the original cavity of piston. It is found in these pictures that the flame spread out widely and the flame propagation is also promoted greatly. But, too much turbulence brought about a bad flame propagation. It was also found that, when the ignition was retarded, the mixture built up before the ignition burnt suddenly and resulted in the combustion pressure vibration. Generally speaking, the flow becomes hastier as the engine speed increases. But the engine speed did not affect the promotion of flame propagation as largely as expected. It was also observed that an appropriate intensity of squashing helped the reduction of NOx formation. In addition, EGR and cold hydrogen injection also decreased the amount of NOx in the exhaust gas [31].   (Shoichi Furuhama)

7.3.2. Spray combustion of heavy oil fuels

(a) Introduction

To improve the burning condition of the residual fuel in the combustion chamber of a diesel engine, visualization and combustion diagnosis of the residual fuel spray were carried out using a specially designed single compression machine (Fig. 7.17) [32,33]. Pyrex-glass was fitted in both the cylinder head and the top of the piston. Two kinds of photographic techniques, "back diffused light" and "shadowgraph" were used for the visualization tests. Only the liquid part of the spray can be observed before ignition with the former technique, and the outline of the spray including the gaseous part can be photographed with the latter technique.

Six kinds of fuels whose properties are summarized in Table 7.2 were prepared for the study. Distributions of distillate components of the fuels are shown in Fig. 7.18. In this figure, the column marked "500" shows the residual component whose distillation temperature is higher than 500°C. The marine diesel oil was injected without heating. Other five fuels were heated and injected with the same viscosity of 20mm/s.

(b) Combustion characteristic of marine diesel oil and bunker fuel oil

Figure 7.19 shows the visualized fuel sprays of the marine diesel oil and the bunker fuel oil. The injection nozzle hole diameter and the injection pressure were set at 0.3mm and 100MPa respectively. The sprays were injected in the swirling air and in the downward direction to impinge on the piston wall (Pyrex-glass) as shown in the figure. The profile of swirl velocities is also shown in the figure. In the comparison of the two fuels before ignition, at 1.6ms after the start of injection, a clear difference can be observed especially with the back diffused light technique. The liquid part of the bunker fuel spray is much larger than that of the marine diesel oil. At 2.15ms, it is observed that the liquid core of the bunker fuel oil is bent and some of the liquid part is dispersed downstream by the swirl and ignitions occur in that dispersed region, rather near to the injection nozzle.

The effects of spray impingement on the piston wall and of air swirl on the combustion of bunker fuel spray were examined with the heat release rate and $Q/Gf \cdot Hu$ in Fig. 7.20. $Q/Gf \cdot Hu$ (Q: net released heat excluding unburnt portion and heat loss, Gf: quantity of fuel injected per cycle, Hu: lower calorific value of fuel) means the ratio of the net released heat to the energy of the fuel injected per cycle. In this figure, (a) shows the data obtained in the case with neither impingement nor swirl, (b) shows the data with impingement, and (c) shows the data with both impingement and swirl. The effect of impingement is clearly seen in the increase of heat release rate in the diffusion combustion region (the second peak of heat release rate) from (a) to (b). The effect of swirl is also seen in the increase of heat release rate from (b) to (c). However, $Q/GfHu$ after the combustion is rather less in case (c). It is considered, therefore, that the swirl increases the heat loss to the combustion chamber wall.

From such photos and data, it has become clear that the combination of impingement of spray and swirl is an effective means to take off the gaseous part and small droplets from the liquid core and supply the spray with fresh air, at the same time sweeping off the burnt gas from the flame area. However, the difference of combustion condition

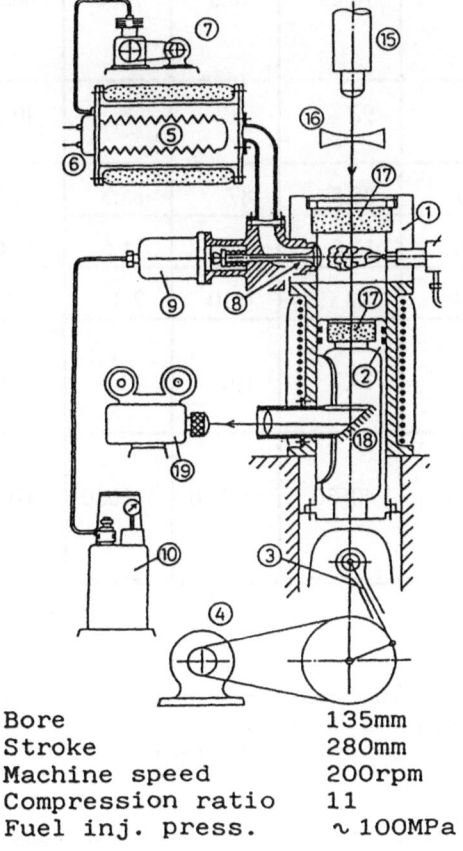

Bore	135mm
Stroke	280mm
Machine speed	200rpm
Compression ratio	11
Fuel inj. press.	∿100MPa

Fig. 7.17   Single compression machine, from [33] with permission

Table 7.2 Fuel Properties, from [33] with permission

		Marine Diesel Oil	Bunker Fuel Oil	Sample Fuels			
				(1)	(2)	(3)	(4)
Density (15°C) kg/m³		843	974	867	935	931	930
Kinematic Viscosity (50°C) mm²/s		2.5	165.5	18.5	74.6	66.9	68.9
Flash Point °C		72	92	212	68	40	42
Wt %	C	86.1	85.5	87	85.3	85.1	85.7
	H	13.0	11.0	12.9	11.6	11.6	11.7
	S	0.84	3.3	0.10	2.8	3.0	2.5
Residual Carbon Wt%		0.07	13.7	0.0	12.2	14.2	13.9
Lower Calorific Value kJ/kg		42500	40530	42790	41030	41030	41240

between the two fuels still remains even after such means to improve combustion have been tried. As it is considered that the poor dispersion of bunker fuel spray is caused by poor evaporation, some means to improve the evaporation rate are necessary for the bunker fuel spray.

(c) Influence of air temperature and distribution of distillate components

As it is very difficult to improve the evaporation rate of residual components themselves in advance of ignition when the compression air temperature is lower than the distillation temperature of the residual components, it is considered as a realistic means to accelerate

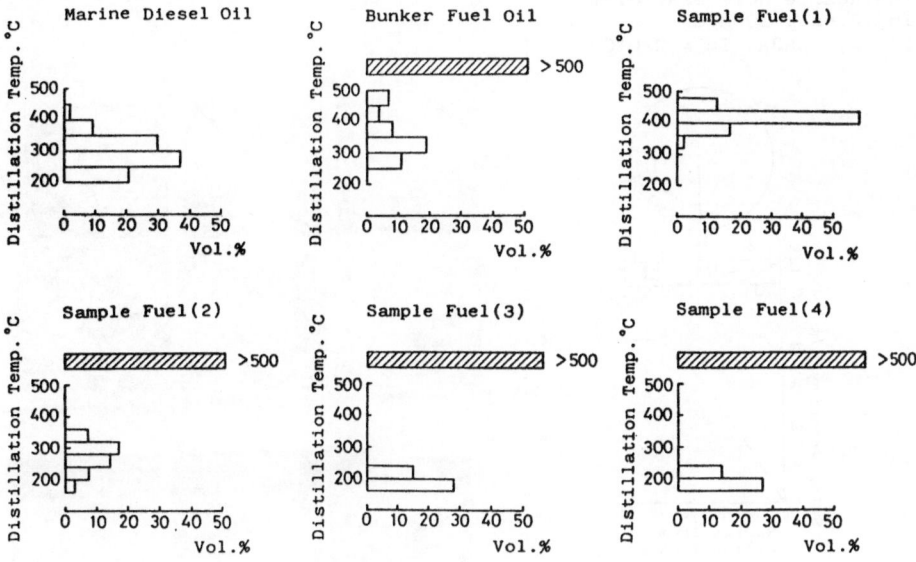

Fig. 7.18 Distributions of distillates in test fuels,
from [33] with permission

the evaporation of distillate components in the residual fuel and promote the combustion of residual components. In this study, influences of compression air temperature and distribution of distillate components on the fuel spray combustion were investigated with four kinds of sample fuels shown in Table 7.2 and Fig. 7.18. The air pressure in the combustion chamber was also changed with air temperature keeping the air density constant.

First, the ignition delays of sample fuels were examined as shown in Fig. 7.21. According to the figure, sample fuel (1), whose distillation temperature is limited in the range of 320–480°C, shows the shortest ignition delay. And, sample fuels (3) and (4), composed of residual components and distillate components of lower distillation temperature than 240°C, show the longest. However, in the range over 550°C of air temperature, ignition delays of all fuels concentrate on about 1.0ms.

Combustion characteristics of the sample fuels were examined with the heat release rate and $Q/Gf \cdot Hu$. Fig. 7.22 shows the change in maximum value of $Q/Gf \cdot Hu$ with the compression air temperature for sample fuels (1)~(3). From which it is recognized that the increase of $Qmax/Gf\ Hu$ with the rise of the compression air temperature is the largest with sample fuel (3) and the smallest with (1), though (1) doesn't include any residual components.

According to the results, it is considered that the evaporation of the components of lower distillation temperature was accelerated by higher air temperature and promoted the combustion of the residual components.

Figure 7.23 shows the change in $(Q/Gf \cdot Hu)$, $Q/Gf \cdot Hu$ value at the end of the diffusion combustion period. In this case, $(Q/Gf \cdot Hu)$ increases with the increase of compression air temperature, and the effects of the impingement of spray and the air

Inj.Nozzle Hole Dia. 0.3mm
Inj.Press. 100MPa
(Pc = 4.4MPa, Tc = 420°C)

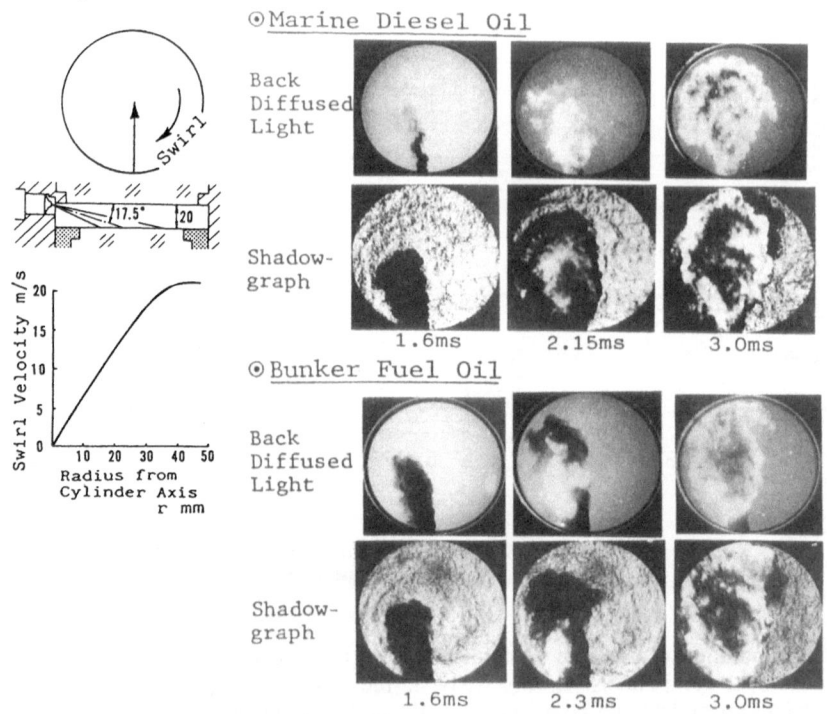

Fig. 7.19 Visualized sprays in the single compression machine,
from [33] with permission

Inj.Nozzle Hole Dia. 0.3mm
Inj.Press. 100MPa
(Pc = 4.6MPa, Tc = 450°C)
Fuel ···· Bunker Fuel Oil

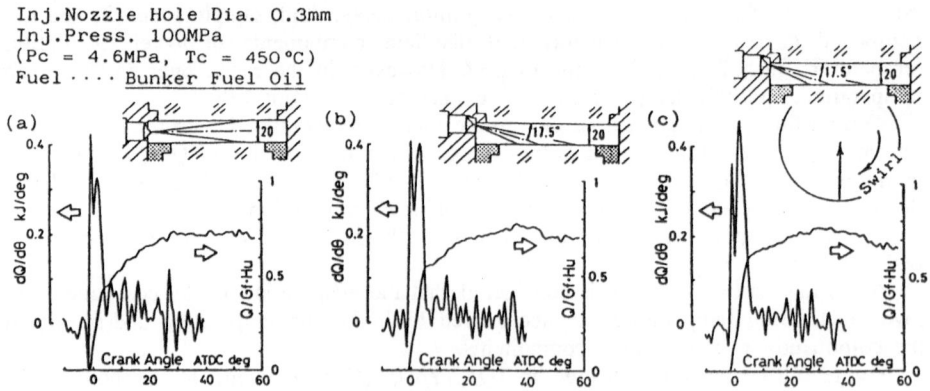

Fig. 7.20 Heat release rates of bunker fuel oil,
from [33] with permission

swirl can both be seen as mentioned previously. However, attention should be paid to the fact that the ($Q/Gf·Hu$) value of residual fuels goes down extremely with the increase of ignition delay in the low temperature region.

The cause of this can be explained as follows. In the case where the distillation temperature of residual component is higher than the compression air temperature, the residual component can not evaporate in advance of ignition. Therefore, it is considered that a great deal of residual component of liquid phase clings onto the piston wall during the ignition delay, and a greater part of the clinging portion remains on the piston wall without being evaporated. On the other hand, in the case of such fuels as marine diesel

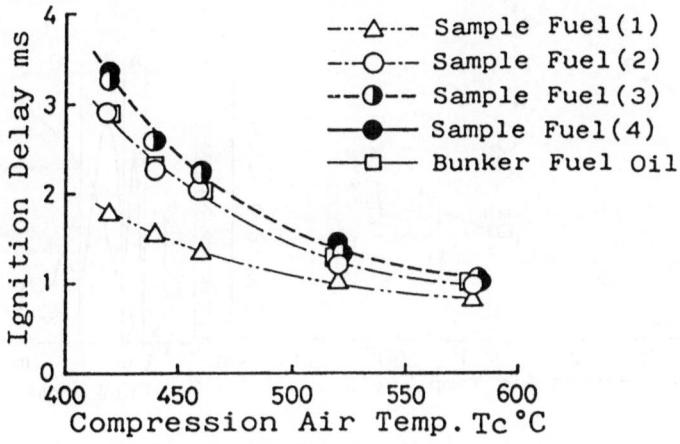

Fig. 7.21 Ignition delays of sample fuels,
from [33] with permission

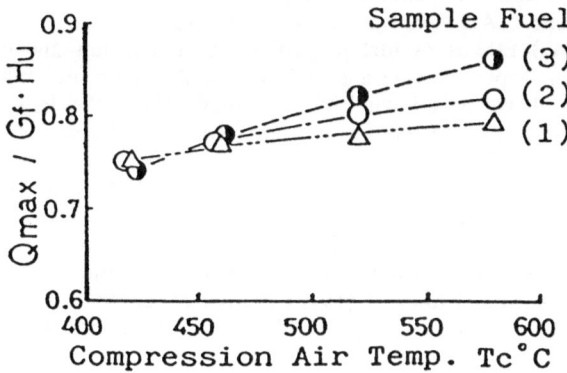

Fig. 7.22 Change in maximum $Q/Gf·Hu$ with compression air temperature,
from [33] with permission

oil and sample fuel (1) which are composed only of distillate components whose distillation temperatures are lower than the compression air temperature, the fuel can easily evaporate before the ignition.   (Yutaro Wakuri and Koji Takasaki)

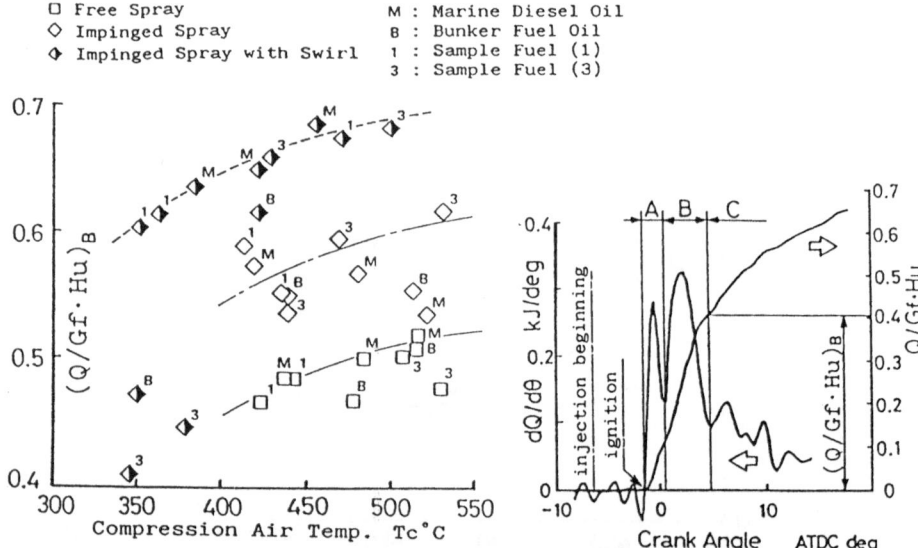

Fig. 7.23 Change in $(Q/Gf{\cdot}Hu)_B$ with compression air temperature, from [33] with permission

References

[1] Tosaka S, Fujiwara Y and Murayama T (1989) The effect of fuel properties in diesel engine exhaust particulate. SAE technical paper series 890421

[2] Shimokawa K (1989) Effects of fuel properties on combustion and exhaust gas emission, Sympo.:Recent topics on fuel and emission of diesel engine, JSME

[3] Ohi A, Nakamura S, Aoyama H, Suzuki M and Yamaki N (1979) The effect of fuel aromatic content on hydrocarbon composition from vehicular exhaust gas. Pollutions 14–3

[4] Parachristos M, Priestman GH, Swithenbank J and Lois E (1988) Fuel structure effects on the performance and emissions of spark ignition engines. I Mech E, C63/88

[5] Onuma Y, and Ogasawara M (1974) Studies on the structure of a spray combustion flame, 15th Sympo. (intl.) on combustion, The Combustion Institute

[6] Onuma Y, Ogasawara M and Inoue T (1976) Further experiments on the structure of a spray combustion flame. 16th Sympo. (intl.) on combustion, The Combustion Institute

[7] Zeelenberg AP, et al. (1983) The ignition performance of fuel oils in marine diesel engines. Proc. 15th International congress on combustion engines. CIMAC, Paris, D–13,2.

[8] Dexter SG (1989) Residual fuels–combustion, emissions and particulates. Proc.

I.Mech.E. Seminar Diesel fuel injection systems. 31

[9] Nomura H. et al. (1987) Indices for estimating the ignition and combustion properties of marine diesel fuels. Proc. 17th International congress on combustion engines. CIMAC, Warsaw.

[10] Nakano H, et al. (1987) Experimental studies on low grade future fuels. Proc. 3rd International congress on combustion engines. D-59.

[11] Groth K. et al. (1988) Brennstoffe fur dieselmotoren Heute und Morgen. Expert Verlag.

[12] Chikahisa T, Murayama T and Takenawa K (1988) Similarity in jet diffusion flame structure for different fuels. 26th Combustion Symposium in Japan, p287

[13] Chikahisa T, et. al.(1990) Similarity in jet diffusion flame structure for different fuels; 2nd report. 28th Combustion Symposium in Japan,p158

[14] Chikahisa T, et. al.(1991) Observation of eddy structure and boundary layer shape in turbulent jet diffusion flame. 29th Combustion Symposium in Japan, p58

[15] Abramovich (1963) The Theory of Turbulent Jets. MIT Press, p345

[16] Chikahisa T, Kikuta K and Murayama T (1992) Combustion similarity for different size diesel engines: Theoretical prediction and experimental results. SAE technical paper series 920465

[17] Shundoh S and Tsujimuya K (1990) Combustion improvement and exhaust emission decrease of a diesel engine by high pressure fuel injection. Proc. of Annual Meeting, No.900 p.14

[18] Knuth HW and Garthe H (1988) Future diesel fuel compositions-Their influence on particulate. SAE technical paper series 881173

[19] Miyamoto N, Ogawa H and Shibuya M (1991) Distinguishing the effects of aromatic content and ignitability of fuels in diesel combustion and emissions. SAE technical paper series 912355

[20] Miyamoto N, Ogawa H, Goto N and Sasaki H (1988) Observation and analysis of the formation of diesel soot with a laser light extinction method Ignition lag and soot formation. Trans. of the JSME, 54-506

[21] Shibuya M, Miyamoto N, Ogawa H and Suda T (1992) The influence of fuel properties on diesel combustion and emissions. Trans. of the JSME, 58-550

[22] Miyamoto N (1989) Effects of sulfur content on diesel combustion and emissions. Symposium Recent topics on fuel and emission of diesel engines, JSAE

[23] Miyamoto N, Hou Z, Harada A, Ogawa H and Murayama T (1987) Characteristics of diesel soot suppression with soluble fuel additives. SAE technical paper series 871612

[24] Miyamoto N, Hou Z, and Ogawa H (1988) Catalytic effects of metallic fuel additives on oxidation characteristics of trapped diesel soot. SAE technical paper series 881224

[25] Haynes BS (1991) Soot and hydrocarbons in combustion. Fossil fuel combustion, John Wiley & Sons

[26] Miyamoto N, Ogawa H, Doi T, Migita H, Arayashiki A and Takeda N (1991) Soot formation process in premixed flame. Proc. of 29th Combustion Symposium, Japan Soc. Comb.

[27] Winter C (1990) Hydrogen and solar energy. Proceeding WHEC 8, 1-3

[28] Peschka W (1990) Cryogenic fuel technology and elements of automotive vehicle propulsion systems. CEC-Conf

[29] Kobayashi K (1989) Studies on ignition methods for hydrogen engines with injection right before TDC. JSME 55-511, B:895

[30] Furuhama S (1991) The power system of computer controlled hydrogen car. IMechE, C430-028:179

[31] Ninomiya Y (1992) NOx control in LH2–pump high pressure hydrogen injection engines. Proceeding WHEC 9

[32] Wakuri Y, et al (1990) Residual fuel sprays – evaporation, dispersion and combustion characteristics. Proc. International Sympo. COMODIA'90:539

[33] Wakuri Y, et al (1991) Studies on combustion of residual fuel sprays in a diesel engine with the aid of visualization method. Proc. 19th International Congress on Combustion Engines(CIMAC,Florence),D–52

[34] Pichainarong P (1990) Studies on heat loss in high pressure hydrogen injection engine. Proc. WHEC 8, Vol. 3, 1275

Chapter 8
New Approaches to Controlling Combustion

8.1 Combustion Control Based on Electrical Aspects

8.1.1 Introduction

The history of the subject of electrical aspects of combustion is a very ancient one. In England, the ability of the flame to discharge an electroscope was demonstrated by W. Gilbert at the court of Queen Elizabeth I as early as AD 1600. The recent topicality of the subject owes much to its potential applicability to the many problems in combustion which have arisen, not least in response to current concerns about diminishing fuel resources and increasing levels of pollution. A review published in the late 1960s [1], whilst dealing largely with academic aspects, already drew attention to the considerable practical potential of these techniques. By 1986, it was appropriate to include separate sections on plasma jets in combustion and the effect of large electric fields on flames in a book on "Advanced Combustion Methods" [2]. The effect of magnetic fields on flames is another area of potential interest (see pp 336 of [1]). Here again, the pioneering work may be traced to classical researches, in this instance those of Faraday [3]. More recently, however, the main practical interest has developed around systems of very fast gas flows – e.g. in magneto–hydro–dynamic generation of electrical power (e.g. pp 259 of [1]). What follows is intended to draw attention to some of the interesting new and practically important directions the subject has taken since the publication of the above–mentioned texts in two particular areas; injection of plasmas into combustion systems, and, complementarily, high voltage, low current, electric field studies applied to flames.

Plasma jets, in addition to producing very high temperatures, provide the most direct method for injecting specific radicals into combustion processes. Major progress has been made in studies on the use of both, pulsed plasma jets for enhanced ignition and continuous flow plasma jets for flame stabilisation. In western countries, unfortunately, the drive towards using the greatly enhanced ignition and flame propagation induced by pulsed plasma jets for lean–burn internal combustion engines has been attenuated (along with combustion – driven jets and other means of producing radical – rich initiation) in favour of stoichiometric combustion with added catalytic clean up of combustion – generated pollutants. However, these decisions pre–date concern about carbon dioxide emission and, hopefully, may be reconsidered in view of the additional costs in fuel used up and greenhouse gases produced. In the meantime we have learned a great deal, using new laser–based diagnostic methods about the manner in which free radicals from plasma jets promote ignition and how centrifugal fields concentrate them in vortices which can be projected by the igniter [4,5].

Considerable improvements have also been achieved in converting the much more energetic surface discharge igniters used for jet engine combustors into plasma jets, particularly for dealing with the critical problem of relight following extinction at high altitudes [6,7]. It emerges that this kind of radical initiation works well even in a spray of aviation fuel. In the meantime, continuous plasma jets have been studied for their potential as flame holders in scram jets both in the USA and in Japan [8-10]. A few years ago it seemed highly unlikely that the same principle might be applied in conventional jet propulsion. However, the pressure for emission reduction from aircraft is changing the industrial outlook. Jet engine combustor design for the next century is being based on leaner combustion, often employing multiple injectors with the aim of varying thrust by shutting down some of them; designs with plasma jet pilots are being put forward [11] because of the hazard posed by flame – out in such systems. Based on this fundamental research, plasma jets suitable for aviation gas turbine use are now commercially available from Russia [12]. A somewhat similar field of study concerns ignition and flameholding in pulverised coal burning furnaces. A six year development of this principle for furnaces used for power generation in Australia has yielded most promising results [13].

Plasma jets offer at least three further interesting areas for study. The radicals they generate have been shown capable of destroying certain pollutants – notably soot and NO (the latter by the addition of nitrogen atoms – e.g. see [2]). More light has been thrown on the details of this process recently [14], though the mechanism is likely to depend critically on the type of plasma jet used. Moreover, sulphur dioxide has been efficiently converted to SO_3 by this means [15]. Current work is concentrating on treating a cocktail of pollutants and on using higher electrical potentials. Another interesting new scheme is based on using plasma jets as generators of nuclei to increase radiation in natural gas burner flames [16]. In addition to previously observed increases in flame stability, the soot so formed increases heat transfer and reduces flame temperature – and hence NO_x formation. Finally, plasma jets offer a potential and much neglected tool for the study of fundamental kinetics. Instead of the enormous number of different radicals produced in flames, they are capable of providing a steady flow of just one or two labile species with appreciable life-times.

Electric fields applied to flames have also yielded several novel developments promising well for the future. It has been shown that the force exercised on neutral gas by flame ions in electric fields can be used to increase flame stability against blow–off substantially on industrial burners [17]. This allows not only larger throughputs for a given equivalence ratio but the use of leaner mixtures with concomitant reductions in NO_x emissions. The same principle has been used for controlling flames in the absence of earth gravity [18,19]. Under normal gravity, we rely on natural or forced convection to replenish reactants and direct hot products to the surface to be heated. Forced convection requires a compressor for keeping fuel under pressure in cylinders, and thus entails a considerable weight penalty. Natural convection is not available under micro-gravity conditions, making heat transfer from flames difficult to achieve. Diffusion flames, indeed, will tend to become spherical and will eventually extinguish as a result of blanketing by their own products. It has been shown [18] that controlling flames in zero gravity by the application of electric fields which can achieve the equivalent of 800 g using compact, light weight, equipment can provide very efficient and economical heating of specified areas by small fames, making optimum use of the oxygen available in the working environment and using liquid or solid fuel requiring no pressurised containment.

As regards transposing charge carriers, numerous attempts have been made to use ion currents in the exhaust products of IC engines to monitor the combustion process with the ultimate aim of feedback control. Another area which has recently resurfaced ([20], but see also [21]) is the manipulation by electric fields of charged refractory particles in order to control their deposition and size in manufacturing processes. Soot continues to be of interest (e.g. [22]), but the technique may well prove important in future also in conjunction with the recent development of new combustion systems which lead to the production of valuable refractories and metals in particulate form.

Japan has been in the forefront of many of the new research developments in these fields and it gives me particular pleasure to introduce contributions from several distinguished Japanese workers, to this volume.   (Felix J. Weinberg)

8.1.2 Plasma jet ignition fundamentals

Among the problems related to ignition, the spark ignition in internal combustion engines is the one noticed from the early age of combustion research. For the ignition in flowing type combustors, successive spark methods are also used widely. It has been shown relatively recently that plasma jets in continuous mode and in pulse mode work effectively on the ignition, on the promotion of combustion, or on the flame stabilization in several combustion experiments [23–27]. In a plasma jet, electric energy is given to the feed stock gas through an arc discharge, and the high temperature gas jet produced is used for the ignition of combustible gas mixtures, although in the case of spark ignition the electric discharge is carried out directly in the combustible gas mixtures.

The ignition phenomenon in a combustible mixture caused by a plasma jet shows various aspects, according to the situation of the mixtures (quiescent mixtures or flowing mixtures, kind of fuels and stoichiometric ratio), the characteristics of plasma jets (continuous type or pulse type, power level or pulse energy, reactive or inert feed stock, flow rate of the feed stock), and especially in the case of diffusion type combustion, the relative position of plasma jet injection for fuel jets.

(a) The composition of high temperature gas produced by plasma jets

In electric discharges, the source energy is first given to electrons through electric field applied, and then the kinetic energy of electrons is transferred to heavy particles (molecules and atoms). In this case, the energy transfer rate into the forms of vibration, dissociation and ionization is large compared to that into the forms of translation and rotation. The states of the high temperature gas produced by electric discharges can be divided roughly in two classes, one is in thermal and chemical equilibrium, and the other, in nonequilibrium. In the case of nonequilibrium, the number density of dissociated and ionized particles is large and the electron temperature is higher than the heavy particle temperature. The magnitude of deviation from equilibrium in an electric discharge depends on the value of E/N in the discharge region (E:electric field strength, N:number density of molecules).

In plasma jets used for combustion, usually, the pressure in the discharge region is the order of 1 atm and the produced high temperature gas is ejected from outlet port through a passage with some length. Under these situations, the composition of the high temperature jet ejected can be assumed thermal and chemical equilibrium, approximately.

(b) Chemical and thermal effects of plasma jets

Plasma jets contain specific radicals such as H, O and N, according to the feed socks used. The effectiveness of plasma jets in igniting combustible gas mixtures is mainly due to the existence of high temperature radicals. The rate constants of the reactions related to such radicals are very large, and furthermore there is the possibility that the effect of high temperature is kept during following several chain reactions. Plasma jets contain also vibrationally excited molecules in the case of molecular feed stocks. It is known that some of the vibrationally excited molecules have fast routes for the production of radicals when they are introduced into the combustible mixtures [23]. Ionized species are also involved in plasma jet, of which reaction rates are one order larger than those of usual radicals, but usually the number density of them is not large.

Argon is used widely as feed stocks in plasma jets because of easy establishment of stable electric discharges, although argon plasma jets are not effective for ignition compared to the plasma jets using molecular species as feed stocks. The ineffectualness of argon plasma jets is due to the fact that the effect is mainly thermal; argon has the mode of energy distribution in only translational form, and the route of production of radicals is relatively slow even at high temperature, because of the large values of activation energies.

Nitrogen is generally assumed inert for the progress of main combustion reaction, however, nitrogen plasma jet is effective for ignition, because nitrogen atoms produced in electric discharge have the following route to produce oxygen atoms, $N+O_2 \rightarrow NO+O$, and nitrogen molecule also has the mode of energy distribution in the form of vibration, which is easily transferred to fuel or oxidizer molecules involved in combustible mixtures through v–v energy transfer.

(c) Aerodynamic effects of plasma jets on combustion process

For the interaction of a plasma jet with a quiescent or low–speed mixture, aerodynamic effects of plasma jet, such as penetration into the mixture and production of swirl or turbulence, can play an important role for the progress of combustion, although the chemical contribution of electrically produced species is an essential one for ignition. In the cases of high speed flow, it is expected that the relative importance of the aerodynamic effects may be decreased, although in supersonic flows, shock waves appear locally.

(d) Direct application of continuous arc discharges

Direct application of a continuous arc discharge seems also to be a powerful ignition source for flowing combustible mixtures. The composition of high temperature gas in the arc column, which participates in reaction after diffusion into the combustible gas mixtures, shows the nature of nonequilibrium, and this matter causes usually a favorable effect on reaction promotion. In flowing gases usually an electrical discharge is unstable, so for the utilization of a direct DC discharge, the cathode spot must be located at the stagnation region of the flow field.   (Itsuro Kimura)

8.1.3 Plasma jet effect on steady diffusion flames

(a) Plasma jet utilization for diffusion flames

As described in the preceding section, plasma jets exhibit chemical and thermal effects, and aerodynamic effects in the interaction with combustion process. As in the cases of premixed type combustions, plasma jets can also be used for the ignition and the combustion controls in diffusion flames, such as the flame–stabilization, the promotion of combustion and the suppression of soot or NO_x.

In the case of diffusion type of combustions, it must be noted that the promotion of mixing of fuel and oxidizer (air) is involved additionally in the aerodynamic effects of plasma jets, and that to demonstrate those plasma jet effects markedly, the relative position of plasma jet injection to fuel jets must be carefully determined.

Here, some experimental results reported so far, which show successful application of plasma jets for the ignition and the combustion controls in diffusion flames, are described.

Figure 8.1 shows the importance of plasma jet injection position relative to the hydrogen fuel jet [28]. In the case of plasma jet injection I_{25} (25 mm downstream of the fuel injection), I_{50} and I_{75}, the ignition, flame–stabilization and spread of combustion are performed successfully.

A detailed diagnostic of the flame region showed that for I_{25} the progress of combustion is 68 % at the section 10 cm downstream of plasma jet injection. In the other cases (I_5 and the plasma jet injection at the upstream of the fuel injection), the ignition and small flames are observed, but the spreads of combustion are limited.

Sub–scale scramjet (supersonic combustion ramjet) engine tests in combination with plasma jet (plasma torch) igniters have been conducted in co–operation of combustion

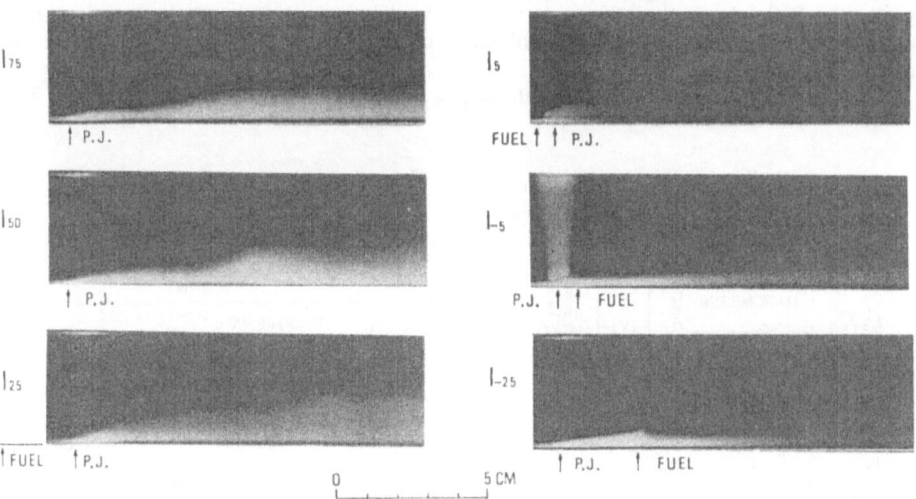

Fig. 8.1   Hydrogen flame photographs (M=2.1, static temperature=154 K, static pressure=1.0 atm, hydrogen plasma jet of 4.7 kW), from [28] with permission

research groups in the United States [29].

The effect of plasma jets on suppression of soot was investigated using a model of a gas turbine combustor [24]. The fuels used are propane, or toluene injected with propane. It was confirmed that a nitrogen–containing plasma removes soot completely, although an argon plasma is also effective for suppression of soot. The effectiveness of argon plasma suggests that for the soot suppression, the thermal effect or the aerodynamic effect of plasma jets plays an important part also. In Reference 24, the result of a preliminary experiment on the suppression of NO_x formation is also involved.

(b) Studies on the mechanism of ignition and flame–stabilization by plasma jets in high speed air streams

Studies on the mechanism of ignition and flame–stabilization by upstream–injection plasma jets were performed for H_2 diffusion flames in high speed air streams (Fig.8.2). The vitiated air that simulates high–temperature air in scramjet combustors was provided by a hydrogen–oxygen–air burner. A newly developed uncooled plasma jet was operated with argon–hydrogen mixtures. It was seen this type of plasma jet is very effective for the ignition and has a wide operating power range 630–2590 W. The recent combustor design decided after extensive investigations, incorporated small upstream pilot fuel injectors, a step for recirculation, and primary fuel injectors downstream of the recirculation region, and the plasma jet was located in the recirculation region. Both unconfined and ducted combustion tests were conducted successfully in a Mach 2 flow, under simulated scramjet combustor conditions.

Recently, the combustion research groups in Japan also have conducted sub–scale scramjet engine tests in combination with plasma jet igniters, in co–operation [9,10].In these tests, some practically important features on the ignition and flame–stabilization by upstream–injection plasma–jet–igniters were made clear, and also the durability of a plasma jet operated by oxygen–containing feed stocks were confirmed. In the both–walls injection experiments (Fig. 8.3), a plasma jet on one side wall could not ignite the fuel jets on the other side wall; after elaborate experimental investigations, a successful ignition method, by combination of a plasma jet and a step attached at the top wall, is suggested.

The plasma jets were operated with Ar gas involving active species such as O_2, N_2,

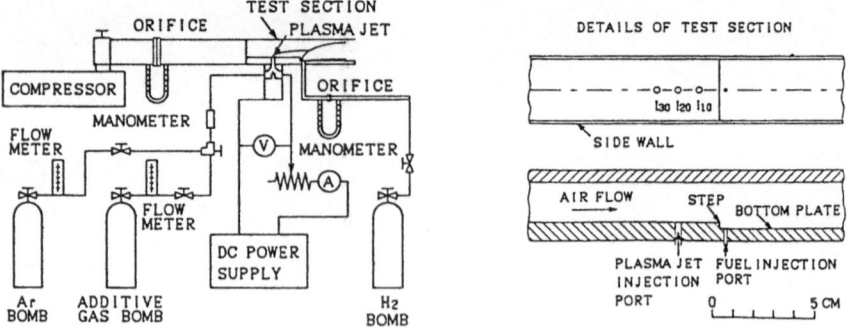

Fig. 8.2   Experimental apparatus

H_2. It was observed that for the ignition and flame–stabilization, the concentration of active species in the feedstock is more important than the plasma jet power (Fig. 8.4). This fact shows that the radical concentration supplied to the ignition point of a fuel jet is the controlling factor for the establishment of ignition. It was also confirmed experimentally that the effectiveness of additives is in the order, $O_2 \geq N_2 > H_2$ (Figs. 8.4 and 5), and the results of numerical simulation on combustion–reaction promotion by the radicals added agreed qualitatively with experimental ones(Fig. 8.6).

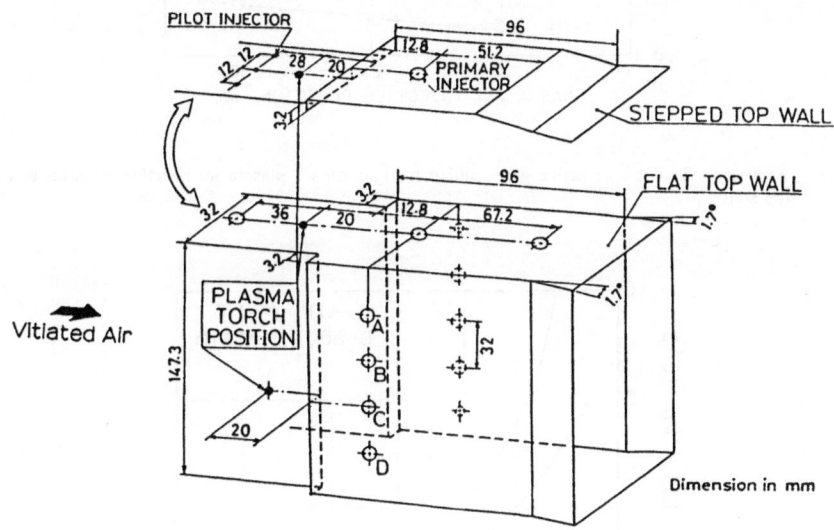

Fig. 8.3   Sub–scale scramjet engine combustor,
from [10] with permission

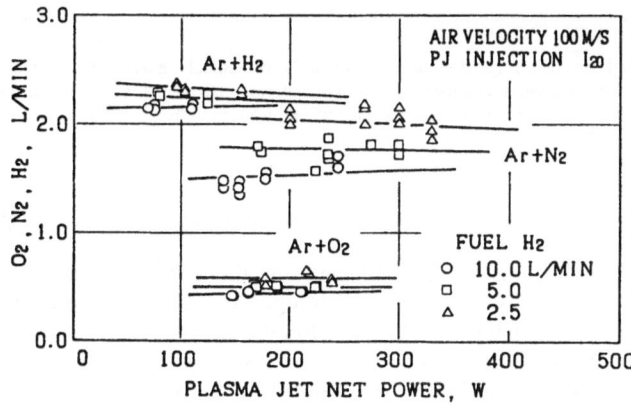

Fig. 8.4   Necessary flow rate of active gases added to argon for ignition vs. plasma jet net power

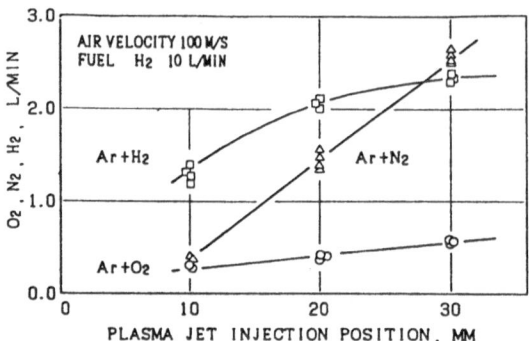

Fig. 8.5   Necessary flow rate of active gases added for ignition vs. plasma jet injection position upstream
the step

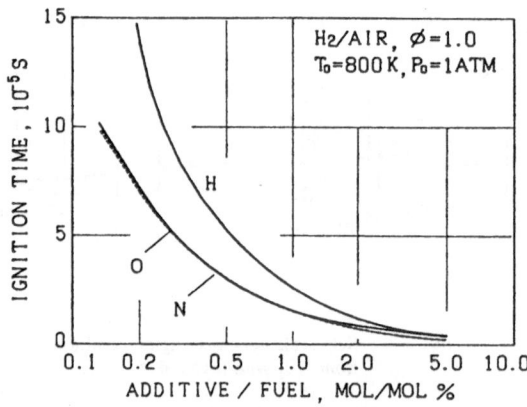

Fig. 8.6   Calculated results of ignition time vs. quantity of radicals introduced (dotted line: N). These
calculated results agree qualitatively with the experimental results, Fig. 8.5, showing that when the
distance between step and plasma jet injection is short, the effectiveness of additive gases is in the
order $O_2 \geq N_2 > H_2$

(c) Studies on the suppression of soot formation in diffusion flames by plasma jets

Experimental investigations were conducted for the control of soot formation in a
propane diffusion flame, by application of nitrogen plasma jets. The flame was formed
in an air stream of low velocity along the bottom plate of test section, by transverse
injection of the fuel at one point on the axial center line of the plate. The quantity of
soot formed can be evaluated by the brightness of the direct photographs of flames (Fig.
8.7).

It was found that for the suppression of soot by the chemical effect of the nitrogen

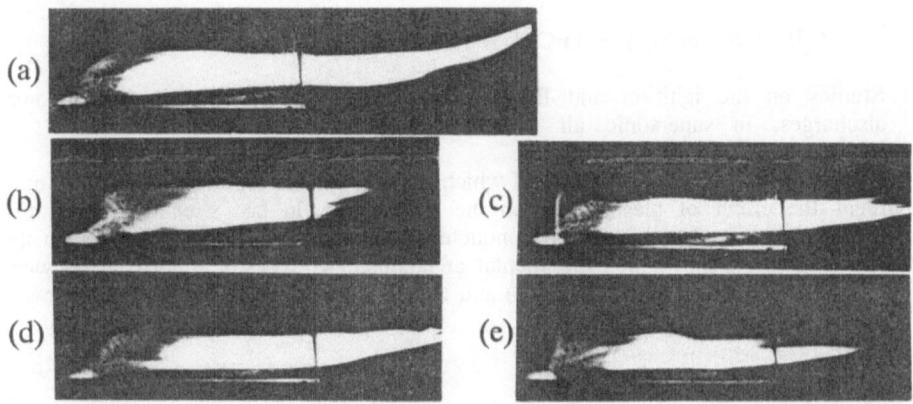

Fig. 8.7   Soot control by plasma jets. (a) original diffusion flame, (b) nitrogen jet injection and (c) nitrogen
plasma jet injection on the axial center line of the bottom plate at the section 2 cm downstream
from the fuel injection, (d) nitrogen jet injection and (e) nitrogen plasma jet injection on the point
0.5 cm off transversely from the axial center line at the section 2 cm downstream from the fuel
injection

plasma jet, it must be injected through the upstream reaction zone of the flame (for
example,   at the point 0.5 cm off from the center line on the section of 2 cm
downstream from the injection point of the fuel); when the plasma jet is injected at a
point on the center line and it contacts with a thick fuel gas directly, the soot formation
is enhanced.

Here, it must be noted that the jet injection of nitrogen gas or nitrogen plasma into
the diffusion flame has the tendency to decrease the formation of soot by the
non–chemical effects, the promotion of mixing of the fuel gas with air and the
rarefaction of the fuel gas with the jet gases. For the mechanism of the suppression of
soot formation when nitrogen plasma was injected into the reaction zone, an enhanced
oxidation of hydrocarbon by the oxygen atoms, produced through the following reactions
related to high temperature   nitrogen atom N^* or excited nitrogen molecule N_2^*, is
suggested.

$$O_2 + N^* \text{ (or } N_2^*) \rightarrow O + O + N \text{ (or } N_2) \qquad (1)$$

For the mechanism of enhancement of soot formation when nitrogen plasma was injected
into the thick fuel region, the following C_2H_4 and C_2H_2 production reactions are
suggested.

$$C_3H_8 + N^* \text{ (or } N_2^*) \rightarrow CH_3 + C_2H_5 + N \text{ (or } N_2) \qquad (2)$$

$$C_2H_5 + N^* \text{ (or } N_2^*) \rightarrow C_2H_4 + H + N \text{ (or } N_2) \qquad (3)$$

$$C_2H_4 + N^* \text{ (or } N_2^*) \rightarrow C_2H_2 + H_2 + N \text{ (or } N_2) \qquad (4)$$

The Reaction (2) is faster than the following reaction, because the bond energy C–C is smaller than that of C–H.

$$C_3H_8 + N^* \text{ (or } N_2^*) \rightarrow i,nC_3H_7 + H + N \text{ (or } N_2) \qquad (5)$$

(d) Studies on the ignition and flame–stabilization by direct application of arc discharges, in supersonic air streams

Direct application of arc discharges, of which electric powers are relatively small, may augment the effect of plasma jets on the combustion in fast streams. Supersonic combustion experiments have been conducted with dc arc discharges exposed in the streams. Figure 8.8 shows the experimental apparatus. Two types of arc discharges were applied: one is perpendicular (Fig. 8.8) and the other is parallel (Fig. 8.9) to the main

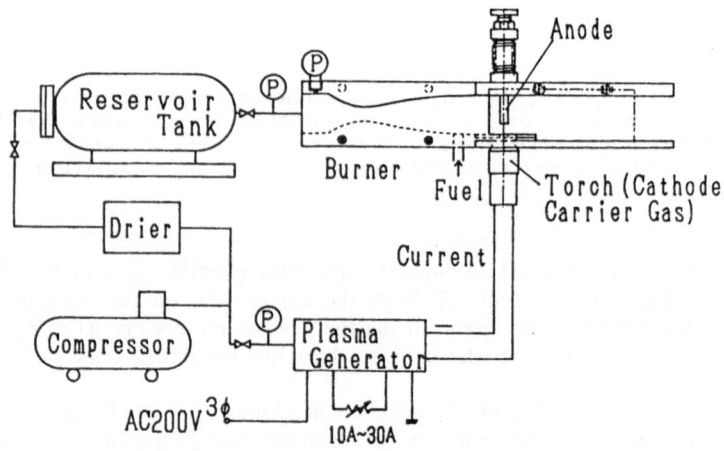

Fig. 8.8 Experimental apparatus (the case of transverse arc discharge)

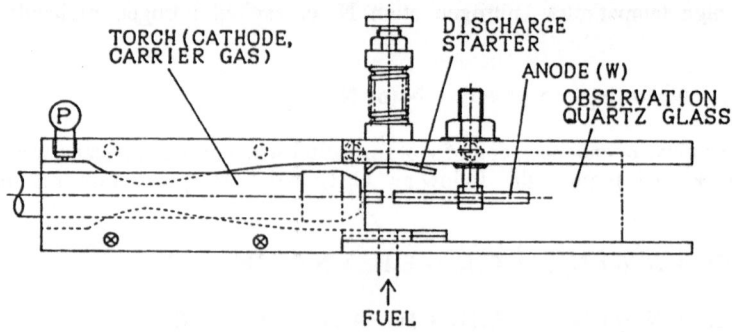

Fig. 8.9 Combustor in the case of parallel arc discharge

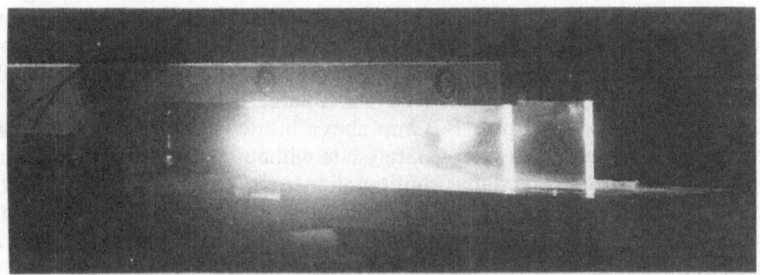

Fig. 8.10   Methane diffusion flame stabilized in a supersonic air stream

air flow. Air or oxygen seemed to be more appropriate than argon, nitrogen or helium
as the carrier gas supplied at the cathode, from viewpoints of effectiveness and practical
use. Fuel used is methane. At this moment stable combustion has been observed up to
M=1.6 with total temperature and total pressure, To = 300 K, Po = 250 kPa, and
equivalence ratio = 0.2 (Fig. 8.10). Combustion under higher Mach number will be
possible with the selection of more fitting configuration in the fuel injection and the arc
discharge.

(e) Concluding remarks

Plasma jet igniters, which exhibit excellent ability for ignition and flame–stabilization,
will be applied to practical scramjet engines and play an important role for the increase
of flexibility in the operation (Mach number in the flow of combustor, the combustion
intensity or the thrust magnitude, kind of fuels used, etc.), and also to the increase of
reliability of the engines. As for the effective application of plasma jets for the ignition
and combustion control, the complexity in chemical and physical processes involved will
require further detailed and extensive studies. (Itsuro Kimura and Takeshi Tachibana)

8.1.4 Plasma jet effects on unsteady diffusion flames

This section is concerned with the utilization of plasma jet to control unsteady diffusion
combustion and to reduce pollutant emissions, especially soot in it. Owing to the two
main properties, i.e. injection of chemically active species and enhancement of
turbulence[2], plasma jet has potential to improve diffusion combustion. However, the
investigations into the effects of plasma jet on soot formation and / or oxidation
processes are scarcely known even in the case of steady diffusion flame, and much less
in that of unsteady one. Therefore this section begins with the presentation of the
investigation into the effects of plasma jet on soot distribution in steady diffusion flames
using continuous plasma jet [31], and then moves to the discussion of the possibility of
the promotion and control of unsteady diffusion flame using pulsed plasma jet[32].

(a) Effects of plasma jet on soot distribution in steady diffusion flame

Figure 8.11 shows the schematic view of the continuous plasma jet generator used in the experiments. Working gas of plasma media is either argon or argon / nitrogen mixture, and is introduced into the generator tangentially. Steady electric discharge is maintained between copper anode and tungsten cathode, and both electrodes are water cooled. The typical results are shown in Fig. 8.12 as the contour maps of soot and OH radical emission distributions inside acetylene diffusion flame, with or without plasma injection. The field of observation is up to 140 mm above burner nozzle tip, which corresponds to half height of the flame length at natural state without plasma injection. It can be seen that with plasma injection the flame configuration changes greatly and concentrates into the vicinity of the plasma jet, which suggests highly enhanced combustion in the region of plasma injection. As for soot distribution, dense cloud disappears and the area of existence decreases greatly. Only dilute soot cloud exists near the plasma jet, and seems to be extinguished before post–flame zone. On the other hand, OH radical emission increases greatly near the plasma jet, which also suggests that combustion is very much promoted by plasma injection.

The parameters of plasma jet controllable in this study are plasma media, charge energy for plasma jet, and flow rate or momentum of working gas. The effects of each parameter is shown in Figs.8.13, 14, and 17 respectively. Each figure shows the effects on the total soot amount inside the field of observation plotted in relative value to that of the same flame without plasma injection. Fig.13 shows the effects of plasma media. As can be seen, nitrogen addition is effective, which can be attributed mainly to the superiority in chemical activity of nitrogen to argon. This is similar to the case of the promotion of the premixed combustion by plasma jet[2].

Figure 8.14 shows the effects of charge energy for plasma jet on the total soot amount. With increasing charge energy, the total soot amount decreases rapidly towards one tenth of that of natural flame at the charge energy of 700 W, which corresponds to one eighth of the heat release of the flame. The reduction of soot amount shows saturating tendency near that value. For more details, Figs. 8.15 and 16 show the vertical distributions of soot and OH radical emission, also plotted in relative value against the distance from burner nozzle tip. With increasing charge energy, the area of soot existence becomes more limited near the plasma jet, and both total amount and peak concentration decrease greatly. On the other hand, OH radical emission increases greatly with increasing charge energy, especially near the region of plasma injection. Since the

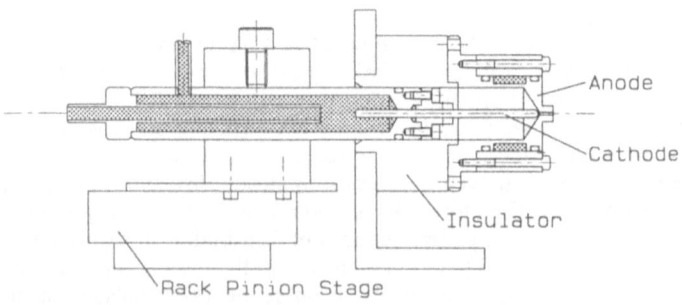

Fig. 8.11   Schematic view of the continuous plasma jet generator

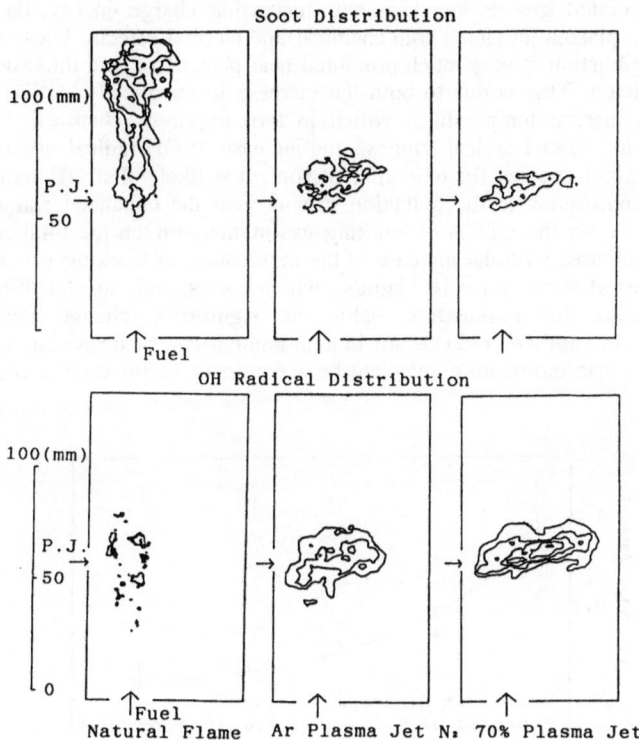

Fig. 8.12 Soot and OH radical distributions in acetylene diffusion flame, with or without plasma injection. Reynolds number of fuel jet is 1.0×10^4, and flame heat release is 5.7 kW. Plasma media is pure argon or argon/nitrogen mixture with the ratio 3:7, momentum of working gas is 2.0×10^{-3} kgm/s^2, and charge energy for plasma jet is 400 W.

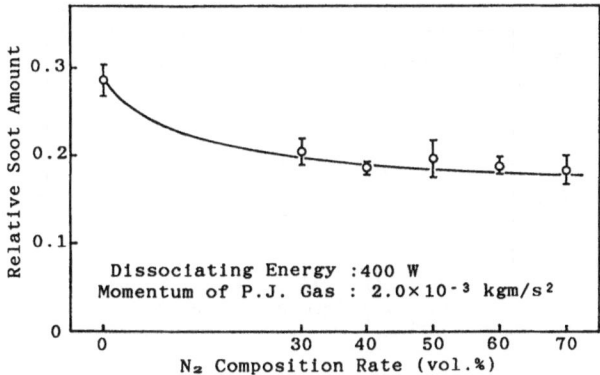

Fig. 8.13 Effects of plasma media on total soot amount

amount of dissociated species increases with increasing charge energy, the increase of charge energy of plasma jet yields both chemical and thermal effects. These three figures suggest that combustion is very much promoted near plasma jet, and thus soot formation process is restricted. This is due to both the increase in chemically active species and the increase in plasma temperature, which in turn improves chemical activity. The increased emission of OH radical suggests the increase of OH radical amount, which is said to be important species for oxidation of soot. It is likely that OH radical induced by plasma jet contributes to the reduction of soot near the enhanced flame region.

Figure 8.17 shows the effects of working gas momentum on the total soot amount. Soot amount decreases with the increase of the momentum of working gas, but becomes saturated already at about 2.0×10^{-3} kgm/s², which corresponds to one fifth of fuel gas momentum. From this momentum value on, significant change neither in the distributions of soot and OH radical, nor in total amount can be observed. This suggests that the working gas momentum may not be a dominant factor over a certain value.

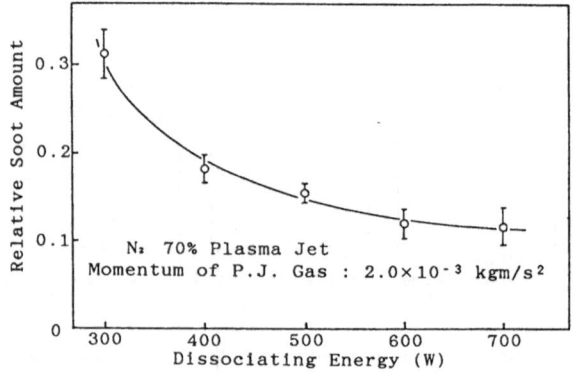

Fig. 8.14   Effects of charge energy for plasma jet on total soot amount

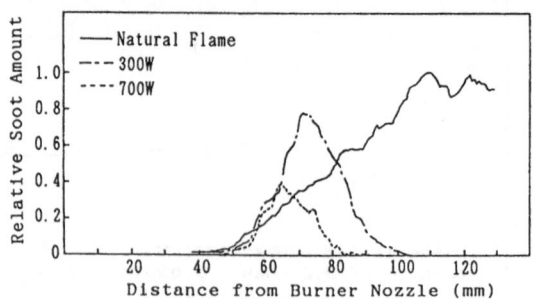

Fig. 8.15   Vertical distribution of soot. Plasma media is argon/nitrogen mixture with the ratio 3:7. The momentum of working gas is 2.0×10^{-3} kgm/s².

From the results above it is shown that, due to both the introduction of chemically active species and the enhancement of turbulence, plasma injection can promote diffusion combustion well and at the same time reduce soot emission greatly.

(b) Effects of plasma jet on soot distribution in unsteady diffusion flames

Figure 8.18 shows schematic view of pulsed plasma jet injector used in the experiments in this subsection. It is a modified spark ignition plug with small cavity (67×10^{-9} m^3) and orifice (ϕ2.0 mm) anode, which is a similar type of plasma igniter normally used

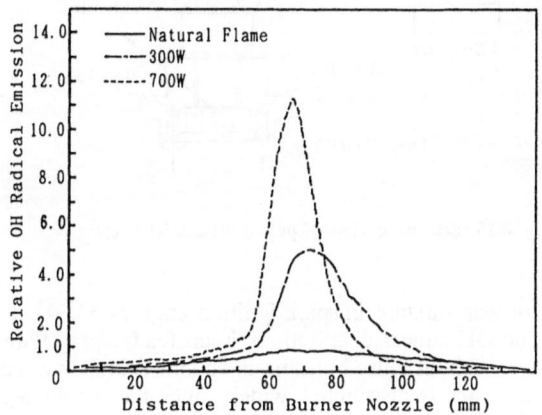

Fig. 8.16 Vertical distribution of OH radical emission. The experimental conditions are identical with those of Fig. 8.15.

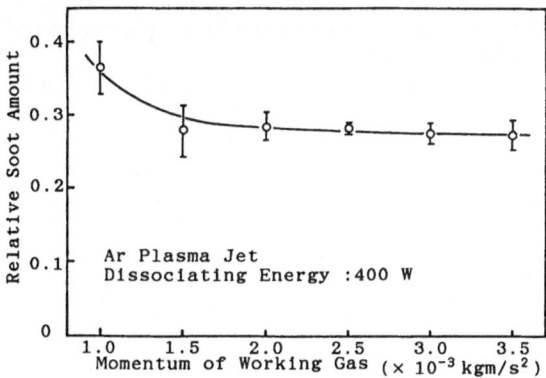

Fig. 8.17   Effects of working gas momentum on total soot amount

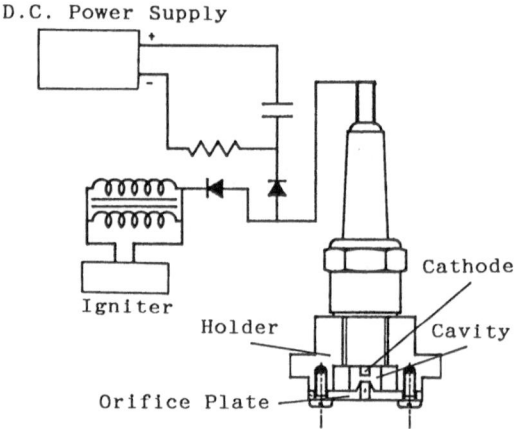

Fig. 8.18 Schematic view of pulsed plasma jet injector

in the study of ignition of lean mixture in spark ignition engines[33,34]. Fig. 8.19 shows
contour maps of soot and OH radical distributions in unsteady spray flames at 3.5 msec
after fuel injection with or without pulsed plasma injection. It is shown that, in the flame
without plasma injection, soot is distributed widely over the whole region, and very
dense soot cloud is observed in the spray tip region of the flame. On the other hand, OH
radical emission has only weak intensity level. The close comparison of twocontour
maps leads to the observation that the dense soot cloud appears in the region outside the
main combustion area. However, in the flames with plasma injection, peak spots with
very high emission intensity appear in OH radical contours, and correspondingly
concentration and total amount of soot decrease slightly. Especially as for OH radical
emission, the enhancement by plasma injection is spread over most of the flame. This
can be attributed to both the improvement of mixing by the plasma jet and the extension
of flammable limit by induced radical species[35].

   For more details, Figs. 8.20 and 21 show the changes of total soot amount and OH
radical emission integrated over the field of observation. These figures are plotted in
arbitrary unit against the time from the start of fuel injection. In the case of natural
unsteady spray flame (without plasma jet), the dense soot cloud has been observed to
go out of the field of observation at about 4.0 msec, and the rapid decrease in total soot
amount can be attributed to this fact. However the main region of the flames with
plasma injection remains in the field of observation throughout the period of
investigation, which may be attributed to fluid–mechanical blocking effects of plasma
jet. Judging from Fig. 8.21, plasma injection promotes the initial stage of diffusion
combustion. Here the charge energy for pulsed plasma jet corresponds approximately to
1% of the heat release of the flame. The efficiency of such plasma jet injector is said
to be not more than 10%[2], thus the energy given to the flame by plasma jet is
supposed to be less than 0.1% of the heat release of the flame. Even by this small
energy addition, the flame has been changed significantly. Therefore, with more efficient
plasma jet injector or effective way of giving sufficient energy to plasma jet, there are

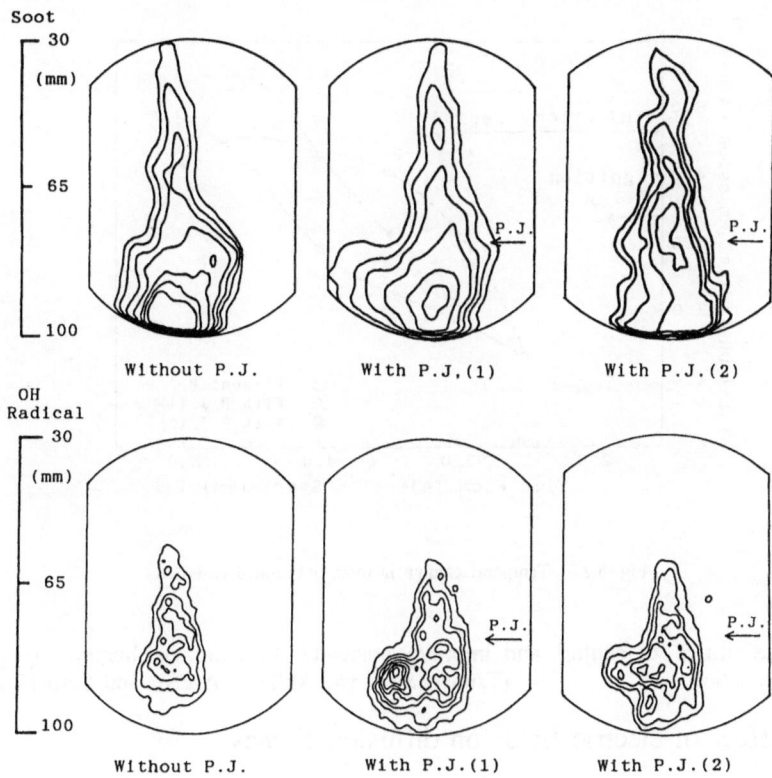

Fig. 8.19 Soot and OH radical distributions in unsteady spray flame, with or without plasma injection at 3.5 msec after fuel injection start. Fuel is JIS #2 diesel oil, amount of fuel injected is 39 mm^3, and charge energy for plasma jet is 12.5 J . Timings of plasma injection are 2.5 msec after fuel injection (P.J.(1)) and 3.0 msec after fuel injection (P.J.(2)) respectively.

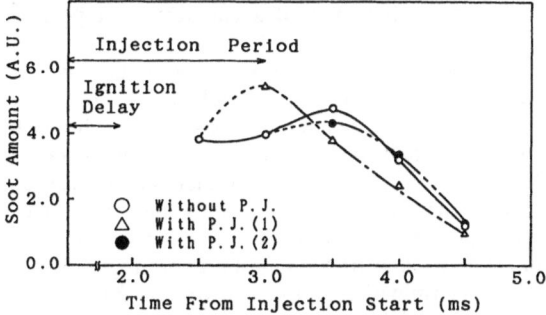

Fig. 8.20 Temporal change in total soot amount

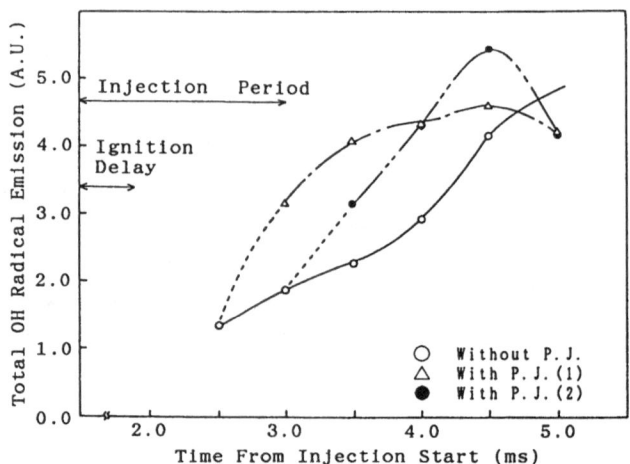

Fig. 8.21   Temporal change in total OH radical emission

good possibilities to control and improve unsteady diffusion combustion by pulsed plasma injection.   (Tsuneo Someya, Akihiko Azetsu and Satoshi Dodo)

8.1.5 Effect of electric fields on diffusion flames

The application of electric fields to flames has been found to affect both the soot formation and flame structure[1,2,36–40]. Electric fields have been reported to alter the luminosity of both diffusion and premixed acetylene flames[36–38], to reduce the soot yield in the exhaust gases of an opposed–jet diffusion flame[1], as well as to change the flame temperature and the intensity of the OH emission in a flame[1,40]. However, flame temperatures, luminosities and species concentrations in such flames have yet to be established. In this section, effects of electric fields on diffusion flames is described on the basis of the study obtained by use of a counterflow type acetylene diffusion flames.

The porous cylinder burner of 60 mm in diameter and 50 mm in length was mounted in the test section of 200 mm × 50 mm in the wind tunnel. A counterflow diffusion flame was stabilized near the forward stagnation region of the cylinder burner[41]. Two types of diffusion flame were used: one sooting, the other non–sooting. To apply an electric field to the flame region, the porous cylinder surface served as one of two electrodes. The gauze was located at a distance of 10 mm under the porous cylinder surface. The experimental apparatus and procedures for measurements of distribution of flame luminosity and soot volume fraction in the flame region were similar to those described previously[38]. Determination of the soot volume fraction, its number density and its average size was made by the laser scattering and extinction method[38,42]. The complex refraction index of the soot particles was assumed to be 1.99–0.56i. The temperature measurements in the flame region used the two–color temperature method based on Wien's law for the sooting pyrolysis zone of the flame and the LIFS of OH for the reaction zone.

(a) Effect of electric fields on flame position

Figure 8.22 shows the changes in the luminosity profiles due to the application of electric fields on the flame. Each curve of the luminosity distribution of the sooting flames can be divided into a large left part and a small right part by a step on the curve. The former corresponds to the yellowish pyrolysis zone, the latter to the blue reaction zone. The step corresponds to the dark zone. For 327 kHz frequency as shown in Fig. 8.22 (a), with increasing voltage of the electric field, the luminosity distributions of the pyrolysis zone expand and slightly move away from the burner surface, and then contract and shift towards the burner surface together with those of the reaction zone. However, the luminosity distributions for DC of both polarities expand and only shift away from the burner surface under the present experimental condition, as shown in Figs. 8.22 (b) and (c). The behaviors of the luminosities for the low frequencies (3.27 kHz and 50 Hz) were similar to those for the DC. Figs. 8.22 (d) to (g) show the luminosity distributions of the non–sooting flames with AC and DC fields. As can be seen in the figures for 327 kHz frequency, the luminosity distributions shift to the burner surface. This feature is consistent with that of the reaction zone of the sooting flame with the field voltages higher than 2400 V, but differs from that with the field voltages lower than 1600 V. This means that the shift of the sooting flame away from the burner

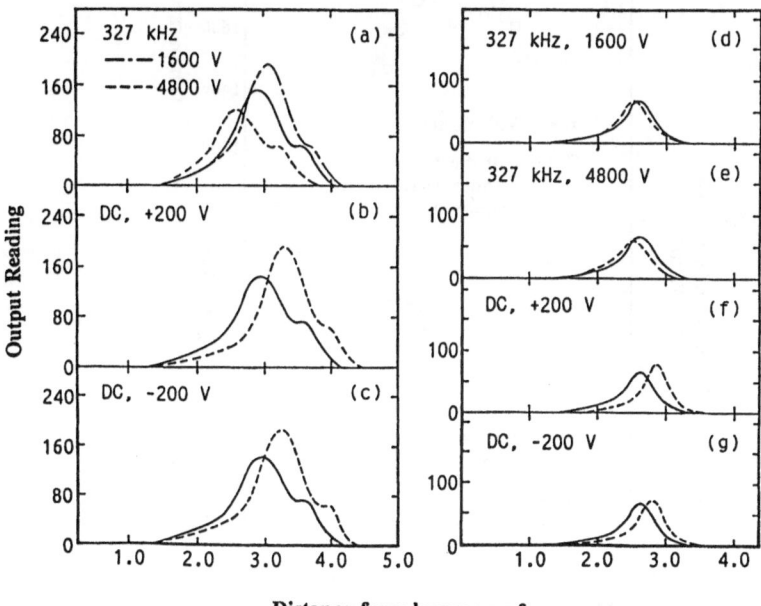

Fig. 8.22 Luminosity distributions in sooting and non–sooting flame
(Solid line : without electric field)

surface is contributed to by the soot particle movement caused by electric fields, while
the shift in the inverse direction is due to the movement of the reaction zone against the
soot particles in the pyrolysis zone. For the low frequencies and DC as shown in Figs.
8.22 (f) and (g), the luminosity distributions shift only away from the burner surface,
like the reaction zone of the sooting flame. Therefore, it is suggested that the expanded
luminosity distribution in Figs. 8.22 (b) and (c) is caused mainly by the movement of
their reaction zones. From this results, it is clear that a simulated result [43] based on
the ionic wind theory is contrary to the present experimental result: the flame shifts
towards the anode, as shown in Figs. 8.22 (b) and (f), and therefore the ionic wind
theory cannot account for the movement of the flame in the present study.

(b) Effect of electric fields on flame structure

Figure 8.23 shows the effect of the electric fields of 1.48 MHz frequency and 4800 V
on temperature and OH concentration in the reaction zone of a sooting flame. The
figure also illustrates the intensity of light scattered by the soot particles. The values of
OH concentration and those intensities were expressed as the ratios to those with no
electric field at 3.7 mm from the burner surface. The figure reveals that in the absence

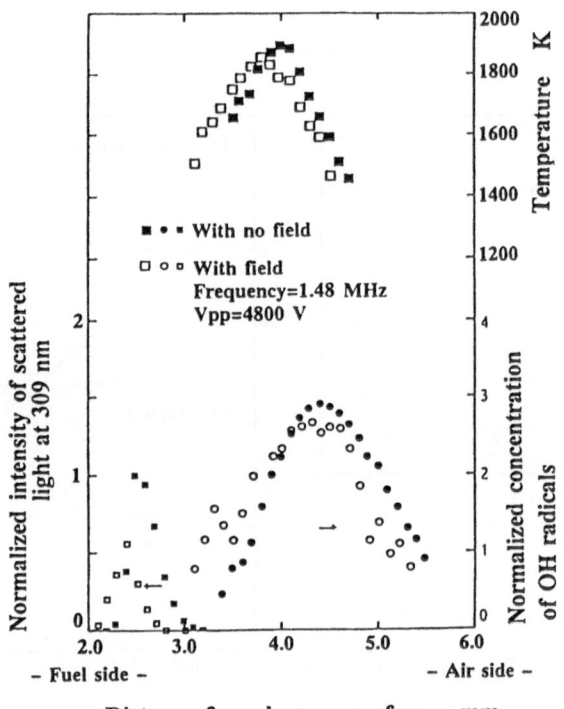

Fig. 8.23 Distributions of temperature, OH radical concentration
and scattered light intensity in sooting flame

of an electric field the maximum temperature of the flame is about 1800 K. The electric fields reduced the maximum value of the flame temperatures by about 40 K and the OH concentrations by about 10%. The decrease in OH concentration is considered to be due to the decrease in the temperature. The figure also shows that, in the OH concentration distribution for the electric field, the second peak appears in or close to the central dark zone between the pyrolysis zone and the reaction zone. The second peak arises only in the case where the soot volume fraction was reduced.

(c) Effect of electric fields on soot formation

The soot formation in flame was suggested in the previous studies to be dependent on the flame temperature in the absence of an electric field[44,45]. In order to examine the temperature dependence of the soot volume fraction in electric fields, the maximum values of the soot volume fraction have been plotted against the reciprocal maximum temperatures of the flames, as shown in Fig. 8.24. This figure shows two separate trends in the relationship between the maximum volume fractions and the maximum temperatures. When the maximum values of the volume fraction are larger than those which occur in the absence of an electric field, they increase with increasing maximum temperatures (line A), whereas they decrease (line B) with increasing maximum

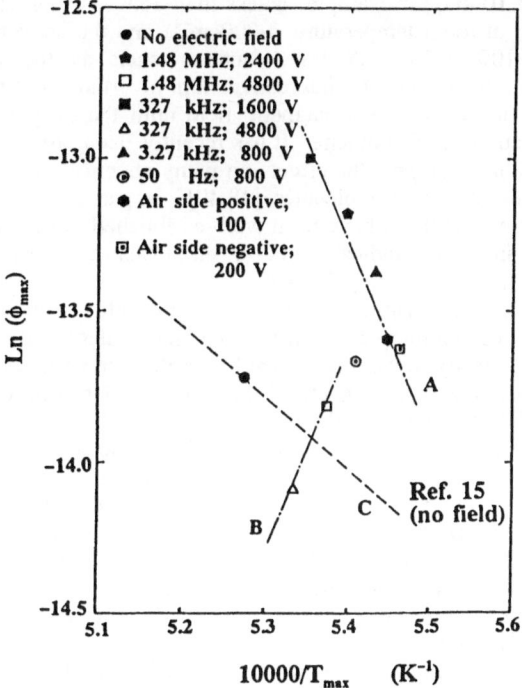

Fig. 8.24 Dependence of maximum value of soot volume fraction
on maximum flame temperature

temperatures. Here the former case is described as the case A, and the latter as the case B. The activation energies obtained is 460 kJ/mole in the case A, and −570 kJ/mole in the case B.

Tesner et al.[46,47] measured the maximum rate of the nucleation in the thermal decomposition of acetylene in diffusion flames, and determined the activation energy for the surface growth to be about 210 kJ/mole and for the nucleation to be about 580 kJ/mole. Clearly, the activation energy for the nucleation is close to the value derived for the case A in Fig. 8.24. The nucleation of soot particles increases the formation of surface area. The area is available for surface growth and the consequent increase in the volume fraction. Therefore, there is a possibility that the production of the incipient particles is the dominant process controlling the soot volume fraction when the electric fields are applied.   (Michikata Kono, Lin Xie and Takeyuki Kishi)

8.2 Magnetic Field Effect on Combustion Reactions

8.2.1 Introduction

Does a magnetic field give any effect on chemical reactions? From the viewpoint of energetics, the answer is negative because the electronic Zeeman energy of doublet intermediate radicals at 10 kG, for example, is less than $1cm^{-1}$, which is much smaller than the thermal energy at room temperature ($\sim 200cm^{-1}$) and the activation energy for chemical reactions ($\sim 1000cm^{-1}$). A few reactions such as the photo−induced predissociation of iodine vapor and the interconversion of ortho− and para−hydrogen have been known to be influenced by a magnetic field until the early 1970s, and have been recognized as exceptional phenomena. It has recently been proved, however, that magnetic fields below 20kG can give the effects on many chemical reactions as well as energy transfer processes of excited molecules [48,49]. In particular, the mechanisms of the magnetic field effects (MFEs) have been well−established for the photochemical reactions in solutions in which radical pairs or biradicals are formed as reaction intermediates with anomalous spin populations.

In the gas phase, MFEs on dynamics of the excited molecules have been detected as the intensity changes of luminescence from the molecules and from reaction intermediates in flames. Many researches on MFEs on the luminescence intensities of small molecules have revealed that magnetic fields can modulate (accelerate, in most case) the radiationless decays from excited molecules [48−50]. Therefore one can expect that a chemical reaction through the field sensitive excited state will show the magnetic field dependence of the reaction yield. This idea has yielded only a few examples [49]. This is a marked contrast to the case of reactions in solutions. This difference is mainly due to the following two; (i) the fast energy relaxation induced by solute−solvent interactions which defines the spin state of the reaction precursors occurs in solution but not in the gas phase, and (ii) the "cage" effect which restricts the rapid diffusion of the spin correlation exists in solution but not in the gas phase. For example, even under a drastic condition such as atmospheric flames, the population distribution and the polarization of laser−induced fluorescence of OH A state in the flames have shown to be far from the complete randomization by collision [51]. Consequently the origin of MFE in the gas phase should be different from that in solution.

At the present stage, the studies of MFEs on the photochemical reactions in solutions and on the radiationless processes of the excited molecules in the gas phase have the definite bases of the "guiding principle" obtained both experimentally and theoretically but those on the reactions in the gas phase remain "phenomena" waiting for further investigations.

8.2.2 Review of MFEs on the gas phase reactions

As the first report of MFE on the reaction of combustion process, Hayashi [52] observed the magnetically induced increase of the emission intensity of the A–X transition of OH radical in a premixed oxygen–propane or oxygen–hydrogen atmospheric flame. The intensity of the emission peak at 310nm was found to increase by 10% with a magnetic field of 15kG. With higher spectral resolution [53], he could show that this increase was particularly large for the N'=4–6 rotational levels of the F_1 spin sublevels. This indicates that the MFE was attributed to some magnetically induced enhancement in the formation process of the particular spin–rotational levels in the flames. Wakayama et al. [54] found the Na D lines in oxygen–hydrogen flames were increased by more than twice in intensity by a magnetic field of 18kG and this increase was dependent in a complicated way on the combustion conditions. These observations indicate that there are some magnetically sensitive processes in the activation to Na atoms from heat bath and/or the de–activation of the excited Na atoms by intermolecular interactions. They [55] also found that the emission intensities of HPO and SnH in oxygen–hydrogen flames were partially quenched by a magnetic field below 18kG. In the case of HPO A–X emission, this quenching did not obey the monotonic dependence on the field strength but the oscillative one. Using this phenomena, they have demonstrated that a phase sensitive detection for the magnetically modulated intensity is very useful as a high sensitive quantitative analysis of P–containing substances.

The above–mentioned observations in the atmospheric flames give the clear evidences to show that there are MFEs on some combustion processes. The detailed mechanisms, however, have not yet been elucidated. What is the elementary process responsible for the MFE and what is the mechanism working upon it? High pressure and high temperature condition in the atmospheric flame may prevent one from separating a particular process from the successive branched chain reactions in it. Obviously the next step for further investigations is the observation of MFE under simpler condition than the atmospheric flame, approaching to the pure elementary process of the reaction. Following this extension of research, Fukuda et al. [56] have studied MFE on some afterglows using the discharge–flow method in the pressure range of a few Torr and at room temperature. They observed the MFE as a remarkable quenching of the NO (B–X) $\beta(0,v'')$ emission in the afterglow produced by a microwave discharge of N_2–O_2 mixtures. At present, this is the most prominent example of MFE in the gas phase. It is known that NO ^{2}B state is produced by the recombination of $N(^4S)$ and $O(^3P)$ atoms including third body followed by collision induced intersystem crossing;

$$N(^4S)+O(^3P)+M \rightarrow NO(^4a)+M \rightarrow NO(^4b)+M \rightarrow NO(^2B)+M.$$

Recently Sumitani et al. [57] studied this system in more detail and indicated that the last step, i.e. the collision induced intersystem crossing is a magnetic sensitive process. From the kinetics of the NO ^{2}B state itself, Cosby and Slanger [58] reached similar conclusion.

Fukuda et al.[59] also found that in a low pressure C_2H_2/O_2 flame produced by a microwave discharge of O_2, the intensity of the CH(A–X) emission was reduced to 95% of that at zero field by a magnetic field of 18kG.

8.2.3 MFEs in low pressure diffusion flames

As a extension of the research of MFEs on the low pressure diffusion flames, we constructed a crossed effusive beam apparatus [60], in which two different microwave discharged flows are crossed between electromagnet poles and the activated reaction intermediates are detected as chemiluminescence from electronically excited species and laser induced fluorescence (LIF). The following is a summary of the results in combination with those of the previous studies.

(a) OH(A–X) emission

(1) In the atmospheric C_3H_8/O_2 and H_2/O_2 flames, the emission intensities from the N'=4–6 rotational levels of the F_1 spin sublevels in the A(v'=0) state was found to be increased by 20–35% in a magnetic field of 18kG. On the other hand, the emission from the F_2 sublevels showed no MFE. The measurements under various reaction conditions indicate that the formation process of F_1 sublevels of N'=4–6 would be promoted by the field [52,53].
(2) No MFE was observed in the low pressure $C_2H_2/O/O_2$ and $C_3H_8/O/O_2$ flames [59].
(3) No MFE was observed in the low pressure $C_2H_2/O/N_2$ flame which was free from O_2 molecules [60].
(4) The LIF intensity from OH A(v'=0) state showed no MFE[61]. This means that the nonradiative process (including collision) from the state is not affected by a magnetic field, which does not contradict the conclusion of (1).
These results, especially the difference between the result of (1) and those of (2) and (3), indicate that a magnetic field promotes some steps of the formation process of the OH A state under high pressure and high temperature conditions. Considering that MFE was observed even in the H_2/O_2 flame, the possible process [62] is;

(i) chemical excitation under high pressure conditions,

$$H + OH + OH \rightarrow H_2O + OH(A) \qquad \text{or}$$

(ii) inverse–predissociation involving third bodies,

$$O + H + M \rightarrow OH(^4\Sigma^-) + M \rightarrow OH(A) + M.$$

(b) CH(A–X) emission

(1) No MFE was observed in the atmospheric C_3H_8/O_2 flame [52].
(2) In the low pressure $C_2H_2/O/O_2$ flame, the emission was decreased by 5% in a magnetic field of 18kG [59].
(3) The emission in the low pressure $C_2H_2/O/N_2$ flame showed no MFE [60].
It has been known [62] that the main formation processes of CH(A,B) states in the hydrocarbon/oxygen flames are

(i) $C_2 + OH \rightarrow CO + CH(A,B)$ \qquad and

(ii) $C_2H + O_2 \rightarrow CO_2 + CH(A,B)$.

The difference in (2) and (3), i.e. the existence of O_2 molecules indicates the process (ii) would be magnetic sensitive.

(c) C_2(d–a) and CN(B–X) emissions

While no MFE was observed in the atmospheric C_3H_8/O_2 flame [52], the emission intensities of C_2(d–a) were found to be reduced by 5–18% in the low pressure $C_2H_2/N/N_2$, $C_2H_2/O/N_2$ and $C_2H_2/O/NO/N_2$ flames in a magnetic field of 12kG [60]. No MFE on the LIF intensity and lifetime of the C_2(d) state indicates the decay process itself is not magnetic sensitive. Therefore a magnetic field would hamper the formation process of C_2(d) state (species). From the study using ^{13}C isotope, it is known that in the $C_2H_2/O/O_2$ system, the C_2(d) state is produced by the decomposition of poly–acetylene such as

$$C_8H_2 \rightarrow C_6 + C_2(d) + H_2 \quad [62].$$

The CN(B–X) emission intensity was also found to be decreased by 32% in the $C_2H_2/O/NO/N_2$ flames in a magnetic field of 12kG [60]. The decrease showed the similar dependence on the field strength to that of the C_2(d–a) emission in this reaction system. This and the resultant vibrational temperature of CN(B) indicate that the magnetic field gives an influence on the process such as

$$C_2 + NO \rightarrow CN(B) + CO \quad [63].$$

8.2.4 SO_2 afterglow and LIF intensity of SO_2 C state

For the study of the low pressure diffusion flames, a burner–type reaction system has been placed between the magnet poles, as shown in Fig. 8.25. Addition of NO to the

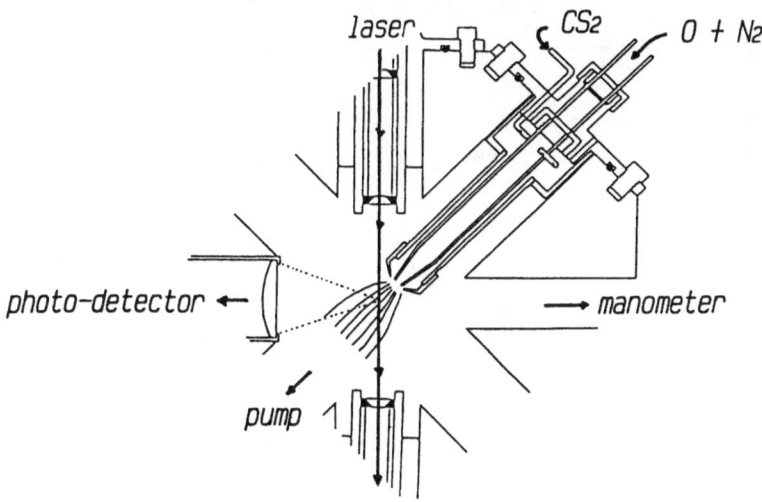

Fig. 8.25   A burner–type reaction system for the study of MFE on the low pressure diffusion flames

discharged N_2 flow produced atomic oxygen without chemically active molecular oxygen by a very fast reaction, $N + NO \rightarrow O(^3P) + N_2$. When this O/N_2 flow encounters a fuel at the center of the poles, a diffusion flame was formed. In the case of CS_2 as a fuel, bright luminescence was emitted from the flame when $[CS_2] \ll [O]$. This broadband emission is shown in Fig. 8.26 and is known as SO_2 afterglow [64,65] produced by the reactions,

$$CS_2 + O \rightarrow CS + SO \text{ , and} \qquad (1)$$

$$SO + O \rightarrow SO_2^* + N_2. \qquad (2)$$

As shown in Fig. 8.26, a magnetic field of 12kG decreased the emission intensity by more than 40% almost uniformly over the whole spectral region. In Fig. 8.27 are shown the magnetic field dependences of the emission intensities at 313nm under various pressure conditions. In a low pressure region (<1Torr), the intensity increased with increasing H until 1kG but decreased with increasing H above 1kG. At higher total pressure (>1Torr), the magnetic enhancement was diminished and the quenching became predominant.

The starting point of the blue side of the emission is ~42000cm^{-1} (see Fig. 8.25), which indicates SO_2^* is the third singlet excited state C $^1B_2$ of SO_2. This may be supported by the facts that the inverse process of (2) (without N_2) is the predissociation of SO_2, which is known to occur above 54500cm^{-1} in the C state [66] and that the afterglow emission (Fig. 8.25) can be interpreted as the overlap of many single vibronic fluorescence from the C levels of SO_2^* [66,67]. Therefore we started to investigate the MFE on the radiationless decay processes of the SO_2 C state in a collision free condition [68]. Fig. 8.28 shows the fluorescence excitation spectra of the $(v_1 v_2 v_3) = (132)$ vibrational

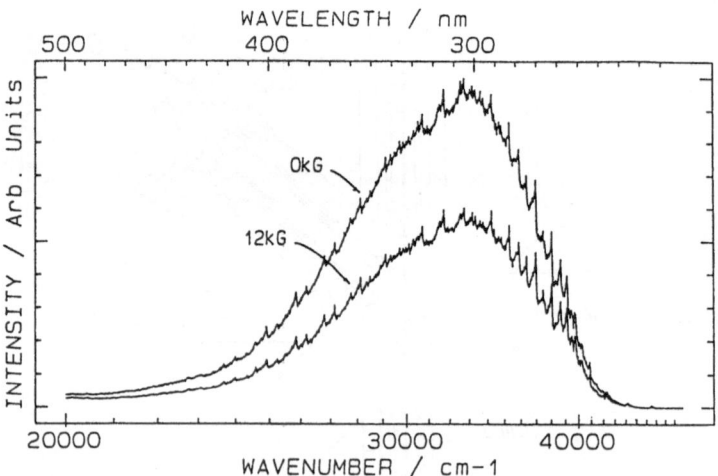

Fig. 8.26   Chemiluminescence spectra of $SO + O + N_2 \rightarrow SO_2^* + N_2$ system at 1.464 Torr under magnetic fields of 0 and 12kG. Flow speed; NO=6.2, CS_2=1.6, N_2=400ccm.

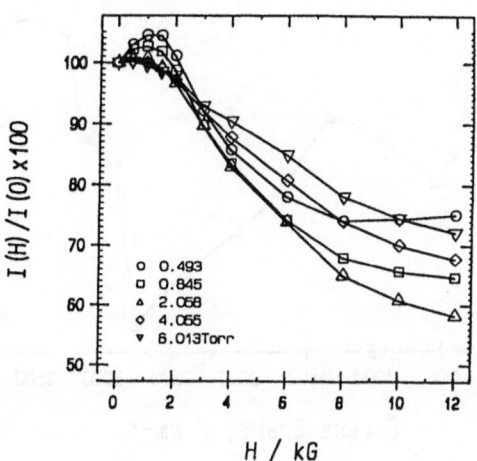

Fig. 8.27 Magnetic field dependences of chemiluminescence intensities at 313nm of SO_2 afterglow at various pressure conditions. Flow speed; same as Fig. 8.26.

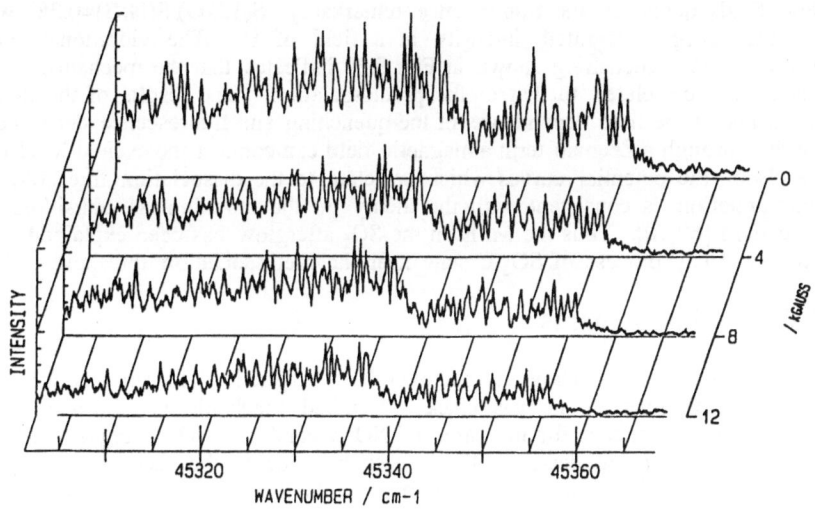

Fig. 8.28 Fluorescence excitation spectra of (132) level of SO_2 C state under magnetic fields of 0, 4, 8, and 12kG. SO_2 pressure was 6 mTorr.

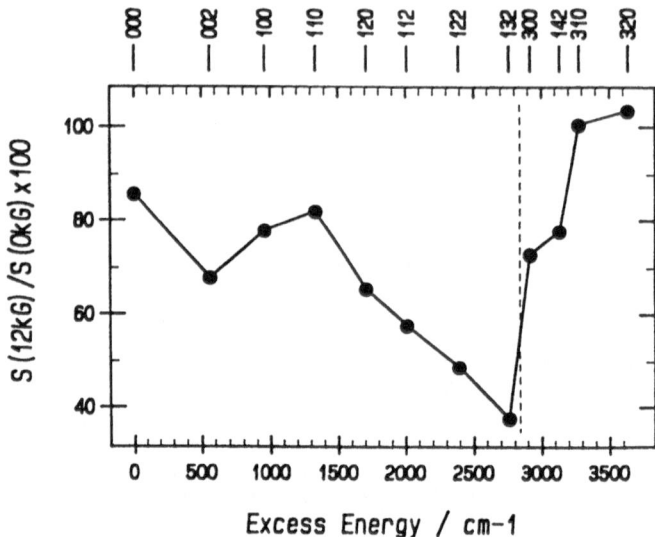

Fig. 8.29   Vibrational energy dependence of the magnetic quenching of fluorescence from $(v_1v_2v_3)$ levels of SO$_2$ C state.   Vertical broken line represents the position of predissociation limit

levels, which is located just below the dissociation threshold. As can be seen, external magnetic fields quenched the fluorescence remarkably; S(12kG)/S(0kG)=0.38, where S(H) is the energy–integrated   intensity at a field of H.   The vibrational energy dependence of the quenching shown in Fig. 8.29 indicates that the mechanism of the MFE is closely correlated to that of the predissociation.   The results of the detailed measurements of the field dependence of the quenching and fluorescence decay curves suggest that through a Zeeman term a magnetic field can connect the excited level to the dark levels on the potential curves which correlate to the dissociation threshold [68]. This interpretation is consistent with the theory on the radiationless transitions in a magnetic field [69,70]. Thus the MFE on the SO$_2$ afterglow has been explained by the MFE on the decay process of SO$_2$ C state near the predissociation threshold.

8.2.5 Concluding remarks

Many examples mentioned above have shown that a magnetic field certainly influences the chemical reactions in the gas phase.   Detailed mechanisms, however, remain unknown at present except for the cases of NO β band and SO$_2$ afterglow, which is described above in some detail.   This is probably due to the fact that a magnetic sensitive process is usually a very small portion of the whole successive chain reaction in the gas phase. Thus it is not easy to investigate separately what elementary process is affected by the field.   Nevertheless we believe that further researches on the MFEs under the simpler reaction conditions are valuable and are the key steps to the dream of scientists; "magnetic control of chemical reaction" [48].

(Haruo Abe and Hisaharu Hayashi)

8.3 Catalytic Combustion – Roles of Catalyst

8.3.1 Introduction

Catalytic combustion is a flameless combustion of fuel–air mixture made possible by the function of solid catalytic materials. The catalytic combustion is utilized either for the production of heat, for the emission control or for both of them. Applications may also be classified by the temperature range of the combustion into low–, medium– and high–temperature combustion: Low–temperature (<300°C); air–cleaning in kitchen, refrigerator, or automobile cabin, etc., medium–temperature (300–900°C); automotive catalysts, catalytic heater, etc., and high–temperature (>900°C); gas turbine, boiler, etc.

Catalytic combustion at a high temperature range is the subject of this section. As pointed out in earlier reviews [71–73], the advantages of catalytic combustion are due to the capability of catalysts to oxidize (i) uniformly (ii) at relatively low temperatures (iii) the low–concentration fuels (iv) without significant formation of nitrogen oxides, carbon monoxide or soot. Therefore, this catalytic combustion is an environmentally friendly technology for the future to convert fuel into heat, and R & D activities are now very high. The problems to be solved are to increase the density and stability of heat production and to improve thermal properties of catalysts.

Several new ideas have been developed for the catalytic combustor as well as for the catalyst materials. As for the combustor, hybrid combustion has been developed by Toshiba [74], in which fuel is supplied before and after the catalyst zone, so that the temperature of catalyst does not exceed the temperature that damages the catalyst, but becomes high enough to combust the fuel injected at the post–catalyst zone. In another combustor called NONOx catalytic combustion system developed by Tanaka Kikinzoku and Catalytica [75], whole fuel is supplied before the catalyst zone at about 350°C, but the portion of combustion that proceeds in the catalyst zone is controlled to keep the temperature below about 900°C. The rest of combustion takes place in the post–catalyst zone to reach the adiabatic flame temperature of about 1300°C.

As for the catalyst materials, high stability at high temperature and against thermal shock is needed. In addition, a highly active catalyst is required to initiate combustion at low temperature. Layered aluminas containing La [76] and Ba [77] have been proposed for thermally stable catalyst supports. The Ba–alumina containing Mn is also active catalytically.

Although some were tested and demonstrated to be effective with medium–scale combustors, further improvements are still required for practical applications. For the design of improved catalysts, it is necessary to understand the mechanism of combustion in more detail, since catalytic combustion is a complex reaction system in which surface and gas–phase reactions are coupled through several mass and heat transfer phenomena.

In this section, the mechanism of catalytic combustion is briefly reviewed and then discussed on the basis of our own work in which experimental data of methane combustion was compared with the simulation analysis. Here, stress will be placed on the effects of catalyst properties such as catalytic activity, surface area and operation conditions.

8.3.2 Mechanism of catalytic combustion

Essential characteristics of catalytic combustion (adiabatic) is understood by a schematic illustration given in Fig. 8.30 [71–73]. In this figure, the solid line shows the rate of

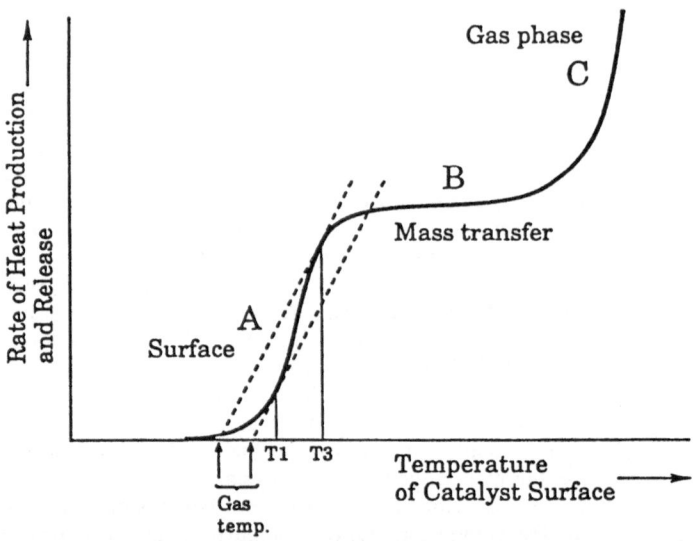

Fig. 8.30   The rate of heat production by combustion and heat release at catalyst surface.

reaction (expressed by rate of heat production) as a function of the temperature of catalyst surface. Broken lines indicate the rate of heat release from the surface to gas phase.

At a low temperature region (region A), the combustion starts on the catalyst surface. The reaction takes place exclusively on the catalyst surface and the temperature of the catalyst is close to that of the fluid. As the temperature reaches T1, catalytic ignition temperature, the rate jumps to the mass–transfer controlled rate. Above T1, the rate depends only slightly on the temperature (region B), because of the mass transfer limit.

At a much higher temperature region (region C), homogeneous gas–phase combustion prevails and the rate increases exponentially. T3 is the extinction temperature, which is observed when the temperature is lowered. In the region B, the concentration of fuel near the catalyst surface is very low and the catalyst temperature is significantly higher than the temperature of bulk fluid.

The catalyst is thought to function in two opposite ways, that is, acceleration by the production of heat and active species and retardation of combustion by the deactivation of active species coming from gas phase. It may be obvious that the combustion starts at a low temperature due to the acceleration effect, but the functions of catalysts at high temperatures are still controversial.

The specific catalytic activity is essential at low temperatures (particularly for the initiation of combustion), but is regarded to be less important at high temperature regions.   A similar consideration may be applied for the catalyst surface area. According to Pfefferle and Pfefferle [71], the catalyst surface needed is little higher than that of catalyst geometry. Hence the pore structure is not important and active ceramics could be useful as monolith catalysts. This is presumably true for micro– and mesopore structure, but macropore structure seems influential under certain conditions as described below. Thus, the effects of catalytic properties such as oxidation activity, surface area, and pore structure are not clear yet and the optimum catalytic properties for catalytic combustion are to be made clearer. Experimental and simulation studies on the initial process of catalytic combustion of methane at a lower temperature range (600–800°C) revealed that the rate of reaction strongly depends on the kind of catalyst and that the chain carriers are HO_2, OH and H radicals [78]. The concentrations of OH and O radical have already been measured in catalytic combustor to elucidate the role of catalyst [71].

8.3.3 Catalytic combustion of methane

As reviewed above, there have been several experimental as well as simulation studies regarding the catalytic combustion. However, few catalytic chemical studies have been reported, e.g., the influences of the chemical and physical properties of catalysts, while those influences are essential in the ordinary catalytic processes, e.g., organic synthesis and automotive catalysts. We carried out catalytic combustion of methane with a series of well–characterized perovskite–type catalysts and analyzed the results by computer simulation based on a one–dimensional steady state model.

(a) Experimental and simulation methods

Methane (5%) was oxidized with air (95%) in the 3–stage (preheating, catalytic, and homogeneous) packed bed flow reactor. Diameter of quartz tube reactor was 6 mm. Powder of perovskite–type catalysts ($La(Sr)CoO_3$, $LaCrO_3$, etc.), of which the catalytic properties were thoroughly characterized [79,80], was used. The temperature of the reactor wall was controlled isothermally by external heating and measured by thermocouples installed before and after the catalyst bed. Only CO_2 and H_2O were detected by GC in the effluents.

Simulation was carried out base on a one–dimensional steady state model [81,82]. Mass and heat balances in the gas phase and catalytic surface, as well as the energy balance in the solid phase and momentum balance in the gas phase were considered. In addition to the ordinary initial and boundary conditions, the temperature gradient at the exit of the catalyst zone was assumed to be zero. Kinetic parameters (preexponential factors and rate constants) for catalytic and gas–phase reactions were determined by using the data of combustions with and without catalysts.

(b) Experimental results

Examples of the temperature dependency of the methane conversion obtained experimentally are shown in Fig. 8.31 [83]. The dependency can be divided into three regions (A, B, and C), as in Fig. 8.30. In the region A, surface reaction is rate–determining. Reaction rate changed little in the region B. Gas–phase reaction prevails in the region C. In the region A (below 650°C), the reaction rate, the dependency of the rate on the mass flow rate (conversion showed a linear correlation with contact time), and the activation energy (80 kJ/mol independently of the mass flow rate) agreed with those of the ordinary catalytic oxidation over the same catalysts.

In the region B, there were no changes upon repeated cycles of reaction temperature (up and down). Therefore, the low dependency of the rate on temperature in the region B was not due to the catalyst deactivation, but really caused by the mass transfer

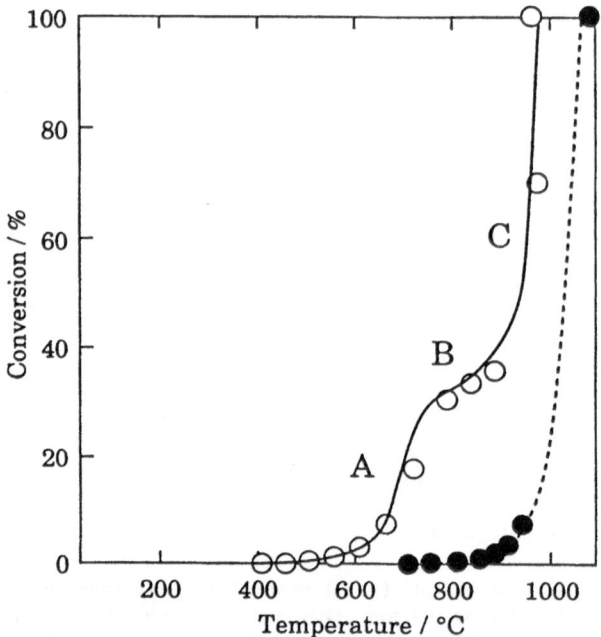

Fig. 8.31 Temperature dependency of the methane conversion (experiment and simulation)
o : Experiment with catalyst, ●: Experiment with an empty tube. Solid line; simulation.

limitation. Methane conversion showed approximately linear correlation with the surface area of the catalyst which was varied by changing the calcination temperature from 850 to 1150°C. The apparent activation energy in this region was low (12 kJ/mol), as expected.

(c) Simulation

The results of simulation are also shown in Fig. 8.31 [81]. The A, B, and C regions were well reproduced by the simulation. Thus it is confirmed again that the region B appeared because of the mass transfer control. The present simulation does not provide evidence of the presence of catalyst–promoted gas–phase reactions in the region C, since the simulation well reproduced the results without considering its presence.

The axial distributions of the gas temperature and methane conversion are simulated as shown in Fig. 8.32 [82]. In the region A, the reaction occurs only on the catalyst surface. The regions B and C show only small differences in the catalyst zone, indicating that mass transfer is the rate–determining step in both regions. A slightly higher temperature for the region C at the end of the catalyst zone causes the gas reaction to start in the post–catalyst zone, while it is low and further reaction does not occur for the region B.

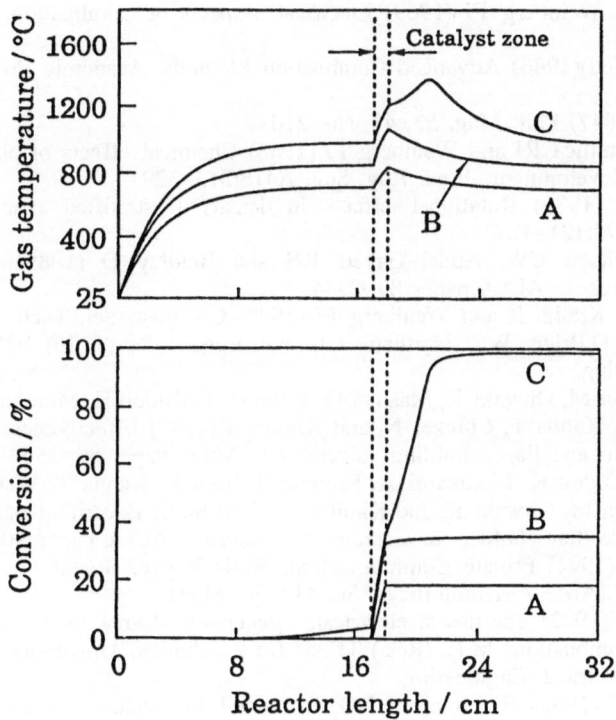

Fig. 8.32 Axial distributions of the gas temperature and methane conversion (isothermal condition).

The advantages of the catalytic combustion may be shown by the following simulation [82]. The simulations for the presence and absence of catalyst were carried out in such a way that 100 % conversion is reached at the exit of the reactor. The reactor wall after the catalyst zone was assumed to be adiabatic. In the catalytic combustion, 1590°C was reached at the exit of the reactor, while the temperature at the exit was 1790°C in the homogeneous gas– phase reaction without catalyst. Thermal NOx is known to be formed above about 1700°C. Therefore, the formation of NOx can be reduced by the catalytic combustion with appropriate preheating and post–catalyst gas–phase combustion (a hybrid system). Since the preheating temperature can be lowered by using catalyst, the temperature at the exit of the catalyst zone becomes lower. It is also shown that the temperature of the catalyst surface can be maintained below 1090°C in the hybrid system. (Makoto Misono, Kwan Y. Lee, and Masayuki Asami)

References

[1] Lawton J and Weinberg FJ (1969) Electrical Aspects of Combustion. Clarendon Press, Oxford

[2] Weinberg FJ (ed)(1986) Advanced Combustion Methods. Academic Press, London and New York

[3] Faraday M (1847) Phil. Mag. 53, 41, No. 210

[4] Vosen SR, Cattolica RJ and Weinberg FJ (1988) Chemical effects of plasma gases on flame kernel development. Proc. Roy. Soc. A418:313-329

[5] Weinberg FJ (1989) Rotational effects in density – stratified reacting gases. Combust. Flame 78:121-127

[6] Low HC, Wilson CW, Abdel-Gayed RG and Bradley D (1989) 25'th Joint Propulsion Conference. AIAA paper 89-2944

[7] Cheriyan GK, Krallis K and Weinberg FJ (1990) Combust. Sci. Tech. 70:171

[8] Wagner PC, O'Brien WF, Northam GB and Eggars JN (1986) 23'rd JANAF Combustion Meeting.

[9] Sato Y, Sayama M, Ohwaki K, Masuya G, Komuro T, Kudou K, Murakami A, Tani K, Wakamatsu Y, Kanda T, Chinzei N, and Kimura I (1989) Effectiveness of plasma torches for ignition and flame-holding in scramjet. AIAA Paper 89-2564

[10] Masuya G, Kudou K, Murakami A, Komuro T, Tani K, Kanda T, Wakamatsu Y, Chinzei N, Sayama M, Ohwaki K, and Kimura I (1990) Some governing parameters of plasma torch igniter/flameholder in a scramjet combustor. AIAA Paper 90-2098

[11] Wilson CW (1991) Private Communication. Rolls Royce, Bristol

[12] Aviaexport CIAM2 Aviamotornaya St., 111250, Moscow.

[13] Foreman CL (1992) The use of electrically generated plasma for the initiation of pulverised coal combustion. M.E. (Res.) Thesis (in 2 volumes) University of Sydney, Department of Electrical Engineering.

[14] Morimune T (1992) Removal of NO_x contained in exhaust gas by N_2 plasma injection. 24'th Symp. (Intl.) on Combust., The Combustion Institute, Pittsburgh, poster no. 16

[15] Carabine MD, Clay PG and Itauma C (1990) SO_2 Control Symposium, EPA/EPR, New Orleans.

[16] Hilliard JC, Strassburger R, Velosa J and Ruiz R (1989) Nitric oxide reduction by radiative cooling of natural gas flames seeded by plasma-generated carbon particles. Fossil Fuels Combustion Symposium PD25, ASME, pp 95-99

[17] Calcote HF and Berman CH (1989) Increased methane-air stability limits by a DC electric field. Fossil Fuels Combustion Symposium PD25, ASME, pp 25-31

[18] Carleton FB and Weinberg FJ (1987) Electric field-induced flame convection in the absence of gravity. Nature 330:635-636

[19] Weinberg FJ (1992) Electrically driven convection. 1'st International Symposium on Fluids in Space, ESA SP-353, pp 123

[20] Katz JF and Hung CH (1991) Initial studies of electric field effects on ceramic powder formation in flames. 23'rd Symp. (Intl.) on Combust., The Combustion Institute, Pittsburgh, pp 1733-1738

[21] Hardesty DR and Weinberg FJ (1973) Electrical control of particulate pollutants from flames. 14'th Symp. (Intl.) on Combust., The Combustion Institute, Pittsburgh, pp 907-918

[22] Xie L, Kishi T and Kono M (1992) Investigation on the effect of electric fields on soot formation and flame structure of diffusion flames. 24'th Symp. (Intl.) on Combust., The Combustion Institute, Pittsburgh, paper no. 121

[23] Kimura I (1988) Promotion of combustion by electric discharges. JSME Int. Journal, Series II. 31:376–386

[24] Hilliard JC and Weinberg FJ (1976) Effect of nitrogen–containing plasmas on stability, NO Formation and sooting of flames. Nature. 259:556–557

[25] Warris M and Weinberg FJ (1984) Ignition and flame stabilization by plasma jets in fast gas streams. 20'th Symp. (Intl.) on Combust., The Combustion Institute, Pittsburgh, pp 1825–1831

[26] Dale JD, Smy PR, and Clements RM (1978) The effects of a coaxial spark igniter on the performance of and the emissions from an internal combustion engine. Combust. and Flame. 31:173–185

[27] Murase E, Ono S, Hanada K, and Nakahara S (1989) Combustion of turbulent lean mixtures by plasma jet ignition. Trans. Japan Soc. of Mech. Engr. B55:1234–1240

[28] Kimura I, Aoki H, and Kato M (1981) The use of a plasma jet for flame stabilization and promotion of combustion in supersonic air flows. Combust. and Flame. 42:297–305

[29] Barbi E, Mahan JR, O'Brien WF, and Wagner TC (1989) Operating characteristics of a hydrogen–argon plasma porch for supersonic combustion applications. J. Propulsion 5:129–133

[30] Horisawa H,Kimura I and Sayama M (1992) Studies on the mechanism of ignition and flameholding by plasma jets. J. Japan Soc. Aero–space Engr.(in Japanese) 40:285–294

[31] Dodo S, Azetsu A ,and Someya T (1991) The effects of continuous plasma jet on steady diffusion flame. Proc. 29th Symp. on Combust. (Japan) :556–558

[32] Someya T and Azetsu A (1992) The control of unsteady combustion using plasma jet. Final Reports of Grant–in–Aid for Scientific Research on Priority Areas "Exploration of Combustion Mechanism":169–176

[33] Oppenheim AK, Teichman K, Hom K, and Stewart HE (1978) Jet ignition of an ultra–lean mixture. SAE paper 780637

[34] Kupe J, Wilhelmi H, and Adams W (1987) Operational characteristics of a lean-burn SI–engine: Comparison between plasma–jet and conventional ignition system, SAE paper 870608

[35] Vince IM, Vovelle C, and Weinberg FJ (1984) The effect of plasma jet ignition on flame propagation and sooting at the rich limit of flammability. Combust. Flame 56:105–112

[36] Kono M, Iinuma K, and Kumagai S (1981) The effect of DC to 10 MHz electric field on flame luminosity and carbon formation. 18'th Symp. (Intl.) on Combust., The Combustion Institute, Pittsburgh, pp 1167–1174

[37] Kono M, Carleton FB, Jones AR, and Weinberg FJ (1989) The effect of nonsteady electric fields on sooting flames. Combust. Flame. 78:357–364

[38] Xie L, Sugiyama G, Tamura K, and Kono M (1990) Investigation on the effect of electric fields on soot formation and flame structure of diffusion flames. First Asian–Pacific International Symposium on Combustion and Energy Utilization, pp 63–68, CAS, Beijing

[39] Weinberg FJ (1981) Electrical Intervention in the sooting of flames, NATO Workshop on Sooting Combustion Systems and its Toxic Properties, NATO, pp 1–15

[40] Heinsohn RJ, Thillard SV, and Becker PM (1969) Temperature profiles of an opposed–jet diffusion flame subjected to an electric field. Combust. Flame. 13:442–445

[41] Tsuji H and Yamaoka I (1967) The counterflow diffusion flame in the forward stagnation of a porous cylinder. 11'th Symp. (Intl.) on Combust., The Combustion Institute, Pittsburgh, pp 979–984

[42] D'Allesio A (1981) Laser light scattering and fluorescence diagnostics of rich flames produced by gaseous and liquid fuels. In: Siegla DC and Smith GW (ed) Particulate Carbon. Plenum Press, New York – London

[43] Jones FL, Becker PM, and Heinsohn RJ (1972) A mathematical model of the opposed–jet diffusion flame: effect of an electric field on concentration and temperature profiles. Combust. Flame. 19:351–362

[44] Sugiyama G and Kono M (1989) The controlling factor of soot formation rate in diffusion flames. Joint. Int. Conf., The Combustion Institute, Pittsburgh, pp 88–90

[45] Vandsburger U, Kennedy I, and Glassman I (1984) Sooting counterflow diffusion flames with varying oxygen index. Combust. Sci. and Tech. 39:263–285

[46] Tesner PA (1974) Formation of soot particles. Faraday Symposium, Chem. Soc., 7:104–108

[47] Tesner PA, Tsygankova EI, Guilazetdinov LP, Zuyev VP, and Loshakova GV (1971) The formation of soot from aromatic hydrocarbons in diffusion flames of hydrocarbon–hydrogen mixtures. Combust. Flame 17:279–285

[48] Hayashi H (1990) Magnetic field effects on dynamic behavior and chemical reactions of excited molecules. In: Rabek JF (ed) Photochemistry and Photophysics. vol. 1 CRC Press, Boca Raton

[49] Steiner UE and Uirich T (1989) Magnetic field effects in chemical kinetics and related phenomena. Chem. Rev. 89:51–147

[50] Lin SH and Fujimura Y (1979) Effect of magnetic field on molecular luminescence, in Excited States. vol. 4. Chap. 3. Lim EC (ed), Academic Press, New York

[51] Doherty PM and Crosley DR (1984) Polarization of laser–induced fluorescence in OH in an atmospheric pressure flame. Appl. Opt. 23:713–721

[52] Hayashi H (1982) The external magnetic field effect on the emission intensity of the A $^2\Sigma^-$ → $X_2\Pi$(0–0) transition of the OH radical in flames. Chem. Phys. Lett. 87:113–116

[53] Hayashi H (1984) Magnetic field effect on combustion –– emission intensity of the OH radicals in flames. Nippon Kagaku Kaishi. pp 1753–1758

[54] Wakayama NI, Ogasawara I, and Hayashi H (1984) The external magnetic field effect on the emission intensity of the Na D line in hydrogen–oxygen flames. Chem. Phys. Lett. 105:209–213

[55] Wakayama NI, Ogasawara I, Nishikawa T, Ohyagi Y, and Hayashi H (1984) Magnetic quenching of the emission intensities of HPO and SnH in hydrogen–oxygen flames. Chem Phys. Lett. 107:207–211

[56] Fukuda Y, Hayashi H, and Nagakura S (1985) External magnetic field effects on the emission intensities of the NO ($B^2\Pi_r$–$X^2\Pi_r$) bands in an afterglow produced by a microwave discharge. Chem. Phys. Lett. 119:480–483

[57] Sumitani M, Abe H, and Nagakura S (1991) External magnetic field effects on the β band emission of NO. J. Chem. Phys. 94:1923–1928

[58] Cosby PC and Slanger TG (1991) External magnetic field effects in NO($B^2\Pi$),$B^2\Pi$/$a^4\Pi$ coupling, and $b^4\Sigma^-$ and $a^4\Pi$ level positions. J. Chem. Phys. 95:2203–2205

[59] Fukuda Y, Abe H, Hayashi H, Imamura T, and Nagakura S (1986) External magnetic field effects on the emission intensities of the OH(A–X) and CH(A–X) bands in low pressure C_2H_2/O_2 and C_3H_8/O_2 flames. Chem. Lett. pp 777–780

[60] Abe H and Hayashi H (1992) External magnetic field effects on chemiluminescence intensities from C_2(d) and CN(B) states in low pressure C_2H_2/N_2O flames. Chem. Phys. 162:225–234

[61] Abe H and Hayashi H, to be appeared.

[62] Gaydon AG (1974) The spectroscopy of flames. Chapman and Hall, London; Gaydon AG and Wolfhard HG (1979) Flames. Chapman and Hall, London

[63] Krause HF (1979) A carbon reaction studied by crossed molecular beams. J. Chem. Phys. 70:3871–3880

[64] Herman L, Akriche J, and Grenat H (1962) Spectres continus d'absorption et de phosphorescence associes a la dissociation de SO_2 et a la recombinaison de SO et O. J. Quant. Radiat. Transfer. 2:215–224

[65] Halstead CJ and Thrush BA (1966) The kinetics of elementary reactions involving the oxides of sulphur. Proc. Roy. Soc. London. A295:363–374

[66] Ebata T, Nakazawa O, and Ito M (1988) Rovibrational dependences of the predissociation in the C^1B_2 state of SO_2. Chem. Phys. Lett. 143:31–37

[67] Yamanouchi K, Takeuchi S, and Tsuchiya S (1989) Spectroscopic studies of vibrational chaos in small polyatomic molecules. Progr. Theor. Phys. Suppl. 98:420–429

[68] Abe H and Hayashi H (1991) External magnetic field effects on SO_2 fluorescence emitted from levels of the C^1B_2 state near predissociation limit. Chem. Phys. Lett. 187:227–232

[69] Stannard PR (1978) Radiationless transitions in a magnetic field. J. Chem. Phys. 68:3932–3939

[70] Matsuzaki A and Nagakura S (1978) On the mechanism of magnetic quenching of fluorescence in gaseous state. Helv. Chim. Acta. 61:675–684

[71] Pfefferle LD and Pfefferle WC (1987) Catalysis in combustion. Catal. Rev. –Sci. Eng. 29:219–267

[72] Trimm DL (1983) Catalytic combustion. Appl. Catal. 7:249–282

[73] Prasad R, Kennedy LA, and Ruckenstein E (1984) Catalytic combustion. Catal. Rev. –Sci. Eng. 26:1–58

[74] Furuya T, Yamanaka S, Hayata T, Koezuka J, Yoshine T, and Ohkoshi A (1987) Hybrid catalytic combustion for stationary gas turbine. ASME paper 87–GT–99

[75] Tsurumi K, Shoji T, Kawaguchi S, and Dalla Betta RA (1992) Development of low NOx catalytic combustion. Preprint of 13rd Symposium on Catalytic Combustion, Tokyo, pp 1

[76] Matsuda S, Kato A, Mizumoto M, and Yamashita H (1984) A new support material for catalytic combustion above 1000°C. Proc. 8th Intern. Congr. Catal. Verlag Chemie. 4:879

[77] Machida M, Eguchi K, and Arai H (1987) Effects of additives on the surface area of oxide supports for catalytic combustion. J. Catal. 103:385–393; Arai H and Machida M (1992) Catalyst materials resistent to high temperature. petrotech. 15:606–611

[78] Kuwabara H (1990) Bachelor's Thesis. University of Tokyo, 1990

[79] Misono M and Lombardo EA (ed) (1990) Perovskites. catalysis today. Elseviervol vol. 8, no. 2

[80] Misono M (1990) Design of practical catalyst at atomic/molecular levels. In: Future Opportunities in Catalytic and Separation Technology. Misono, Moro–oka and Kimura (ed) Elsevier. pp 13

[81] Lee KY, Ogasawara K, Asami M, Mizuno N, and Misono M (1990) Mechanism of Catalytic Combustion (II). 66th Catalysis Symposium, Hiroshima, 4L420

[82] Asami M, Lee KY, and Misono M (1991) Mechanism of Catalytic Combustion (III). 68th Catalysis symposium, Sapporo, 4G413

[83] Ogasawara K and Misono M (1990) Mechanism of Catalytic Combustion (I). 59th Natl. Mtg. of Chem. Soc. Jpn., Tokyo, 3A501

List of Authors

Abe, Haruo, Dr.
Molecular Photochemistry Laboratory, The Institute of Physical and Chemical Research
Hirosawa, Wako, Saitama 351-01 Japan

Aoki, Hideyuki, Dr.
Faculty of Engineering, Tohoku University
Aramaki, Sendai 980 Japan

Asami, Masayuki, Dr.
Faculty of Engineering, The University of Tokyo
Hongo, Bunkyo-ku Tokyo 113 Japan

Azetsu, Akihiko, Professor
Faculty of Engineering, The University of Tokyo
Hongo, Bunkyo-ku Tokyo 113 Japan

Chikahisa, Takemi, Professor
Faculty of Engineering, Hokkaido University
N13, W8, Sapporo 060 Japan

Daisho, Yasuhiro, Professor
Faculty of Science and Engineering, Waseda University
Okubo, Shinjuku-ku Tokyo 169 Japan

D'Alessio, Antonio, Professor
Dipartimento di Ingegneria Chimica, University "Federico II"
Piazzale V, Tecchio-80125 Napoli, Italy

Dodo, Satoshi, Dr.
Faculty of Engineering, The University of Tokyo
Hongo, Bunkyo-ku Tokyo 113 Japan

Durst, Franz, Professor
Lehrstuhl für Strömungsmechanik, Universität Erlangen-Nürnberg
Cauerstrasse 4, D-8520 Erlangen, Germany

Eisfeld, Fritz, Professor
Fachbereich Maschinenwesen, Universität Kaiserslautern
Erwin-Schrôdinger-Strasse 6750 Kaiserslautern, Germany

Enomoto, Yoshiteru, Professor
Faculty of Engineering, Musashi Institute of Technology
Tamatsutsumi, Setagaya-ku Tokyo 158 Japan

Frenklach, Michael, Professor
Department of Materials Science and Engineering, The Pennsylvania State University
University Park, PA 16802, USA

Fujii, Nobuyuki, Professor
Department of Chemistry, Nagaoka University of Technology
Kamitomioka, Nagaoka 940-21 Japan

Fujimoto, Hajime, Professor
Faculty of Engineering, Doshisha University
Karasuma-Imadegawa, Kamigyo-ku Kyoto 602 Japan

Fujiwara, Yasuhiro, Professor
Hokkaido Institute of Technology
Teine-Maeda, Teine-ku Sapporo 006 Japan

Furuhama, Shoichi, Professor
Faculty of Engineering, Musashi Institute of Technology
Tamazutsumi, Setagaya-ku Tokyo 158 Japan

Hayashi, Hisaharu, Dr.
Molecular Photochemistry Laboratory, The Institute of Physical and Chemical Research
Hirosawa, Wako, Saitama 351-01 Japan

Hirai, Shuichirou, Dr.
Faculty of Engineering, Osaka University
Yamada-oka, Suita, Osaka 565 Japan

Hiroyasu, Hiroyuki, Professor
Faculty of Engineering, University of Hiroshima
Kagamiyama, Higashi Hiroshima 724 Japan

Iida, Norimasa, Professor
Faculty of Science and Engineering, Keio University
Hiyoshi, Kohoku-ku Yokohama 223 Japan

Ikegami, Makoto, Professor
Faculty of Engineering, Kyoto University
Yoshida Honnmachi, Sakyo-ku Kyoto 606-01 Japan

Ito, Kenichi, Professor
Faculty of Engineering, Hokkaido University
N13, W8, Sapporo 060 Japan

Kadota, Toshikazu, Professor
Faculty of Engineering, University of Osaka Prefecture

Gakuen-cho, Sakai, Osaka 593 Japan

Kamimoto, Takeyuki, Professor
Faculty of Engineering, Tokyo Institute of Technology
Ookayama, Meguro-ku Tokyo 152 Japan

Katsuki, Masashi, Professor
Faculty of Engineering, Osaka University
Yamada-oka, Suita Osaka 565 Japan

Kimura, Itsuro, Professor
Faculty of Engineering, University of Tokai
Kitakaname, Hiratsuka Kanagawa 259 Japan

Kishi, Takeyuki, Dr.
Faculty of Engineering, The University of Tokyo
Hongo, Bunkyo-ku Tokyo 113 Japan

Koda, Seiichiro, Professor
Faculty of Engineering, The Unviersity of Tokyo
Hongo, Bunkyo-ku Tokyo 113 Japan

Komori, Satoru, Professor
Faculty of Engineering, Kyushu University
Hakozaki, Fukuoka 812 Japan

Kono, Michikata, Professor
Faculty of Engineering, The University of Tokyo
Hongo, Bunkyo-ku Tokyo 113 Japan

Lee, Kwan Y., Dr.
Faculty of Engineering, The University of Tokyo
Hongo, Bunkyo-ku Tokyo 113 Japan

Liao, Chihong, Dr.
Faculty of Engineering, Yokohama National University
Tokiwadai, Hodogaya-ku Yokohama 240 Japan

Massoli, Patrizio, Dr.
Instituto Motori CNR
Piazza Barsanti e Matteucci, 80125 Napoli, Italy

Matsui, Hiroyuki, Professor
Faculty of Engineering, The University of Tokyo
Hongo, Bukyo-ku Tokyo 113 Japan

Misono, Makoto, Professor
Faculty of Engineering, The University of Tokyo
Hongo, Bunkyo-ku Tokyo 113 Japan

Miura, Takatoshi, Professor
Faculty of Engineering, Tohoku University
Aramaki, Sendai 980 Japan

Miwa, Kei, Professor
Faculty of Engineering, Tokushima University
Minami–josanjima–chou, Tokushima 770 Japan

Miyamoto, Noboru, Professor
Faculty of Engineering, Hokkaido University
N13, W8, Sapporo 060 Japan

Mizutani, Yukio, Professor
Faculty of Engineering, Osaka University
Yamada–oka, Suita, Osaka 565 Japan

Murayama, Tadashi, Professor
Faculty of Engineering, Hokkaido University
N13, W8, Sapporo 060 Japan

Naqwi, Amir A., Dr.
Lehrstuhl für Strömungsmechanik, Universität Erlangen–Nürnberg
Cauerstrasse 4, D–8520 Erlangen, Germany

Nishiwaki, Kazuie, Professor
Faculty of Science and Engineering, Ritsumeikan University
Toji–in Kitamachi, Kita–Ku Kyoto 603 Japan

Ogawa, Hideyuki, Professor
Faculty of Engineering, Hokkaido University
N13, W8, Sapporo 060 Japan

Ohtake, Kazutomo, Professor
Faculty of Engineering, Toyohashi University of Technology
Hibariga–oka, Tempaku, Toyohashi 441 Japan

Okajima, Satoshi, Professor
Faculty of Engineering, Hosei University
Kajino–cho, Koganei Tokyo 184 Japan

Okamoto, Tatsuyuki, Professor
Faculty of Engineering, Osaka University
Yamada–oka, Suita, Osaka 565 Japan

Onuma, Yoshiaki, Professor
Faculty of Engineering, Toyohashi University of Technology
Hibariga–oka, Tempaku, Toyohashi 441 Japan

Peters, Norbert, Professor
Institute für Technische Mechanik, RWTH–Aachen

D5100, Aachen Germany

Pitz, William J., Dr.
Lawrence Livermore National Laboratory
P.O.Box 808, Livermore, California 94550 USA

Sadakata, Masayoshi, Professor
Faculty of Engineering, The University of Tokyo
Hongo, Bunkyo-ku Tokyo 113 Japan

Sano, Taeko, Professor
Faculty of Engineering, Tokai University
Kitakaname, Hiratsuka, Kanagawa 259-12 Japan

Senda, Jiro, Professor
Faculty of Engineering, Doshisha University
Karasuma-Imadegawa, Kamigyo-ku Kyoto 602 Japan

Shimamoto, Yuzuru, Profeesor
Faculty of Engineering, Kyoto University
Yoshida Honnmachi, Sakyo-ku Kyoto 606-01 Japan

Someya, Tsuneo, Professor
Faculty of Engineering, Musashi Institute of Technology
Tamazutsumi, Setagaya-ku Tokyo 158 Japan

Sugiyama, Gen, Dr.
Japan Automobile Research Institute,Inc.
2530, Karima, Tsukuba 305 Japan

Tachibana, Takeshi, Professor
Department of Mechanical Engineering, Kyushu Institute of Technology
Sensui-cho, Tobata-ku Kitakyushu 804 Japan

Takagi, Toshimi, Professor
Faculty of Engineering, Osaka University
Yamada-oka, Suita, Osaka 565 Japan

Takasaki, Koji, Professor
Faculty of Engineering, Kyushu University
Hakozaki, Fukuoka 812 Japan

Takeno, Tadao, Professor
School of Engineering, Nagoya University
Furo-cho, Chikusa-ku Nagoya 464-01 Japan

Tanabe, Hideaki, Professor
Department of Mechanical System Engineering, Kanazawa Institute of Technology
Oogiga-oka, Nonoichi, Ishikawa 921 Japan

Terao, Kunio, Professor
Faculty of Engineering, Yokohama National University
Tokiwadai, Hodogaya–ku Yokohama 240 Japan

Wakisaka, Tomoyuki, Professor
Faculty of Engineering, Kyoto University
Yoshida Honnmachi, Sakyo–ku Kyoto 606–01 Japan

Wakuri, Yutaro, Professor
Faculty of Engineering, Kyushu University
Hakozaki, Fukuoka 812 Japan

Wang, Hai, Professor
Department of Materials Science and Engineering, The Pennsylvania State University
University Park, PA 16802, USA

Weinberg, Felix J., Professor
Department of Chemical Engineering and Chemical Technology,
Imperial College of Science and Technology, London SW7 U.K.

Westbrook, Charles K., Dr.
Lawrence Livermore National Laboratory
P.O.Box 808 Livermore, California 94550 USA

Xie, Lin, Dr.
National Aerospace Laboratory
Jindaijihigashi–machi, Chofu–shi, Tokyo 182 Japan

Yoshihara, Yoshinobu, Professor
Faculty of Science and Engineering, Ritumeikan University
Tojiin–Kitamachi, Kita–ku Kyoto 603 Japan

Yoshizawa, Yoshio, Professor
Research Laboratory for Nuclear Reactor, Tokyo Institute of Technology
Ookayama, Meguro–ku Tokyo 152 Japan

Index